Bioelectric Recording Techniques

PART C

Receptor and Effector Processes

METHODS IN PHYSIOLOGICAL PSYCHOLOGY

EDITOR: Richard F. Thompson
DEPARTMENT OF PSYCHOBIOLOGY
UNIVERSITY OF CALIFORNIA, IRVINE

Vol. I. BIOELECTRIC RECORDING TECHNIQUES,
Richard F. Thompson and Michael M. Patterson (Eds.)

Part A. Cellular Processes and Brain Potentials, 1973
Part B. Electroencephalography and Human Brain Potentials, 1974
Part C. Receptor and Effector Processes, 1974

Bioelectric Recording Techniques

PART C

Receptor and Effector Processes

Edited by

RICHARD F. THOMPSON

MICHAEL M. PATTERSON

Department of Psychobiology
University of California
Irvine, California

ACADEMIC PRESS New York and London   1974
A Subsidiary of Harcourt Brace Jovanovich, Publishers

COPYRIGHT © 1974, BY ACADEMIC PRESS, INC.
ALL RIGHTS RESERVED.
NO PART OF THIS PUBLICATION MAY BE REPRODUCED OR
TRANSMITTED IN ANY FORM OR BY ANY MEANS, ELECTRONIC
OR MECHANICAL, INCLUDING PHOTOCOPY, RECORDING, OR ANY
INFORMATION STORAGE AND RETRIEVAL SYSTEM, WITHOUT
PERMISSION IN WRITING FROM THE PUBLISHER.

ACADEMIC PRESS, INC.
111 Fifth Avenue, New York, New York 10003

United Kingdom Edition published by
ACADEMIC PRESS, INC. (LONDON) LTD.
24/28 Oval Road, London NW1

Library of Congress Cataloging in Publication Data

Thompson, Richard Frederick, Date
    Receptor and effector processes.

    (His Bioelectric recording techniques, pt. C)
(Methods in physiological psychology)
    Bibliography: p.
    1. Electrophysiology–Technique. 2. Psychology,
Physiological–Technique. I. Patterson, Michael M.,
joint author. II. Title. III. Series.
QP341.T48 pt.c [ ]   591.1'9'127   72-88357
ISBN 0–12–689403–5

PRINTED IN THE UNITED STATES OF AMERICA

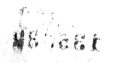

Contents

List of Contributors ix
General Preface xi
Preface to Part C xiii
Contents of Parts A and B xiv

RECEPTOR POTENTIALS

Chapter 1   Electrophysiology of the Cochlea
Jack Vernon and Mary Meikle

I. Types of Electrical Potentials Recorded from the Cochlea 4
II. Characteristics of the a.c. Cochlear Potential 8
III. Types of Cochlear Potential Data 10
IV. Recording the a.c. Cochlear Potential 14
Appendix 58
References 60

Chapter 2   The Electroretinogram
Harold Koopowitz

I. Introduction 64
II. History of the Techniques 64
III. Recording Techniques 66
IV. The Stimulus 73
V. Kinds of Preparations 76
VI. Measurements and Interpretation 83
References 85

Chapter 3 Recording of Bioelectric Activity: The Electro-Olfactogram
David S. Phillips

I. Introduction	87
II. Technique	92
References	94

EFFECTOR PROCESSES

Chapter 4 Recording of Human Eye Movements
Bernard Tursky

I. Introduction	100
II. Neuromuscular Control Mechanisms	100
III. Eye Movement Recording Methods	104
IV. History of Electro-Oculography	113
V. Electro-Oculography Recording Problems	114
VI. Applications	128
VII. Conclusions	130
References	131

Chapter 5 Electromyography: Single Motor Unit Training
John V. Basmajian

I. Introduction	137
II. General EMG Techniques	140
III. Basic Motor Unit Training	143
IV. Further Psychophysiologic Techniques Employing SMUT	147
References	153

Chapter 6 Electromyography: Human and General
Joseph Germana

I. Introduction	155
II. The Coordination of Behavior and Response Uncertainty	156
III. The Effects of Response Uncertainty on Behavior	158
IV. Autonomic–Somatic Integration	159
V. Conclusion	161
References	162

Chapter 7 Electrocardiogram: Techniques and Analysis
Neil Schneiderman, George W. Dauth, and David H. VanDercar

I. Introduction	166
II. The Heart	167
III. Regulation of Cardiac Responses	172
IV. Basic Instrumentation	179
V. Data Analysis	191
VI. Operant Conditioning and Biofeedback	193
References	197

Chapter 8 The Cardiac Response during Infancy
Michael Lewis

I. The Use of Physiological Recording in Infancy	201
II. Cardiac Response as a Measure of Psychophysiological Responsivity	203
III. Studies of Heart Rate Response in Infants	215
IV. Developmental Issues in Heart Rate Response	219
V. Heart Rate Response and Cognitive Functions	222
VI. Discussion and Summary	225
References	226

Chapter 9 Mechanisms of Electrodermal Activity
Don C. Fowles

I. Introduction	232
II. Skin Conductance Responses	239
III. Skin Conductance Level	251
IV. Skin Potential Responses	253
V. Skin Potential Level	259
VI. Summary and Conclusions	265
References	267

Chapter 10 Recording of Electrodermal Phenomena
William W. Grings

I. Nature of the Physical Measurement	273
II. The Electrode	277
III. Input Signal Conditioning, Amplification, and Registration	281
IV. Measurement Units	286
References	292

Chapter 11   The Electrogastrogram
Daniel Lilie

I. Introduction	297
II. Recording Technique	298
III. Data Analysis	299
IV. The Data	300
V. Clinical Application of the EGG	301
VI. Problems and Caveats	301
VII. Concluding Statement	304
References	304

Bibliography   Methods and Techniques

Cochlear Potential Methods	307
Electroretinography	308
Electro-Olfactogram Methods	309
Eye Movements	309
Electromyography	310
Electrocardiogram	313
Electrodermal Activity	315
Electrogastrogram	316

AUTHOR INDEX	317
SUBJECT INDEX	326

List of Contributors

Numbers in parentheses indicate the pages on which the authors' contributions begin.

JOHN V. BASMAJIAN (137), Emory University Regional Rehabilitation Research and Training Center and Georgia Mental Health Institute, Atlanta, Georgia

GEORGE W. DAUTH* (165), Department of Behavioral Physiology, New York State Psychiatric Institute, and Columbia University, New York, New York

DON C. FOWLES (231), Department of Psychology, The University of Iowa, Iowa City, Iowa

JOSEPH GERMANA (155), Department of Psychology, Virginia Polytechnic Institute and State University, Blacksburg, Virginia

WILLIAM W. GRINGS (273), Department of Psychology, University of Southern California, Los Angeles, California

HAROLD KOOPOWITZ (63), Department of Developmental and Cell Biology, University of California, Irvine, California

MICHAEL LEWIS (201), Educational Testing Service, Princeton, New Jersey

DANIEL LILIE (297), Southern Wisconsin Colony and Training School, Union Grove, Wisconsin

MARY MEIKLE (3), Department of Otolaryngology, Kresge Hearing Research Laboratory, University of Oregon Medical School, Portland, Oregon

*Present address: Department of Neurology, College of Physicians and Surgeons, Columbia University, New York, New York.

DAVID S. PHILLIPS (87), Department of Medical Psychology, University of Oregon Medical School, Portland, Oregon

NEIL SCHNEIDERMAN (165), Department of Psychology and Laboratory for Quantitative Biology, University of Miami, Coral Gables, Florida

BERNARD TURSKY (99), Department of Political Science, State University of New York, Stony Brook, New York

DAVID H. VANDERCAR* (165), Laboratory of Physiological Psychology, Rockefeller University, New York, New York

JACK VERNON (3), Department of Otolaryngology, Kresge Hearing Research Laboratory, University of Oregon Medical School, Portland, Oregon

---

*Present address: Department of Psychology, University of South Florida, Tampa, Florida.

General Preface

The major approaches used to characterize the organization and functions of the brain can be grouped into four categories of techniques—electrophysiology, anatomy, chemistry, and behavior. All these approaches to the study of the brain and its functions will be treated in this series on *Methods in Physiological Psychology*. The series begins with the present treatment of bioelectric recording techniques in three volumes (Parts A, B, and C). Much that has been learned about the brain is due to the development of bioelectric recording techniques. Perhaps more important, the basic processes among neurons that underlie all aspects of brain function and behavior, from simple reflexes to relativity theory, are fundamentally bioelectric in nature—they result in potential and current changes across membranes.

Bioelectric recording techniques currently offer one great advantage over anatomical and chemical procedures—they minimize the "uncertainty" principle. An electrode placed on the surface of the scalp can record human brain activity with minimum perturbation of the system. Activity of the heart can be recorded by placing wires on the arms; the autonomic signs of brain activity and behavior such as the galvanic skin response can similarly be measured by peripheral electrodes. However, we cannot infer central events from peripheral measures without basic knowledge of the cellular processes underlying the generation of bioelectric activity. Part A treats the analytic techniques used to study the basic bioelectric phenomena of neural tissue in animal preparations. Part B deals with electroencephalography and peripheral recording of brain events in man. Part C treats receptor and effector processes.

Brain electrical activity was first recorded by Caton in 1875, and the human EEG was described by Berger in 1929. However, the major developments in bioelectric techniques have come in the past 30 years. Recording techniques and methods of analysis have grown almost exponentially over this period. There is no up-to-date review of the methods of bioelectric recording; hence these three volumes. All of the articles here are original contributions written by experts. Each author is a recognized authority in his or her area, who is actively engaged in the use of the techniques described.

The best and perhaps only way to understand properly a method or technique is to use it; the next best way is for an expert to describe it. The authors of these chapters have succeeded very well in presenting the techniques of bioelectric recording clearly and in sufficient practical detail to be of immediate use. Moreover, their chapters go well beyond details of technique and provide an analysis of experimental and theoretical issues as well. In our opinion, these volumes provide an outstanding synthesis of what we know and where we stand today in our understanding of the bioelectric aspects of brain function and behavior. The authors of these chapters have done their job very well indeed.

We express our very sincere gratitude to the many contributors who have written this work. We also acknowledge the invaluable editorial assistance of Nancy M. Kyle, and the help of the UCLA Brain Information Service, and Information Services, Pacific Southwest Regional Medical Library Service for providing the bibliographies of technique articles listed at the end of each volume.

RICHARD F. THOMPSON

Preface to Part C

Methods for recording the bioelectric signs of receptor and effector activity are treated in Part C. The first section begins with a comprehensive review of current techniques and knowledge of cochlear potentials by Vernon and Meikle. Koopowitz then provides a most interesting review of the electroretinogram in man, vertebrates, and invertebrates. In the last chapter of this section, Phillips presents an overview of the olfactogram.

The second section of Volume III begins with a review of electromyography, a topic of fundamental importance for those interested in the immediate physiological substrates of behavior. In a most comprehensive chapter, Basmajian describes techniques for recording single motor units in man and reviews the intriguing topic of training responses of single motor units. Germana presents a generalized discussion of electromyography in the context of psychophysiology.

Next Schneiderman, Dauth, and VanDercar present an extensive discussion of the electrocardiogram and techniques for recording and analysis. Lewis reviews the methodological issues and techniques for recording the cardiac response during infancy.

The electrodermal response ("galvanic skin response" in earlier literature), the most widely used autonomic measure in psychology, is treated in the next section. Fowles provides a discussion of the mechanisms of electrodermal activity and Grings presents an extensive review of recording techniques and interpretation of electrodermal phenomena.

In the final chapter, Lilie describes an interesting and less familiar peripheral measure of effector activity—the electrogastrogram.

Contents of Parts A and B

Part A: Cellular Processes and Brain Potentials

INTRACELLULAR RECORDING

1. Intracellular Single Unit Recording
    Marc A. Dichter
2. Iontophoretic Injection Techniques
    Albert Globus
3. Microelectrode Recordings in Brain Tissue Cultures
    Stanley M. Crain

EXTRACELLULAR SINGLE UNIT RECORDING

4. Sampling Single Neuron Activity
    Arnold L. Towe
5. Extracellular Unit Recording
    Russell L. De Valois and Paul L. Pease
6. Extracellular Single Unit Recording
    D. Max Snodderly, Jr.
7. Multiple Unit Recordings from Behaving Rats
    James Olds

Multiple Unit Recording

8. Multiple Unit Recording: Technique, Interpretation, and Experimental Applications

    Jennifer S. Buchwald, Solon B. Holstein, and David S. Weber

9. The Study of Neuronal Networks in the Mammalian Brain

    M. Verzeano

10. Generation of Brain Evoked Potentials

    John Schlag

11. Brain Evoked Potentials: Acquisition and Analysis

    E. Roy John

Slow Potentials

12. Slow Potential Changes

    Arnaldo Arduini

13. Cortical Steady Potential Shift in Relation to the Rhythmic Electrocorticogram and Multiple Unit Activity

    Vernon Rowland and George Dines

    Bibliography: Methods and Techniques

Part B: Electroencephalography and Human Brain Potentials

Electroencephalography

1. The Electroencephalogram: Autonomous Electrical Activity in Man and Animals

    Donald B. Lindsley and J. D. Wicke

2. The Electroencephalogram: Human Recordings

    Janice R. Stevens

Evoked Potentials in Man

3. Human Average Evoked Potentials: Procedures for Stimulating and Recording
    William R. Goff
4. The Analysis of Scalp-Recorded Brain Potentials
    Herbert G. Vaughan, Jr.
5. Computer Use in Bioelectric Data Collection and Analysis
    Monte Buchsbaum and Richard Coppola

Contingent Negative Variation

6. The Contingent Negative Variations
    Dale W. McAdam
7. Cerebral Psychophysiology: The Contingent Negative Variation
    Jerome Cohen
8. Methodological Issues in CNV Research
    Steven A. Hillyard

    Bibliography: Methods and Techniques

Receptor Potentials

Chapter 1

Electrophysiology of the Cochlea

Jack Vernon
Mary Meikle

*Department of Otolaryngology
Kresge Hearing Research Laboratory
University of Oregon Medical School
Portland, Oregon*

I. Types of Electrical Potentials Recorded from the Cochlea		4
A. The a.c. Cochlear Potential		4
B. The d.c. Potentials		5
C. Action Potentials of the Auditory Nerve		7
II. Characteristics of the a.c. Cochlear Potential		8
III. Types of Cochlear Potential Data		10
A. Frequency Function (Often Called Sensitivity Function)		10
B. Intensity Function (Often Called Input–Output Function)		13
IV. Recording the a.c. Cochlear Potential		14
A. Anatomical Considerations		14
B. Choice of Experimental Animal		16
C. Surgical Procedures		18
D. Chronic Preparations		23
E. Physiological Variables Affecting the Cochlear Potential		24
F. The Acoustic Stimulus		30
G. Recording Apparatus and Methods		43
H. Quantitative Measurements		51
I. Artifacts		55
Appendix		58
References		60

I. Types of Electrical Potentials Recorded from the Cochlea

The ear is the peripheral organ of the hearing sense; therefore, the electrical activity of that organ during stimulation by sound is of interest in studying the mechanism of hearing. There are several different types of electrical activity which can be recorded from the inner ear (cochlea), depending upon electrode location as well as upon the characteristics of the recording instrumentation. These electrical phenomena include action potentials of the auditory nerve, d.c. potentials originating in various portions of the cochlea, and the alternating or a.c. cochlear potential (often referred to as the "cochlear microphonic").

A. The a.c. Cochlear Potential

The first electrical output of the ear to be discovered was the a.c. cochlear potential. Its cochlear origin was not recognized at first, however. Perhaps because the study of peripheral sensory nerve activity received much attention as a result of Adrian's work in the 1920s, the auditory nerve was an early choice for electrophysiological study. In 1930 Wever and Bray placed an electrode on the portion of this nerve which passes through the internal auditory meatus en route to the brainstem (Fig. 1-1). When they presented acoustic stimulation to their experimental animal, which was a cat, they were very much surprised to record high-fidelity reproduction of acoustic frequencies as high as 4100 Hz. Their surprise resulted from the apparent demonstration that the auditory nerve was capable of following and reproducing such high frequencies, when it was well known that other peripheral nerves could fire no faster than a few hundred times per second. However, it later developed that what they had recorded was not primarily activity of the auditory nerve, but rather an electrical response from the organ of Corti within the cochlea. As Adrian showed (1931), the electrical potential which Wever and Bray had discovered was at its largest when recorded from the round window of the cochlea (see Fig. 1-1).

The question of what to call this electrical response of the ear deserves some consideration. For many years, following comments by Adrian (which he later disavowed), it was thought by some that the potential was artifactual, that is, much like a microphonic response generated by mechanical vibrations within some electronic tubes.[1] Stemming from this misconception, the term "cochlear microphonic" gained wide acceptance.

---

[1] Following is a definition of the term "microphonic" given by H. Petersen (1958): "Microphonics are by definition the spurious electrical signals which are developed in an electrical unit when subjected to mechanical vibrations."

It has now been demonstrated beyond doubt that this cochlear potential is genuinely physiological in origin, and in fact there is good evidence that it is generated by the hair cells of the organ of Corti (Tasaki, Davis, & Eldredge, 1954; Davis, 1961). In view of the inappropriate use of the term "microphonic" as originally applied to this bioelectrical phenomenon, it seems high time to discard this misleading terminology.

Because of its faithful representation of the oscillations which characterize pure tone stimuli, the electrical response discovered by Wever and Bray has been called the "alternating cochlear potential" (Wever, 1966). This designation has the advantage that it distinguishes the Wever–Bray potential from action potentials and from other electrical phenomena of the cochlea, which are d.c. potentials (see below). We would like, therefore, to urge strongly that this potential be called the *alternating cochlear potential* or, more briefly, the *a.c. cochlear potential*.

As will be discussed further in the detailed sections on recording methods, it is possible to record even larger magnitudes of the a.c. cochlear potential if the electrode is inserted through the bone of the cochlea, into one of the fluid-filled scalae (see Fig. 1-2). Once the recording electrode is properly located within the cochlea, it is also possible, with proper amplifying circuitry, to record various d.c. potentials generated within.

B. The d.c. Potentials

1. Endolymphatic Potential

It was first established by von Békésy (1951, 1952) that the scala media has a positive resting potential of some 50–80 mV (relative to a reference in the scala tympani or neck tissue). This potential responds to acoustic stimulation by a decrease in magnitude, which is correlated with the intensity and duration of the stimulating tone. Succeeding observations have indicated that this d.c. potential is generated independently from the a.c. cochlear potential (Rice & Shinabarger, 1961; Small, 1963). Much interesting work has been undertaken to study the origin and nature of the resting potential (for discussions see Grinnell, 1969; Hawkins, 1964). In addition, investigations into its relation to intracochlear ionic concentrations and to other factors such as hypoxia and acoustic trauma have been undertaken (Butler, 1965; Misrahy, Shinabarger, & Arnold, 1958; Suga, Nakashima, & Snow, 1970). The reader who is interested in recording the resting potential is referred for procedural details to the report of Suga, Morimitsu, and Matsuo (1964), as well as to the initial experiments by von Békésy and the work of Butler, cited above.

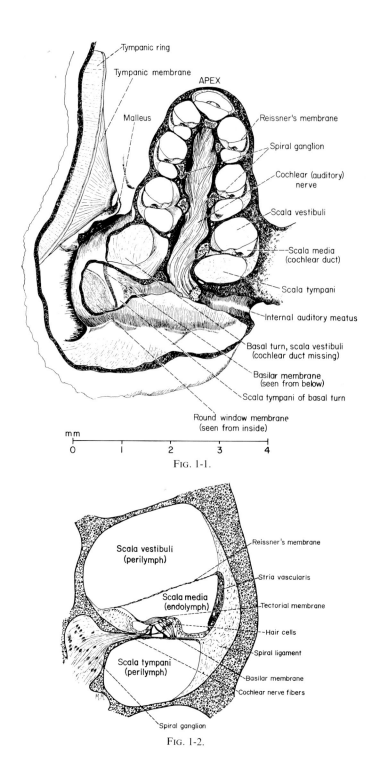

Fig. 1-1.

Fig. 1-2.

2. Summating Potential

Another type of d.c. potential recorded within the cochlea is the summating potential, first observed by Davis, Fernandez, and McAuliffe (1950). This potential is seen as a d.c. bias imposed on the alternating cochlear potential. In other words, the summating potential, unlike the endolymphatic potential, is only observed during sound stimulation. Usually its polarity is seen as negative, when it is measured with an electrode in scala media relative to a reference electrode in scala tympani or in tissues of the neck. However, the summating potential is prone to unexplained variability and capriciousness. Sometimes it has a positive sign, and this seems to be true when conditions are especially favorable—in fresh preparations, for instance (Davis, Deatherage, Eldredge, & Smith, 1958). In order to achieve comparable levels, approximately ten times more sound is required for the summating potential than for the a.c. cochlear potential. The magnitude of the summating potential is directly proportional to the intensity of the stimulus sound, but it does not appear to reach a maximum. The latter observation has led to some doubt that the summating potential represents usual or normal activity of the cochlea. For further details on this potential the reader is referred to Konishi and Yasuno (1963) and to the work by Davis *et al.* (1958).

C. Action Potentials of the Auditory Nerve

No description of electrical events recordable from the vicinity of the cochlea is complete without mention of the neural potentials generated by the auditory nerve. When a brief, moderately intense stimulus such as a

---

FIG. 1-1. The cochlea of the guinea pig. The cochlea is shown here in a cut-away view, with the upper turns in cross section (derived from a camera lucida drawing). Except for the malleus (which is attached to the narrow, funnel-shaped end of the tympanic membrane), the ossicles of the middle ear are not visible as they are hidden behind the basal turn of the cochlea. The basal turn has been opened so as to show the underside of the round window membrane. The outside of this membrane is now the standard electrode site for recording the alternating cochlear potential. This view also shows the internal auditory meatus (with auditory nerve cut away). The auditory nerve at its passage through the meatus was the recording site used by Wever and Bray (1930) when they first discovered the alternating cochlear potential. In this experiment the auditory nerve served as a low resistance path for the passive spread of the cochlear potential to the recording site. The oval window cannot be seen in this view as it lies deeper behind the parts of the basal turn which are visible here.

FIG. 1-2. Cross section view of a single turn of the cochlea. This enlarged view, based on a camera lucida drawing, shows the organ of Corti resting on the basilar membrane. The hair cells of the organ of Corti are aligned in rows which are seen in cross section here. There are three rows of outer hair cells separated from one row of inner hair cells by the tunnel.

click or a high-frequency tone pip is given, the auditory nerve responds with a characteristic and distinctive evoked potential. This response usually contains two well-marked negative waves which have been labeled "N_1" and "N_2." As with the responses of other peripheral nerves to brief stimuli, the amplitude and latency of N_1 and N_2 are dependent on stimulus intensity. N_2 is somewhat more difficult to measure than N_1 as its amplitude is more variable, hence in many experiments N_2 is ignored and attention is focused upon N_1.

The round window is often a convenient place to record N_1, as it requires much less surgery to get to the round window than to get to the auditory nerve itself as it passes through the internal auditory meatus. The only problem with recording N_1 at the round window is that sometimes it is difficult to distinguish between neural activity and the a.c. cochlear potential. This difficulty is particularly acute when the frequency of the acoustic stimulus is close to 1000 Hz,[2] for the waveforms of N_1 and N_2 have a predominant frequency around 1000 Hz.

To avoid confusion between N_1 and the cochlear potential, one method is to use the differential electrode technique described in Section IV,G. Still another way around the problem is to use only acoustic stimuli of higher frequency (such as 10 kHz) so that cochlear potentials are clearly distinguished because their frequency is too high for any neural response. This method can be used with satisfactory precision, and has yielded much information on the activity of the auditory nerve under various conditions. The reader who wishes to pursue this technique is referred to the report of Ruben, Fisch, and Hudson (1962).

Of the various electrophysiological phenomena which can be recorded in the cochlea, none provides a greater variety and depth of information, or is more closely related to the world of hearing, than the a.c. cochlear potential. This chapter will therefore be concerned primarily with the intriguing phenomenon first heard when Wever and Bray connected the output of the cat's ear to a set of earphones. The procedure of "listening in" on these signals generated by a living bioelectric transducer is still to be highly recommended as a most striking demonstration of the delicacy and fidelity of the ear. It is a procedure which should be easily available to anyone who will follow the steps outlined below.

II. Characteristics of the a.c. Cochlear Potential

The alternating cochlear potential is produced in response to sound energy entering the ear and is an almost exact electrical analog of that

[2] Hz (Hertz) is the presently accepted unit of frequency, replacing older designations of cps (cycles per second) or the symbol ∼. kHz replaces the older kc (kilocycles).

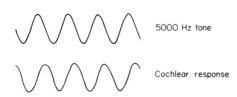

FIG. 1-3. Comparison of an acoustical stimulus with cochlear response. *Upper trace:* the sound stimulus, a 5000 Hz tone as registered by a calibrated microphone in the vicinity of the eardrum. *Lower trace:* the electrical output of the cochlea of a guinea pig in response to the 5000 Hz tone. The similarity of the two waveforms is apparent. The amplitude as well as the phase relationships of each trace were arbitrarily adjusted to provide an easy comparison.

energy. The fidelity of this electrical analog is attested to by the pictures shown in Fig. 1-3, which compare the acoustical stimulus acting upon the ear with the resulting electrical activity produced by the ear. A more qualitative expression of this fidelity is found in the exercise of connecting the amplified output of the ear to a speaker. Now, words spoken into the animal's ear are heard at the speaker with perfect intelligibility. Even an individual person's accent is identifiable.

An advantage of the alternating cochlear potential is that it can be very accurately quantified. That is, it bears a clear and precise relationship to the intensity of the sound stimulus, as can be easily demonstrated by utilizing accurate sound measuring equipment. This relationship will be discussed more fully below.

Another advantage of the cochlear potential[3] is its high degree of replicability which can be demonstrated over long periods of time. The guinea pig is not the easiest animal in which to demonstrate this point. Nevertheless, even this animal will produce a constant output of cochlear potential for periods of 10–12 hr if sufficient care is taken. On the other hand, the cat has an output which is highly reliable over periods of days in acute preparations, and over periods of months in chronic preparations (Rahm, Strother, & Gulick, 1958; Simmons, 1967; Wever, 1966). The constancy of the cochlear output makes it a good monitor for stimulus constancy in a wide variety of experiments. In addition, because there is usually bilateral symmetry between the output of the two ears, one ear can provide baseline data with which to compare the output of the other ear after it is subjected to some experimental manipulation.

The possibility that the cochlear potential may be that event which initiates auditory nerve impulses leading to hearing makes it an interesting phenomenon in its own right, with both theoretical and clinical importance.

[3] Unless otherwise stated, the term "cochlear potential" will be used hereafter for the sake of brevity to mean the a.c. cochlear potential.

Whether or not it is such a generator potential remains an open question, and a crucial one. Nevertheless, regardless of the physiological role of the cochlear potential, it has been shown repeatedly to be a useful indicant of hearing in general or of the condition of the ear in particular. It is its usefulness as a quantitative, precise, and reliable research tool which leads us to advocate it as an important method for studying auditory physiology.

What, then, are the methods for observing the cochlear potential? We will begin by discussing the end product, that is, the data that are to be observed. After that, we will describe the procedures required to obtain such data.

III. Types of Cochlear Potential Data

A. Frequency Function (Often Called Sensitivity Function)

A frequency function is a description of the amount of sound required at each frequency to elicit a constant level of cochlear output. Figure 1-4 illustrates typical frequency functions for several kinds of animals. Note that an arbitrary level of 1 μV was established as the response to be recorded. A level of 1 μV was chosen because it represents a good compromise in that it is usually clearly detectable above background electrical noise, while in almost all animals it would be far below the damaging level.

The 1 μV level is well above the level of cochlear activity associated with the absolute threshold of hearing. The best evidence to date suggests that 1 μV of cochlear potential is about 40 dB above the hearing threshold. Said differently, the cochlear potential at the threshold of hearing would be about 0.01 μV. Such a low level of signal is too small for present-day detection techniques.[4] Whether or not the cochlear potential itself has an absolute threshold has never been established. If there is such a threshold, it is buried even further below the noise limit of present-day electronic equipment.

As Figure 1-4 indicates, the shape of the curve representing the frequency function has been shown to vary between different animal species. In addition, the shape of the frequency function also depends to a certain extent upon location of the recording site. The data in Fig. 1-4 were recorded from round window electrodes.

---

[4] At high frequencies (above, say, 5000 Hz), where electrical noise is much less a problem, the state of the art at present allows readings as low as approximately 0.03 μV. This low a level can be achieved only by means of narrow bandwidth filtering such as a wave analyzer provides. A further improvement in the detection of cochlear signals buried in noise may be possible with signal averaging techniques utilizing computer processing of the data.

1. ELECTROPHYSIOLOGY OF THE COCHLEA

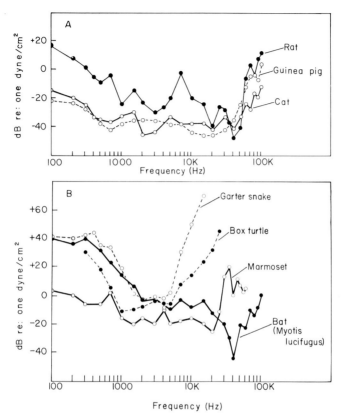

FIG. 1-4. Representative frequency functions for several animal species. In each case the amount of sound required to produce 1 μV of cochlear potential is indicated. Choice of frequencies is arbitrary and is intended to obtain a representative sample across each animal's frequency range. The sound is measured in decibels relative to 1 dyne/cm². The lower the curve the greater the sensitivity. Clearly, individual species differ as to the shape of their frequency functions. The curves shown above were plotted from data reported in the following references: (A) rat: Crowley, Hepp-Raymond, Tabowitz, and Palin (1965); guinea pig: Wever, Vernon, and Peterson (1963); cat: Wever, Vernon, Rahm, and Strother (1958). (B) garter snake: Wever and Vernon (1960; original data measured at 0.1 μV, here adjusted to 1 μV for easier comparison); box turtle: Wever and Vernon (1956); marmoset: Wever and Vernon (1961a); bat: Wever and Vernon (1961b).

Figure 1-5 represents the data obtained when electrodes are inserted in different locations of the cochlea. The idea here is that the closer the electrode is to the locus of activity within the cochlea, the larger will be the resulting recording. It is clear in Fig. 1-5, for example, that a round window electrode does not register as much potential for low frequencies as does an apical electrode. Similarly, a middle turn electrode favors inter-

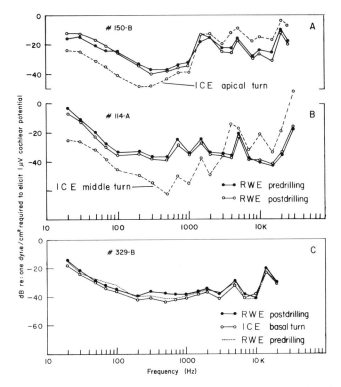

FIG. 1-5. Comparison of frequency functions measured at various intracochlear sites. RWE, round window electrode; ICE, intracochlear electrode. All curves plot the 1 μV level of cochlear potential. Data were obtained from three different guinea pigs. In each case predrilling and postdrilling records from the RWE provide the necessary checks against damage due to drilling. (A) The apical turn ICE favors the low frequencies and is correspondingly poorer for the high frequencies when compared to the RWE. (B) The middle turn ICE is best for intermediate frequencies. (C) The basal turn ICE is in general only slightly better than the RWE. The difference is probably due to the attenuation resulting from the electrical resistance of the round window membrane, which is bypassed when recording from the ICE.

mediate frequencies. On the other hand, a basal turn electrode may offer little advantage over a round window electrode.

The shape of the frequency function is affected by leaving an opening in the bulla (the bulla is shown in Fig. 1-9). The principal effect is at the low frequencies, as shown in Fig. 1-6. In order to obtain these data, an electrode was placed on the round window in the usual manner and the bulla sealed. Then a second hole was drilled in the bulla and fitted with a small plug which could be inserted to provide a sealed bulla and then easily removed to provide an open-bulla reading. The same procedure was

1. ELECTROPHYSIOLOGY OF THE COCHLEA

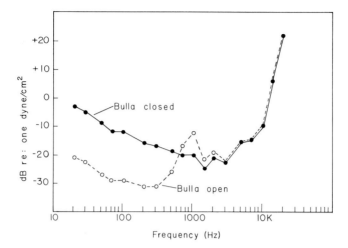

FIG. 1-6. Comparison of frequency functions measured under two conditions: open bulla versus closed bulla. The data shown were obtained from one guinea pig. The following procedure was repeated at each frequency:

(1) The sound intensity required to elicit 1 μV of cochlear potential was measured with the bulla open. (2) The same measurement was carried out with the bulla sealed. (3) The bulla was reopened, and the reading in step (1) verified. An identical procedure was carried out during calibration of the sound system following the data collection. The results, which were highly reliable both within one animal and between different animals, demonstrate that the open bulla condition exaggerates the sensitivity of the cochlea at the low frequencies.

Similar open versus closed bulla data are also available for the cat (Guinan & Peake, 1967) and the kangaroo rat (Vernon, Herman, & Peterson, 1971).

followed during calibration, which was done with a calibrated probe tube inserted into the sound cannula in the ear. It is evident that there exists a difference between the open and closed bulla. Leaving the bulla open appears to exaggerate the ear's sensitivity to low frequencies. Therefore, we conclude that a more faithful recording from the cochlea will be given by a closed bulla. Since it requires considerable effort to achieve proper sealing of the bulla, leaving it open may be the easiest procedure. In addition, an open bulla allows visual checking of the round window electrode placement. The main point here is that each investigator must clearly state his procedure.

B. *Intensity Function (Often Called Input–Output Function)*

It is widely known that mammalian hearing encompasses a very large range of sound intensities, perhaps over a millionfold. It is assumed, although it has not been established, that the peripheral mechanism must be capable of at least the same dynamic range. Technical limitations have thus far prevented measurement of the cochlear potential at very low sound levels. Indeed, it has never been measured at sound levels corresponding to hearing thresholds. Thus, it is not known whether the cochlear potential has an absolute threshold. It is known, however, that the voltage output of the cochlea is directly proportional to stimulus intensity up to the maximum output of the ear, and is almost perfectly linear over most of that range (Fig. 1-7). Beyond the maximum point, the output of the ear decreases with further increases in the sound intensity. This phenomenon is referred to as "bend-over." Excessive or prolonged stimulation at bend-over intensities usually produces damage to the ear. Inability to maintain a high level of output is seen near bend-over, and this fact can be used as an indicator that the ear's capabilities are being strained. In fact, in measuring the input–output function, great care should be exercised in exposing the ear only for brief intervals to the loud sounds. Where it is important to maintain a normally functioning ear, while at the same time measuring intensity functions, after each such determination the absence of injury should be verified by demonstrating that the sound intensity required to produce, say, 1 μV output, has not changed. (Utilizing careful procedure, it is not unusual to find 1 or 2 dB improvement upon rechecking the 1 μV level, for reasons as yet not understood.)

IV. Recording the a.c. Cochlear Potential

A. *Anatomical Considerations*

In order to talk about surgical and other procedures involving the middle and inner ear of our various experimental animals, a little anatomical terminology is useful. The middle ear extends from the tympanic membrane or eardrum to the cochlear walls, thus including the ossicles (malleus, incus, and stapes) the middle ear muscles (tensor tympani and stapedius), and all the space around them. The inner ear is synonymous with the labyrinth, which includes both the vestibular organ and the cochlea. Only the cochlea is auditory in function. The bone of the inner ear is called the bony labyrinth. In some animals (carnivores and rodents, among others), the cochlea projects into an air-filled bony cavity called the bulla. In these animals the part of the bony labyrinth encasing the cochlear structures is called the cochlear capsule, and in small animals, such as the guinea pig, this capsule may be very thin.

The cochlea (the Greek word for snail) is a coiled tube subdivided longitudinally into three major compartments (see Fig. 1-1). The scala tympani

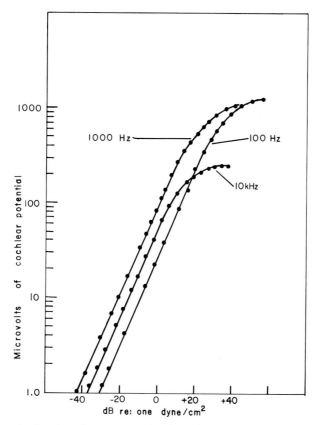

FIG. 1-7. Intensity functions (input–output functions) for guinea pig. The data shown were collected from a round window electrode in one guinea pig. Cochlear output in microvolts is plotted as a function of sound intensity for three different frequencies, one low, one intermediate, and one high. The shape of the intensity function is characteristic, being linear until the region of bend-over is approached. The departure from linearity is accompanied by progressive distortion of the waveform. Further increases in sound intensity would have caused a decrease in cochlear output (the "bend-over" phenomena) but were avoided so as not to damage the ear. The animal from whom these data were collected had a maximum output around 1050 μV; it is not unusual to find guinea pigs capable of putting out 1500–2000 μV. Similar measurements seem to indicate lower maximum outputs for animals higher on the phylogenetic scale.

and the scala vestibuli are filled with perilymph, which is very similar to cerebrospinal fluid. Between these two chambers is the scala media, also called the cochlear duct, which contains endolymph. The ionic composition of endolymph is quite different from that of the perilymph. Perilymph is like extracellular fluid in that is has a relatively high concentration of sodium with relatively little potassium. The situation is reversed for the endolymph, where the concentration of potassium is high and that of sodium is low.

The outline of the cochlea narrows toward the apex. Within the apex, the scala tympani and scala vestibuli communicate with each other through a passage called the helicotrema. The cochlear portion of the bony labyrinth contains two apertures opening into the middle ear cavity: one is the oval window, which leads into scala vestibuli of the basal turn of the cochlea, and which is completely filled by the footplate of the stapes; the other is the round window, which is situated in the scala tympani of the basal turn and is sealed by a thin elastic membrane. It is this round window membrane which provides a convenient site for recording the electrical activity of the cochlea.

The cochlea contains a complex sensory apparatus, the organ of Corti (see Fig. 1-2). Within the organ of Corti there are, in addition to supporting structures, specialized sensory epithelial cells called hair cells. It is thought that these cells are the transducers which convert sound energy (mechanical vibrations) into the form of energy (chemical or electrical) which is appropriate for activating the afferent nerve fibers of the cochlear or auditory nerve. The terminals of the nerve fibers synapse on the hair cells. The mechanism by which these terminals are activated in response to sound is unknown, and this question remains one of the most important and challenging in the field of auditory physiology. However, one possibility which cannot be ruled out, and in fact at present remains untested, is that the alternating cochlear potential may act directly to produce depolarization of the auditory nerve terminals.

The peripheral processes of the auditory nerve fibers run from their synapses with the hair cells to cell bodies in the spiral ganglion, which is contained in the bony spiral lamina. This lamina lies twisted around the central core of the cochlea, the modiolus. The central projections of the ganglion cells extend into the modiolus where they unite to form the cochlear nerve. The cochlear nerve fibers enter the brain stem by way of the internal auditory meatus in the temporal bone.

B. Choice of Experimental Animal

The cochlear potential has been recorded in a very wide range of animals, including not only many different mammals (Dalland, Vernon, & Peterson, 1967; Peterson & Heaton, 1968; Peterson, Heaton, & Wruble, 1969; Peterson, Wruble, & Ponzoli, 1968) but also birds (Schwartzkopff, 1968), reptiles (Wever, 1967), and amphibians (Strother, 1962; Wever & Vernon, 1956). Despite the diversity of animals which have been studied, auditory electrophysiology has tended to concentrate heavily on two species in particular: the guinea pig and the cat. Therefore, our discussion of surgical and other procedures will be concerned with these two animals.

The guinea pig, by virtue of its anatomy, seems to have been designed expressly for the purpose of studying the inner ear. Unlike many other animals, the guinea pig has an extremely accessible cochlea. In primates, for example, this is not the case, for there the cochlea is deeply embedded in the petrous part of the temporal bone (the hardest bone in the body). In the guinea pig, by contrast, a large portion of the cochlea projects clear of this bone and into the auditory bulla, a hollow balloon-shaped cavity of the skull. As the bulla requires the simplest surgery to expose and open it, it is an easy matter to gain access to the middle ear as well as the cochlea of the guinea pig. Another point in the guinea pig's favor is that its cochlea contains $3\frac{1}{2}$–5 turns (with considerable individual variation), unlike the human cochlea which has only $2\frac{1}{2}$ turns. Thus there is a wider variety of intracochlear electrode placements available in the guinea pig. Furthermore, the cochlear bone which is accessible for drilling is very thin (about 20 μ to 100–150 μ, thickest toward the basal turn). Like all rodents, the guinea pig is resistant to postoperative infection, so chronic experiments can be undertaken with a minimum of sterile precautions. One disadvantage of the guinea pig is that his respiratory system is easily compromised under anesthesia. This problem and some possible solutions will be discussed later.

It should not be assumed that the guinea pig is lacking in other problems. Such is far from being the case. For example, an infrequently mentioned fact is the ease with which ruptures in the region of the round and oval windows can be produced. The resulting fluid leaks render good cochlear potential records extremely difficult to obtain.

The cat is the second animal used very frequently for auditory electrophysiology, probably because the cat has been traditionally such a common choice for neurophysiological study. Certainly the cat's convenient size, hardiness under anesthesia, ability to be trained, and durability in long-term studies make it well suited to auditory investigation. Like the guinea pig, the cat possesses an auditory bulla, thus rendering surgical exposure of the round window relatively easy. However, the cochlea in the cat is not as accessible as that of the guinea pig, being enclosed in a considerable amount of dense bone. In addition, the bony capsule of the cochlea itself is significantly thicker in the cat (being measured in millimeters rather than in tens of microns), thus making drilling and insertion of intracochlear electrodes more difficult. Therefore, the cat is not so convenient an animal when placement of numerous intracochlear electrodes is planned.

An animal that perhaps should be used more extensively than it has been so far is the kangaroo rat, *Dipodomys*. It outdoes the guinea pig in respect to accessibility of the bulla and in having a large cochlea with many turns.

In addition, the skull in this animal is so thin as to impose practically no impediment to surgery, and thus it would seem to be a potentially useful animal in which to study other portions of the auditory nervous system. For chronic studies it also has the advantage of resistance to infection, and in addition may be more easily trained than the guinea pig. Incidentally, for the same reasons the chinchilla is currently being used in several laboratories.

C. Surgical Procedures

General anesthesia is achieved using intraperitoneal (I.P.) injection of diallylbarbituric acid with urethane for acute cats and guinea pigs, and pentobarbital sodium for chronic animals.[5] The animal's head and throat regions are shaved. In our laboratory a tracheal cannula is routinely employed in acute experiments and is especially necessary for guinea pigs.

It is extremely convenient, and essential in experiments which involve drilling into the cochlea, to have the animal's head rigidly immobilized. Considering that we must always leave at least one, and often both, of the ears available for auditory stimulation, the choice of a head holder becomes an important decision. We have come to prefer a nontraumatic type, shown in Fig. 1-8, which is adaptable to either acute or chronic preparations, either in guinea pig or cat. The head holder consists of a small piece of metal drilled to accept bolts in either end. The metal piece is attached to the animal's skull, using skull screws and dental acrylic cement. The head holder is then bolted to a rod which can be clamped in any desired position.

1. Guinea Pig

It is easy to palpate the auditory bulla prior to any surgery. It can be felt as a hard, convex, bony structure just posterior to the external meatus and extending inferior to the meatus under the corner of the mandible. For recording from the round window, a postauricular opening of the bulla is used, as described with full details in Fig. 1-9.

When it is desired to drill into the basal turn of the cochlea, in order to insert one or more intracochlear electrodes, it is often possible to do so through a postauricular opening of the bulla, particularly if the opening is enlarged as much as possible. If intracochlear electrodes are to be placed in the middle or apical turns of the cochlea, however, a more ventral approach to the bulla is used, as described in Fig. 1-10.

---

[5]Doses as follows: Pentobarbital sodium (Nembutal^R), 32 mg/kg; diallylbarbituric acid, 100 mg/ml; urethane, 400 mg/ml; dose = 0.6 ml/kg. Dial^R-urethane should not be used in chronic preparations because of its adverse effect on white blood cells. As this anesthetic is no longer available commercially, a formula for its preparation is given in the Appendix.

1. ELECTROPHYSIOLOGY OF THE COCHLEA

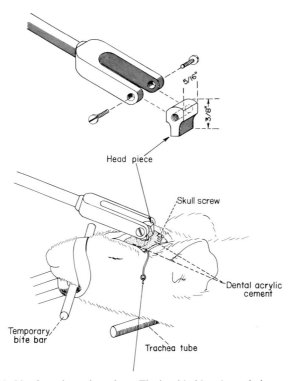

FIG. 1-8. Head holder for guinea pig and cat. The head holder pictured above was made from brass. The small headpiece is affixed to the skull with dental acrylic cement (cold-cure cements are recommended). Two skull screws are first set into holes drilled through the skull. They serve as anchors for the dental acrylic. While the acrylic is hardening it is necessary to keep the head level and immobile; a temporary bite bar held with a rubber band serves this purpose. After the acrylic has set thoroughly, the Y-shaped holder is bolted to the headpiece. Rotation of the head holder around the bolt axis is prevented by tightening the bolts into the threaded holes in the headpiece. The long shaft of the Y bar is fastened into a clamp in the desired location. The bite bar is then no longer needed. In our experience this head holder is well worth the time and effort it requires as it provides very rigid and stable fixation while at the same time allowing great flexibility in arrangement.

Drilling into the cochlea of the guinea pig is a delicate job because the cochlear walls are so thin. Particularly near the apex, care must be taken not to fracture the cochlear capsule. Very small (0.002–0.010 inch) pivot or twist drill bits are convenient for drilling; these can be obtained from manufacturers of jewelers' tools. They can be used in any small pin vise, but it is better to have a long and slender handle in the vicinity of the cochlea, so as not to obscure the view of the spot being drilled, either by a bulky drill handle or by one's own hand. A long slender shaft to hold the drill

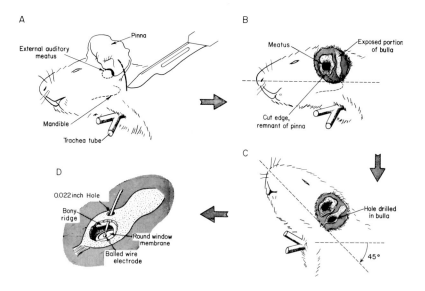

Fig. 1-9.

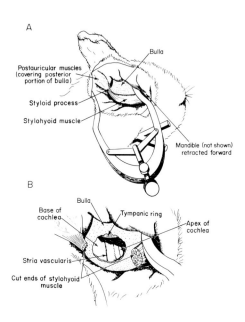

Fig. 1-10.

1. ELECTROPHYSIOLOGY OF THE COCHLEA

Fig. 1-9. Round window surgical approach for electrode in guinea pig—postauricular approach. (A) A curved incision is made in a vertical direction about ½ cm posterior to the base of the pinna. A small hairless area directly behind the pinna is often found in guinea pigs; the incision should run through this bald spot. In acute experiments it is convenient to remove the entire pinna at this stage, in order later to insert the sound tube into the external auditory meatus with optimal visualization. (B) With pinna removed, a very smooth, thin layer of muscle can be seen overlying the bulla. This tissue is cleared away using blunt dissection, revealing the bony surface of the bulla. The bone of the bulla has a characteristic stippled surface texture. (C) An opening into the bulla can easily be made with any sort of drill, using a small burr, or even with the point of a scalpel, rotated by hand. In order to view the round window, the animal must be turned approximately 45° in the horizontal plane. The proper orientation directs the experimenter's view in a line passing over the animal's shoulder and toward the noise. Once the bulla is open, an operating microscope is needed to provide the $10-25 \times$ magnification required for further stages of the work. There are individual variations in internal bony ridges of the bulla which sometimes overhang and obscure the round window. It is desirable to be able to tilt and bend the head and neck at this point and thus obtain the clearest view of the round window membrane. (D) Placing the electrode in contact with the round window membrane requires delicate control, in order not to tear or rupture the fragile membrane, thus releasing perilymph. A micromanipulator is essential for this step. In addition, it is helpful to drill a small hole of about 0.022 inch or so (just large enough to admit the balled ending of the electrode wire) close to the opening through the bulla. The micromanipulator is used to advance the electrode through this hole until it rests gently upon the round window membrane. It is advisable to anchor the electrode in some way to protect it from inadvertent movement. For this purpose dental acrylic can be used to fill the passage hole through the bulla, thus forming a solid plug adhering to the bulla surface and encasing the electrode wire for a short distance. The bone must be dry in order to obtain a good bond.

Fig. 1-10. Surgical approach for intracochlear electrodes in guinea pig—ventrolateral (submandibular) approach. (A) The animal is placed so that the ventral surface of the neck faces the experimenter. An incision is made starting at about the posterior-inferior corner of the pinna downwards toward the midline. The ventral approach is both deeper and more vascular than the postauricular approach; therefore retractors are needed, and there may be major blood vessels to be ligated. The operating microscope greatly facilitates isolation and tying-off of troublesome blood vessels. The mandible is the major landmark. Blunt dissection is directed toward separating tissues over it and posterior to it, always working in a direction aimed at a point about ½ cm inferior to the meatus. Self-retaining retractors are used to force the mandible forward and out of the way, for left in its normal position the mandible tends to cover the target area on the bulla. Sometimes in a large, old animal the mandible cannot be retracted forward sufficiently, and then it is necessary to break and remove a short segment of it. As there are numerous blood vessels around and deep to the mandible, this must be done with caution. Eventually blunt probing will reveal the hard, curved surface of the bulla, about ½ cm inferior to the meatus. The styloid process of the bulla gives rise to the stylohyoid muscle with hyoid cartilage; these must be snipped away in order to uncover the bulla cleanly. Hemostasis should be achieved before opening the bulla. (B) The bulla is opened by drilling. The upper portion of the bulla just below the meatus should be avoided, as the tympanic ring with the attached tympanic membrane lies directly beneath it, and the sound conduction of the ear can no longer be considered normal if the tympanic membrane has been ripped or otherwise damaged, however slightly. After an opening has been made in the bulla, the animal's head may need to be turned even further in a throat-upward direction, in order to bring the cochlea into view. The dark outline of stria vascularis can be seen through the bone of the cochlea. This can be used as a guide when holes are to be drilled into a particular scala. In each turn the stria vascularis separates the scala vestibuli from the scala tympani with the latter being in a posterior direction. (See p. 6.)

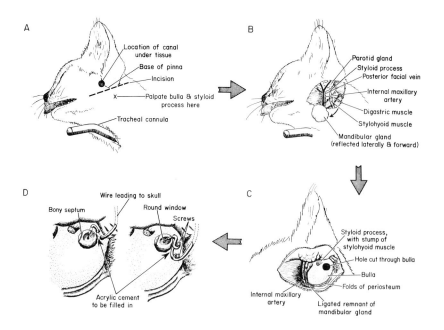

FIG. 1-11. Surgical approach to the round window in the cat. (A) The incision is made on a line drawn from the corner of the mouth to the bottom of the pinna. The bulla and the styloid process (a pointed projection just superior to the curved bone of the bulla) can be palpated before any cut is made. The approximate location of the ear canal is indicated by a shaded circle. In chronic preparations an opening should be cut through the skin and underlying tissue into the cartilaginous canal at this spot, in order to gain clear access to the external auditory meatus for introduction of the sound tube. The cut edges of the wound should be sutured. In acute preparations the entire pinna can be removed for easier access to the ear canal. The head holder is not shown here (see Fig. 1-8) although we use it routinely for cats. (B) Blunt dissection around the mandibular gland frees it for reflection laterally and forward. The posterior facial vein should be ligated and cut in acute preparations. In chronic preparations, loose ligatures can be slipped around it so that it can be pulled anteriorly, out of the way of the surgical exposure. The internal maxillary artery should be carefully avoided. The muscles overlying the bulla are separated and stripped away with further blunt dissection. (C) The mandibular gland has been excised, after ligating its blood supply. The stylohyoid muscle and hyoid bone are cut close to their attachment to the bulla, and a short segment removed. In this view the posterior facial vein has been removed. The periosteum, the tough membrane sheathing the bulla, is slit and pushed to the sides. In order to visualize the round window, a hole is cut through the bulla just posterior to the styloid process. In chronic preparations the cap of bone removed should be saved for later replacement (it should be kept moist, preferably by laying it in an out-of-the-way portion of the surgical exposure where tissue fluids will continue to bathe it). (D) This enlarged view of the bulla shows two ways of immobilizing the round window electrode in chronic preparations. One method is to drill several small holes through which the wire is led and pulled taut. It is essential to maintain the integrity of the wire insulation, therefore we use fine polyethylene tubing as an additional protective sheathing around our Teflon-coated wire. The second method involves threading two stainless steel screws into holes drilled to accept them tightly. The screws form a rigid structure around which the wire is then wound tightly before it is led under the tissue of the skull to the permanent skull mount containing connecting pins. Though not shown here, dental acrylic cement is then

bit can be made from a length of stainless steel hypodermic needle tubing, into which the bit is cemented.[6]

The smaller the hole drilled, the less the likelihood of damage to inner structures of the cochlea, and the less the leakage of perilymph out of the cochlea. Although microelectrodes with correspondingly minute holes may tend to minimize these dangers, and may be especially desirable for drilling into scala media or when a number of holes must be drilled into one cochlea, we have found that holes of 0.004 inch in the apex and as much as 0.008 inch in the basal turn are well tolerated in acute experiments. These hole diameters are chosen to fit snugly around the platinum electrode wire we use for intracochlear recording. Further details regarding insertion of intracochlear electrodes are given below in the section on electrodes.

It is essential that the condition of the cochlea be ascertained before and after each drilling. This is an important routine which we do by first determining a frequency function with a round window electrode, prior to drilling. That function is measured again after the cochlear hole is drilled and the intracochlear electrode inserted. It can be shown that with proper care, drilling need have no deleterious effects (refer again to Fig. 1-5).

2. CAT

Again, it is helpful to palpate the auditory bulla before making an incision. To reach the round window of the cat, the bulla must be approached more ventrally than is the case for the guinea pig. Thus, the head holder should be clamped in such a way that the head and neck of the cat are turned slightly upward (see Fig. 1-11). It is particularly important to remove the pinna when working on acute cats, as the cat's pinna contains intricate folds, making the external canal so tortuous it is difficult to insert the sound tube properly. In order to reduce bleeding and provide additional local anesthesia, 2% xylocaine with epinephrine (1:10,000) can be injected subdermally all around the base of the pinna before excising it.

D. Chronic Preparations

Aseptic surgical procedure is not required with guinea pigs or kangaroo rats, as these animals have unusual resistance to infection. With cats, it is

[6] A power-driven MicroDrill is made by the Circon Co. of Goleta, California.

---

used to fill in all around the wire and the screws at their contact with the bulla. (The bone must be dried thoroughly first.) When intracochlear electrodes are to be used, the bulla must be opened more widely, to allow for drilling into the cochlea, and the bony septum which partially divides the interior of the bulla may have to be removed. In chronic preparations the cap of bone should be replaced as neatly as possible. Surgical adhesive can be used to hold it in place. The periosteum should be pulled back over the bulla if possible, and the muscle layer sutured loosely back together before closing the skin incision with either sutures or wound clips.

not necessary to use strict sterile techniques unless the surgery is very protracted or unless the cochlea is to be drilled. Cats recover very well from rapid implantation of a simple round window electrode, especially when the bulla is open for only a short time. Antibiotics (ampicillin sodium, 250 mg) should be given to cats the day of surgery and the day following. Antibiotics are likely to provoke allergic reactions in guinea pigs and therefore should not be given to these animals (Eyssen, de Somer, & Van Dijck, 1957).

The most demanding feature of a chronic preparation is that great care must be taken to immobilize the electrode inside the ear. One way to achieve this is shown in Fig. 1-11. Another complication in chronic recording is the action of the middle ear muscles, which can reduce the amplitude of the cochlear potential by as much as 20 dB. This action is brief and may be triggered by a variety of stimuli, in particular by loud sounds. In most cases this activity is little more than an annoyance during recording. Nevertheless, when it is desirable to be free of such variability in chronic preparations, it is not a difficult matter to sever the middle ear muscles.

E. Physiological Variables Affecting the Cochlear Potential

It is important to maintain the physiological environment in which the cochlear potential is supported. The two features of that environment most affected by general anesthesia are respiration and body temperature.

1. Cat

Only a minimum of attention is necessary to maintain normal respiration and temperature even under deep anesthesia. Normal body temperature (37.5°C) can be easily maintained with only occasional aid from a heating pad. Careful monitoring via a rectal probe is necessary to ensure against overheating.

Respiration usually offers no problem in the anesthetized cat. Nevertheless, it is a good idea to cannulate the trachea in acute preparations, especially those of long duration, so that aspiration is possible if needed.

2. Guinea Pig

The situation is far different in the guinea pig, which seems compromised by almost everything. In this animal there appears to be a very fragile physiological balance, and the same fragility characterizes his ear. It is not uncommon to see a marked decline in the output of the ear during the course of an experiment, and the causes of the decline are not easily found. However, since the depressant effect of general anesthesia on respiration

1. ELECTROPHYSIOLOGY OF THE COCHLEA

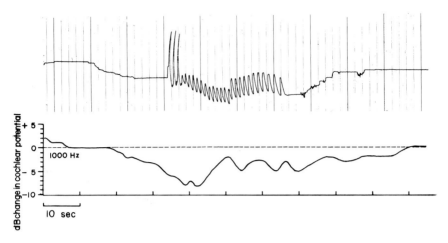

FIG. 1-12. Transient respiratory effects on the cochlear potential. *Upper trace:* polygraph record of unassisted respiration in an anesthetized guinea pig, obtained by fastening a strain gauge to the outside chest wall. This record shows an example of Cheyne–Stokes breathing, a form of depressed respiration characterized by long apneustic periods (in which the trace remains nearly flat) interspersed with short periods of vigorous breathing (see the three initial gasps followed by rapid breathing). *Lower trace:* simultaneous recording of cochlear potential in response to a sustained tone of 1000 Hz. The cochlear potential began to decline about 17 sec before the onset of the breathing, reaching its most depressed level about 9 sec after breathing had begun. The potential then gradually recovered, returning to normal about 33 sec after the cessation of breathing. The coincidence of changes observed in the two records, such as that shown above, was observed repeatedly using many different tonal stimuli.

and temperature regulation is especially severe in the guinea pig, maintenance of these two functions seems the logical place to begin in the attempt to keep the ear from deteriorating.

a. RESPIRATION. It is well known that anoxia can severely reduce the a.c. cochlear potential (Rice & Shinabarger, 1961; Butler, 1965). Our own observations indicate that even minor changes in respiration may affect the magnitude of the cochlear potential. Figure 1-12 shows a case in which we observed Cheyne–Stokes respiration,[7] during recording of the cochlear potential in a guinea pig whose breathing was unassisted. It can be seen that during each breathing cycle the cochlear potential showed a reversible alteration in amplitude of about 10 dB.

If the cochlear potential is capable of such a sensitive response to changing respiratory levels, we must ask how well it can tolerate the respiratory depression produced by anesthesia. We believe that in fact animals last longer, and their ears remain more stable, when artificial respiration is

---

[7] An abnormal respiratory rhythm characterized by periods of rapid breathing alternating with periods of no breathing.

used. Therefore, we use a positive-pressure respiration pump with variable rate and stroke volume, adjustable to different sizes of guinea pigs.[8]

Unfortunately, the use of artificial respiration raises problems of its own. It is very difficult to determine the proper level of ventilation for an animal as small as the guinea pig. Ventilation graphs relating body weight to respiratory minute volume[9] are only moderately useful, as there is no practical way to measure the actual volume of inspired air (tubing lengths vary from one setup to another, thus affecting the dead air space as well as effective pressure), and the pump itself may not be calibrated accurately.

Despite these difficulties, our experience supports the following generalizations: It is probably easier to overventilate than to underventilate, so as a beginning approximation the volume and rate should be set on the low side. If the pump settings are too low, the animal will start to fight the rhythm, and obtain extra air by "double breathing." If the pump is set too high, the animal's own natural breathing reflexes will be suppressed and it will be difficult to get him to resume breathing on his own if he is removed from the pump. With the proper setting the animal will resume natural breathing almost immediately after he is taken off the pump. Continued observation in a succession of animals can thus provide a basis for developing a working approximation of the correct ventilation level.

An alternative approach to the question of how to set the proper respiratory level is provided by an interesting feature peculiar to the guinea pig. When anesthetized, its middle ear muscles contract spontaneously and periodically. These contractions are clearly visible as displacements of the round window membrane and also of the eardrum, and it appears that the contractions are linked in some way to the adequacy of respiration (Meikle, Brummett, & Vernon, in preparation). We have found that we can eliminate the contractions simply by placing the guinea pig on artificial respiration of sufficient rate and volume. A useful procedure therefore is to start the guinea pig at a pump rate and volume low enough for the spontaneous contractions to occur, and then to increase rate and/or volume until the contractions just disappear. In this way the inconvenience of periodic movements of the round window is avoided; at the same time we do not believe there is much risk of hyperventilation.

b. Temperature. We have attempted to analyze carefully the reports already in the literature which show decreases in the cochlear potential when body temperature drops below normal (38.5°C for the guinea pig).

---

[8] Small Animal Respiration Pump, Model 763, Harvard Apparatus, Millis, Massachusetts; or Model V5KG Physiograph, E & M Instrument Co., Houston, Texas.

[9] Such as that of Kleinman, L., and Radford, E. P. Tidal volume versus body weight and rate for laboratory mammals in resting state. Prepared for Harvard Apparatus Co., Dover, Mass.

Although many experiments have shown that N_1 is highly sensitive to cooling (Harrison, 1965; Kahana, Rosenblith, & Galambos, 1950; Fernandez, Singh, & Perlman, 1958), both in amplitude and latency, the cochlear potential data are more inconsistent. In a careful study using a number of guinea pigs, Fernandez, Singh, and Perlman (1958) obtained less than 3 dB decrease in the cochlear output when the animals were cooled as much as 12°C below normal. Since Fernandez *et al.* (1958) did not use artificial respiration, it is possible the effect they observed may have been caused by respiratory depression which is well known to result from hypothermia. Other work by Gulick and Cutt (1960, 1962) has shown much larger decrements in the cochlear potential during cooling. However, Gulick and Cutt also did not use artificial respiration, and since they did not show reversibility of the downward trend except in one instance, it is equally plausible that their results were owing to a deteriorating preparation rather than to the effects of temperature *per se*.

It is apparent that the question of temperature effects on the cochlea is a difficult one to study because of the possibility of an interaction between body temperature and respiration—and as we have already indicated, respiration is difficult to control. Moreover, the question of where to measure "body temperature" adds a further complication. Most experimenters rely on measurements of rectal temperature, yet our own observations indicate that intracochlear temperature is quite independent of rectal temperature (Fig. 1-13). Our data show that if the bulla is sealed, the temperature within the bulla is a good indicator of intracochlear temperature. Therefore, it seems advisable to use a thermal probe in the bulla as a more reliable temperature monitor in cochlear potential experiments.

We consider the question of a temperature effect on the cochlear potential still an open one, even though in our experiments where artificial respiration was provided we found no change in cochlear output down to 28°C. Although we tested the cochlear output at a wide range of frequencies, we have not studied the intensity function during cooling. Our temperature experiments have used the 10 μV level of cochlear output, whereas the temperature effect reported by Fernandez *et al.* (1958) was obtained using cochlear outputs of 600 μV or more. Thus it is possible our results would be different were we to use a higher level of cochlear potential, particularly if the level were close to bendover.

It is clear that a definitive study of temperature effects upon the ear over a range of frequencies and intensities is needed. In the absence of more complete data, the safe course is to maintain head temperature at or near the normal 38.5°C.

c. Accumulation of Fluid in the Bulla. Even where temperature and respiration are adequately controlled, it is not uncommon to see gradual

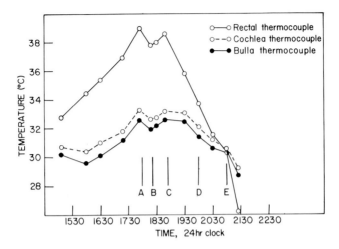

FIG. 1-13. Comparison of body and intraaural temperatures in the guinea pig. Temperatures measured with small (0.008 inch) chromel–constantan thermocouples in one guinea pig. The intracochlear thermocouple was sealed just inside a hole drilled in scala tympani of the basal turn. After insertion of the bulla thermocouple, the hole in the bulla was sealed. The temperature record begins with the animal below normal. Subnormal temperatures are typical of anesthetized guinea pigs if no external heat is supplied. Heating was begun at 1530 and continued until the rectal temperature had reached 39°C. Note that the bulla and cochlear temperatures never rose higher than 33°C. At A, heat was removed. An immediate decline in temperature was observed at all thermocouples. At B heat was restored briefly, demonstrating the rapid following ability at all points. At C the heating pad was turned off but remained covering the animal. At D the animal was uncovered, and at E ice packs were placed around the body. Thereafter a very rapid drop in rectal temperature accompanied a more gradual decline in intraaural temperatures. Three things are evident from these data: (1) Temperatures in the ear tend to fluctuate over a smaller range than does body temperature; (2) intraaural temperatures can be far different from the temperature measured by a rectal probe; (3) the bulla thermocouple provides a good monitor of intracochlear temperature.

deterioration of the cochlear potential in the guinea pig. The single most likely factor in this decreasing output is probably the phenomenon of fluid accumulation in the bulla seen so often in this animal. The fluid typically builds up gradually, with a time course similar to the cochlear down trend. Many investigators take the "mopping up" of fluid for granted, and appear to accept it as a necessary evil. In our experience, mopping out of a flooded bulla can easily improve the observed cochlear potential level by as much as 10–20 dB. There are at least two ways in which fluid in the bulla could be expected to exert a deleterious effect: (1) it might load the ossicular chain and thus decrease the efficiency of sound transmission, or (2) it might partially "short out" the active electrode.

If the loss is due to a loading action, then all electrode placements (intracochlear as well as round window) should suffer the same amount of loss.

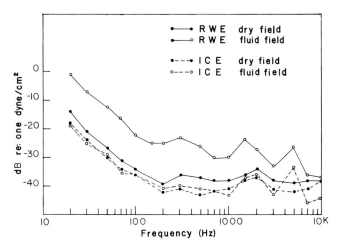

FIG. 1-14. Comparison of frequency functions with and without fluid in the bulla. RWE, Round window electrode; ICE, Intracochlear electrode (basal turn). The RWE was a ball formed on the end of 0.008-inch platinum–iridium wire. The ICE was a 0.004-inch platinum–iridium wire, insulated with Teflon except where its exposed tip was sealed into the hole drilled into the cochlea. As shown above, frequency functions plotting the 1 μV level of cochlear potential were obtained for both electrodes under both conditions (dry bulla versus fluid-filled). Measurements under the fluid-filled condition were made first, after which the interior of the bulla was carefully blotted and data for the dry condition were obtained. The RWE was in a position to be exposed to the fluid buildup and its shorting-out effect. The ICE, on the other hand, was well sealed within the cochlea, where the accumulation of fluid could not make electrical contact with it. The data indicate that in this case, the accumulation of fluids acted to produce depression only in the RWE record—the ICE record appeared unaffected by the presence or absence of fluid. The source of the fluid which accumulated was not determined, beyond ruling out the possibility of a defect in the round window membrane.

Such a case is in fact often seen when fluids have accumulated.

On the other hand, the possibility of shorting-out produced by fluid in the bulla could influence only round window electrodes, since intracochlear electrodes are sealed within the cochlea. Figure 1-14 illustrates the case where shorting out is probably the primary reason for the losses caused by fluid. Note that only the recordings from the round window electrodes are depressed in the fluid-filled condition, while the intracochlear recordings remain unaffected.

Considering the detrimental effects of fluid accumulation, it would be very helpful if we knew how to prevent it. In our opinion the round window membrane is even more easily torn then is usually supposed, and we think a frequent source of the fluid is perilymph leaking out from the cochlea. However, there are times when it is almost certain that no damage has occurred to the round window membrane, and yet fluid comes from somewhere. We have ruled out secretions from salivary glands lying near (outside) the bulla. Other possible candidates are cerebrospinal fluid (perhaps coming from the facial nerve canal), and canalicular fluid oozing from cut surfaces of bone. At this time we can offer no cure for the problem caused by these fluids, other than to emphasize the need for extreme delicacy in touching the round window membrane. Some investigators use a wick or small tube inserted into the bulla in an attempt to provide a drain for fluid. For the most part, however, it should be realized that a certain percentage of cochlear experiments in the guinea pig are doomed to be complicated by the fluid problem.

F. The Acoustic Stimulus

1. SOUND-PRODUCING SYSTEMS

The quantification of the cochlear potential depends upon quantification of the sound stimulus. Although this seems obvious, there are a variety of ways in which quantification of the sound stimulus can go astray if proper care is not taken.

Our comments here will be confined principally to pure tones, for several reasons. First, pure-tone stimuli have proved to be a powerful tool for analysis of the functioning of the ear. There is considerable theoretical basis for this, as pure tones are the "elements" from which complex sounds are formed. The mathematical treatment known as Fourier analysis has demonstrated that all complex sounds can be analyzed and completely described in terms of components which are pure tones of proper frequency and intensity. The second reason for concentrating on pure-tone stimuli is that the techniques for the generation and measurement of pure tones are relatively simple. Complex sounds introduce many difficulties into the measurement of intensity levels, so that it may be extremely difficult to specify a given complex sound precisely. Furthermore, we feel that the state of knowledge regarding the function of the ear is not yet ready to advance beyond the simple cases.

First of all, what is meant by a pure tone? Elementary physics has taught us that the pitch of a tone which sounds "pure" to the ear (such as that generated by a tuning fork) is the sensory correlate of the frequency of oscillation of the tuning fork itself, of the air surrounding it, and of the ear structures finally stimulated. Frequency of oscillation is thus one of the parameters of stimulus tones that is important to specify and measure. These oscillations are of a sinusoidal character, and Fourier analysis has shown that any deviation from a perfectly sinusoidal shape is exactly equivalent to the introduction of additional frequencies. The ear is very sensitive to such adulterations of a pure tone, as can be verified both by listening to

1. ELECTROPHYSIOLOGY OF THE COCHLEA

pure versus complex tones, or by observing the electrical output of a cochlea in response to such stimuli.

In order to verify that a given tonal stimulus is in fact pure, acoustical measurements of the stimulating sound must be carried out. We will reserve discussion of how to do this for the section on sound measurement. However, we wish to emphasize here that it is *not sufficient* to equip oneself with a tone generator (an oscillator, for example) and assume that its nominal frequency setting is accurate and that its output signal remains pure throughout transmission through the sound-generating system. Both parameters, exact frequency and tonal purity, must be verified externally if one wishes to describe the stimulating sound adequately.

2. Equipment to Produce Tones

The following items constitute a tone-generating system: an oscillator to produce the sinusoidal signal, an audio or power amplifier to amplify the signal sufficiently to activate the loudspeaker, attenuators to adjust sound intensity level, and a speaker or other transducer (such as earphones). In addition, an electronic counter should be connected so as to monitor the exact frequency output of the oscillator. For reasons to be discussed (see Section IV,F,3,b), in our laboratory we require that frequency be specified with an accuracy of ±1 Hz. Figure 1-15 schematizes these and several additional elements which we will discuss in detail.

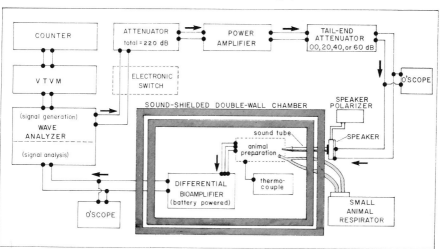

FIG. 1-15. Schematic of stimulating and recording apparatus for typical cochlear potential recording. The function of the oscillator (that of generating the sound signal) is performed by the wave analyzer. As the wave analyzer also performs frequency analysis of the returning output signal from the ear, the "flow pattern" of the experimental setup begins and ends with the wave analyzer.

a. OSCILLATOR. The oscillator must put out those frequencies appropriate to the hearing of the particular animal under study. For human subjects a range of 20 Hz to 20 kHz is sufficient, whereas for bats it is necessary to encompass higher frequencies, up to approximately 100 kHz. Cats are sensitive to frequencies at least up to 60 kHz, and guinea pigs[10] about the same.

In our schema the oscillator is contained in the wave analyzer, as it is a great convenience to have the analyzing circuitry "slaved" to the signal generating source (more will be said about this in Section IV,G). The signal output of the oscillator or wave analyzer is monitored by an external voltmeter, to ensure at the outset that the amplitude of the signal is known and constant.

b. ATTENUATORS. Adjustability of sound intensity levels is achieved by the decade attenuators, which provide convenient 1 and 10 dB steps. As they are not intended to carry heavy loads, electrically speaking, they are placed before the power amplifier.

c. AMPLIFIER. The power amplifier should be (1) relatively "flat" over the frequency range required (i.e., not varying its output with frequency), and (2) powerful enough to activate the desired transducer. Most, if not all, commercially available amplifiers produce large amounts of self-generated noise. This noise can be heard when the attenuators are set very high so that little sinusoidal signal is reaching the amplifier. Amplifier noise sounds like a low but distinct hissing containing a wide band of frequencies. Because even very expensive and high-quality audio amplifiers do this, it is desirable to insert another attenuator into the circuit following the amplifier. In our schema this attenuation is accomplished by a "tail-end attenuator" which can provide 00, 20, 40, or 60 dB of attenuation. This attenuator must be able to tolerate a heavy power load, as it is exposed to the full output of the power amplifier. In calculating the total attenuation applied to the electrical signal, the attenuation settings before and after the power amplifier are simply added together, for it does not matter where in the system the attenuation is applied—it has the same effect on the signal, regardless of its location.[11]

The need for tail-end attenuation may come as a surprise to many people[12] who are used to thinking of audio amplifiers in terms of their high fidelity home sets, and who do not usually listen to an amplifier turned up to its

---

[10] Based on cochlear output, not on tests of overt hearing.

[11] This statement is in general true, but it should be checked in the initial calibration of a sound system, utilizing a wide range of frequencies. In addition, some attenuators may contain elements which introduce capacitative reactance, leading to frequency-dependent variations in attenuation.

[12] Including, it appears, the manufacturers of audio amplifiers, who seem to be unaware of their products' poor performance at high amplification of low signal levels.

maximum output but with no signal applied to it. It is easy and instructive to verify the striking elimination of unwanted noise by the addition of a "tail-ender."[13]

d. TRANSDUCER. Finally, the amplified and attenuated signal is applied to a transducer, which might be a speaker or earphones, or even a condenser microphone driven in reverse. Transducers vary widely in the fidelity with which they convert electrical oscillations into sound energy, and it is here that by far the greatest amount of distortion is likely to be introduced into the sound system. Even with a speaker thought to be "flat," it is instructive to measure the sound output using a calibrated microphone with signals of constant voltage but varying frequencies. Most speakers have peculiarities in regard to frequencies at which they operate more or less efficiently, and all speakers become less efficient as the high and low extremes of their particular frequency range are approached. It is important to know the peculiarities of your transducer before using it as a stimulus source.

A special case of sound stimulation, the use of clicks, deserves some mention here. Clicks are often used to elicit N_1 responses. It is common practice to activate a speaker with a rectangular electrical pulse, which produces an audible "click." The important point is that different speakers produce different clicks when activated in this manner. Because of this idiosyncratic response on the part of the speaker, it is never sufficient to describe a click stimulus by describing the voltage, duration, and shape of the electrical pulse producing it. Instead, the actual sound output of the speaker must be observed using a calibrated microphone and viewing its output on an oscilloscope. When this is done, it will be seen that the speaker's response to an electrical pulse consists of oscillation or "ringing" which is brief enough to be heard as a click. The frequency of ringing is an individual characteristic of speakers. The major frequency and duration of oscillation should therefore be reported, just as the frequency is specified for a pure-tone stimulus, even though the frequency of the ringing gives rise to little sensation of pitch as it is too brief.[14]

Ringing of the speaker can occur in other situations, notably where short tone pulses are used as stimuli. If the signal begins abruptly, it may produce onset transients in the speaker, which simply means the speaker rings at its own frequency before settling down to the signal frequency. In order to avoid this, an electronic switch is inserted between the signal generator and the attenuation and amplification stages (see Fig. 1-15). The electronic

---

[13] It is difficult to find suitable and inexpensive attenuators for this purpose. Therefore, a schematic diagram for an easily constructed tail-end attenuator is given in the Appendix to this chapter.

[14] See the discussion by Licklider (1951), which indicates that for frequencies of 100 Hz or higher, the oscillation must endure at least 3 msec before any sensation of pitch occurs.

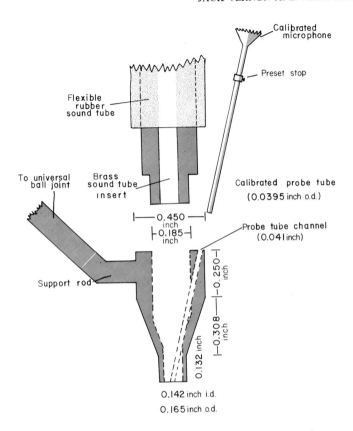

FIG. 1-16. Ear cannula for connecting sound tube to the ear. The sound cannula arrangement shown here consists of two main parts: the actual cannula for insertion into the external auditory meatus, and the removable sound tube insert. The tapered tip at the bottom of the cannula is inserted into the guinea pig's ear (when this cannula is used for cats, a short rubber sleeve is fitted on the tapered end to increase its outside diameter to fit snugly in the ear canal of the cat). The angle of alignment is important, as the hollow bore of the cannula must face the eardrum and not be blind-ended against the canal wall. To facilitate alignment we use the operating microscope to keep the eardrum in view as the cannula is inserted. The universal ball joint in the clamp which holds the support rod makes it easy to obtain any desired alignment. The calibrated microphone-probe tube assembly is then inserted into the probe tube channel. The preset stop prevents the probe tube from entering too far (and thus risk damaging the drum). It is essential to clean the probe tube as well as the probe tube channel prior to each installation. A second clamp on a stand is needed to support the body of the microphone (with its cathode follower) as it is too heavy for the slender probe tube to support. The brass sound tube insert is fitted into place in the cannula when sound stimulation is to be presented. The great advantage in having a removable insert such as we show here, is that it can be removed at any time during the experiment when it is desired to interrupt the sound path to the ear (in checking for "jump-across" artifact, as described in Section IX,I, for instance) or when it is necessary to check the condition of the eardrum or ear canal. With this arrangement, such checks can be made without disturbing the seal between the cannula and the ear. The bright

switch is used to shape the signal so as to generate a gradual onset (usually greater than 1 or 2 msec), and it does this by modulating the amplitude of the signal from zero up to the full amount desired. This is referred to as a gradual signal rise time; it may also be important to consider the decay time, so as not to generate offset transients in the speaker. Once again, the use of a calibrated microphone and oscilloscope display is recommended to observe the actual sound output of the transducer under the stimulus conditions. If there are transients present, they must be noted as part of the accurate description of the stimulus.

The sound system is completed by the conduction of the sound to the ear of the subject. There are two methods of doing this: open and closed. The closed method is the one we recommend and will describe first. The speaker output is led through a tube[15] to the animal's ear, where a tight-fitting cannula connects it to the external auditory meatus. Joints are tightly fitted everywhere, and the sealing of the cannula into the uneven contours of the ear is assured by plastering Vaseline around the juncture so that every gap is filled. The advantages of this method are (1) that external sounds cannot exert much influence on the ear, thus there are less stringent sound-proofing demands on the experimental chamber; and (2) that sound levels can be measured more accurately. Another consideration is that some speakers, such as the often used Western Electric 555, operate more efficiently when facing into a closed system.

The final link in the sound path between the speaker and the ear is formed by the ear cannula. Any rigid tube of the proper diameter can be cut to a short length and used as a cannula inserted into the external auditory meatus, to which the sound tube is then connected. We have improved upon such a simple arrangement by having a cannula that (1) can easily be fixed at almost any position in space, and held there rigidly so as to assure constancy of stimulus conditions, and (2) includes provision for a probe-tube microphone, so that sound calibration can be done without disturbing the connection of the ear to the sound system. (The merits of a calibrated probe-tube microphone are discussed in Section IV,F,3.) The sound cannula we have devised is shown in Fig. 1-16. Another desirable feature of this cannula is

[15] We use flexible soft rubber tubing of $\frac{1}{4}-\frac{1}{2}$ inch inside diameter. No alterations in the configuration of the tubing are permitted between the actual experiment and the measurement of sound intensities during the calibration immediately following each experiment.

---

surfaces of the brass often introduce unwanted reflection of light. A reflection-free black surface for brass may be achieved in the following way: Thoroughly mix $\frac{1}{3}$ lb of copper carbonate in $\frac{1}{3}$ qt of ammonium hydroxide and add 1 qt of water. Heat to 175°F and immerse brass parts until desired color is acquired, usually several minutes. *Caution:* ammonium hydroxide fumes are harmful, thus the entire operation must be conducted under a fume hood. Directions for other chemical coatings for brass may be found in *Product Engineering,* December 1956, 137–138.

that the sound tube connection can be easily lifted out, without disturbing the junction between the cannula and the ear canal, whenever during the experiment it becomes necessary to inspect the eardrum or external canal (for instance, if it is suspected that bleeding has filled and obstructed the canal).

The second method of sound presentation is the open ("free-field") method. The speaker's output is left open to the room; the animal's ear is then placed at a defined point within this general sound field. Keeping in mind that sound energy can be reflected or absorbed by surfaces or objects in its field, it is obvious that in an open system the external surround can influence the sounds actually reaching the ear. This is why experimental control of sound stimuli is usually more difficult with the open presentation than with the closed.

There is, however, one condition in which the open presentation *must* be used: that is when very high frequencies (above about 50 kHz) are used. The reason stems from the fact that wavelength decreases as frequency increases. For very short wavelengths, the standing wave pattern in a closed system is characterized by extremely frequent minima and maxima, which correspond to marked variations in sound intensity. A very minute shift in the position of the ear relative to the speaker can produce a major change in stimulus intensity. Thus, calibration becomes unreliable, and stimulus conditions are difficult to replicate from one experiment to the next. The vulnerability to local variations is much reduced if sound is allowed to travel freely through the air in all directions from the speaker. For further discussion of the technique of high frequency stimulation, see Vernon, Dalland, and Wever (1966).

3. Measurement of Sound

Before describing the procedure used for measuring sound levels, it is appropriate to consider briefly the units used in acoustic measurements. It is conventional to use the decibel (dB) scale for quantifying the intensity dimension of sound. One peculiarity of a decibel scale is that it has no zero. Therefore, all values measured on the scale are relative values only. Some level is arbitrarily chosen as the reference; all other values are then expressed as multiples of this value.

A number of different reference levels are in use, causing a fair amount of confusion. A common reference is "0.0002 dyne/cm^2," which was arbitrarily selected by the American Standards Association as the best representation of the threshold of human hearing at 1000 Hz. Independently of this reference, commercial microphones[16] were calibrated in terms of 1

---

[16] For example, most of the Brüel and Kjaer microphones.

dyne/cm². Then about 1968 there was a world-wide movement to standardize on the mks system of units. Consequently, microphones produced since that time use the reference "1 Newton/m²," which is equivalent to 10 dynes/cm² (difference of 20 dB). The reference we prefer is "1 dyne/cm²." Our reasons for preferring this reference are as follows: (1) it makes computation simple, on account of the prevalence of commercial microphones calibrated in terms of this reference; (2) it is of sufficient amplitude to be registered by microphones as small as ⅛ inch, whereas 0.0002 dyne/cm² is too low a level to cause a discernible response on present-day equipment, even with a 1 inch microphone; (3) considering the wide range of hearing sensitivity among different experimental animals, it seems more appropriate to choose a neutral reference such as 1 dyne/cm² rather than the 0.0002 dyne/cm² which was originally chosen because it represented human threshold at 1000 Hz. To our way of thinking, the reference should not depend upon or be complicated by the psychophysics of hearing.

The concept of adding or subtracting some number of decibels as one does when adjusting attenuation, is confusing to many people new to the area of acoustic measurement. Therefore it is useful to have some quick mental equivalences handy. Constant use will ingrain them until decibel scales come to seem more meaningful to the user. We have found it helpful to recall that:

Addition of :	6 dB =	multiply sound intensity × 2
	10 dB =	multiply sound intensity × 3 (approx.)
	12 dB =	multiply sound intensity × 4
	20 dB =	multiply sound intensity × 10
	40 dB =	multiply sound intensity × 100
	60 dB =	multiply sound intensity × 1000
Subtraction of :	6 dB =	divide sound by 2 (i.e., multiply × ½)
	20 dB =	divide sound by 10
		(etc.)

Incidentally, it may be useful to keep in mind that 0.0002 dyne/cm² is 74 dB below 1 dyne/cm².

a. METHODS OF MEASURING SOUND INTENSITY. Acoustical stimuli can only be specified in terms of acoustical energy. It is never sufficient to describe an auditory stimulus in terms of the electrical voltage applied to a speaker.

Sound stimuli have many physical dimensions, but for most work only two are of major concern: frequency and intensity. Specification of stimulus frequency is easily handled and has already been discussed. The measurement of sound intensity, however, requires much careful consideration. There are many ways in which sound intensity could be measured, but it is generally agreed that measures of sound pressure will be utilized. Fortunately, sound pressure can be measured easily by sensitive calibrated microphones which are commercially available. Here the lion's share of the work has been done for us by the manufacturers. Not only do they provide reliable and stable instruments, but they also provide calibration data for them. With such units one may proceed with confidence of obtaining accurate sound measurement.

In general the calibrated microphone has been used in two ways to measure the intensity of sound stimuli. They are called the substitution method and probe-tube method. A discussion of these two procedures can be better understood if we first briefly consider the action of the calibrated microphone.

The calibrated condenser microphone is a device with a sensitive diaphragm which moves in response to sound. This action produces an electrical voltage which is proportional to the sound pressure exerted upon the diaphragm. Needless to say, this is a delicate instrument requiring protection. The calibration graph for each microphone provides a statement as to how much voltage is generated for specified amounts of sound. Typically the microphone output is expressed in millivolts or microvolts (depending upon its size) when a sound pressure of 1 dyne/cm^2 is acting upon its diaphragm. In addition to having a relatively flat frequency function, these microphones have linear input–output functions. For example, a typical ¼ inch calibrated microphone might generate 90 μV in response to 1 dyne/cm^2 over the range of 10 Hz–100 kHz. If this microphone were activated by 10 dynes/cm^2, it would put out 900 μV; if it were activated by 0.1 dyne/cm^2, then 9μV would result. Incidentally, a convenient and often-used notation for 1 dyne/cm^2 is 1 μbar (microbar).

Now let us consider how calibrated microphones may be used to measure sound intensity.

(*i*) *The Substitution Method.* This is a simple and straightforward procedure whereby the animal's ear is replaced by the calibrated microphone. In the closed sound situation where a sound cannula has been sealed into the ear, the animal is removed from the cannula and the microphone coupled to it with a piece of flexible rubber tubing which provides a tight seal around

1. ELECTROPHYSIOLOGY OF THE COCHLEA

the microphone. The arrangement is such as to place the diaphragm of the microphone at that position previously occupied by the animal's eardrum. Obviously in most cases this placement is an approximation. The ¼ inch and ⅛ inch microphones are most convenient for this purpose. The substitution method is used in the following manner to determine a frequency function: (1) For each frequency, the attenuation of the sound-producing system is adjusted to yield 1 μV of cochlear potential. (2) The microphone is then substituted for the animal. (3) The attenuation of the sound system is adjusted to yield 1 dyne/cm² at each frequency. A typical data sheet might look like that presented in Table 1.1.

TABLE 1.1

Sample Data Collected Using the Substitution Method for Calibration

Stimulus frequency (Hz)	dB Attenuation for 1 μV cochlear potential	dB Attenuation for 1 dyne/cm²	dB re: 1 dyne/cm² for 1 μV cochlear potential
100	92	66	-26
200	101	61	-40
300	106	65	-41
500	109	64	-45
⋮	⋮	⋮	⋮

The frequency function simply plots the last column in graphic form (see Fig. 1-4). To take one example, say 200 Hz—here our sample data page shows that in order for our system to produce 1 dyne/cm², 61 dB of attenuation was needed. (This refers to attenuation of the voltage sent to the speaker; the lower the attenuation setting, the more intense the sound.) At the same frequency, 1 μV of cochlear potential was generated by the animal at 101 dB attenuation. This means that the actual intensity of the sound stimulus was 40 dB *below*, or less than, 1 dyne/cm² (61 − 101 = −40). Expressed differently, 40 dB below 1 dyne/cm² is 0.01 dyne/cm². This then was the intensity of a 200 Hz tone required to elicit a 1 μV cochlear potential in that experiment.

The substitution method has the advantage of convenience but it may be lacking in accuracy. Its disadvantage is that the conditions affecting the sound field during sound measurement differ from what they were during stimulation of the animal. Clearly, such things as sound reflectance will differ for the microphone diaphragm as compared to the animal's eardrum. Because of considerations such as this we have elected to use the probe tube method.

(ii) The Probe Tube Method. The essential feature of the probe tube method is that the sound field is measured within the animal's ear canal, so that conditions during calibration are the same as during recording. The probe tube we use was illustrated previously in Fig. 1-16. The body of the sound cannula, held by the support rod, is sealed into the external auditory meatus. This is best done with the aid of a microscope to sight down the cannula, for misalignment is easy. Once the probe tube is in its place, the sound tube is then connected to the sound cannula. With this arrangement all conditions are as nearly identical as possible for both stimulation of the animal and measurement of sound intensity. That is to say, the probe tube and its effects are present when the ear is being stimulated, and likewise the ear and its effects are present when the sound intensity is being measured.

The probe tube length and diameter depend upon the experimental setup. Each probe tube is therefore an individual and must be calibrated individually. Commercial probe tube calibration kits are available. In brief, they provide a small sound chamber arrangement as indicated in Fig. 1-17. The probe tube is calibrated in the following way: the speaker is adjusted so as to produce 1 dyne/cm^2 in the small chamber (measured by calibrated microphone I). The sound reaches microphone II through the probe tube. The output of microphone II under these conditions provides a calibration for the probe tube. Again, the procedure must be repeated for each frequency to be utilized.

The chamber is very small by design, for if such a chamber is sufficiently small the sound field within it will be uniform for most frequencies. The requirement of uniformity cannot be met for some high frequencies, however. Thus for these frequencies calibration may not be possible as there may be insufficient pressure exerted through the probe tube upon the diaphragm of microphone II. In our probe tube calibration arrangement, 14 kHz and 20 kHz were easily calibrated, but 15 kHz did not produce sufficient sound pressure in the probe tube to be registered at microphone II.

One final check on the probe tube is essential. One must be sure that the sound energy which activates microphone II has actually arrived there by way of the air column inside the probe tube (i.e., not by travelling along the metal walls of the tube). This is checked by blocking the probe tube and repeating the calibration procedure. The resulting output should decrease, in our experience by 30–40 dB or more.

An example of a probe tube calibration is presented in Table 1.2. Note that the output of the probe tube varies with frequency. Each probe tube has its own characteristic pattern of variations. The output data cited above have been replicated several times over the past few months, and have thus demonstrated the stability over time of this particular probe tube. Although we expect the probe tube output to remain stable, nevertheless we recommend repeated calibration of a probe tube at regular intervals.

1. ELECTROPHYSIOLOGY OF THE COCHLEA

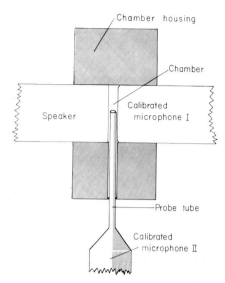

FIG. 1-17. Probe tube calibration chamber. The diagram above shows the essential features of the probe tube calibration kit supplied by Brüel & Kjaer (Model #UA 0040). The small speaker, similar to those used in hearing-aid ear pieces, is the major limiting factor in determining available sound pressure levels at different frequencies. The probe tube can be inserted to any depth within the chamber bounded by Calibrated Microphone I (it can be verified that its location does not affect the sound pressure level at either microphone I or microphone II). The port through which the probe tube enters the chamber housing must, however, provide a snug fit around the probe tube.

In order to compare the two methods of calibration described above, we determined the frequency function for a guinea pig using first one method and then the other. The results are presented in Fig. 1-18. It can be seen that differences do result from using the two techniques. These differences are generally greater at the higher frequencies, as might be expected. In sum-

TABLE 1.2

CALIBRATION OF A TYPICAL PROBE TUBE

Frequency (Hz)	Microphone—probe tube output for 1 dyne/cm^2 (μV)	Frequency (Hz)	Microphone—probe tube output for 1 dyne/cm^2 (μV)
100	1210	2000	320
200	1300	3000	310
310	1500	5000	130
500	2100	7000	155
700	2300	10,000	70
1000	1000	14,000	33
1500	440	20,000	99

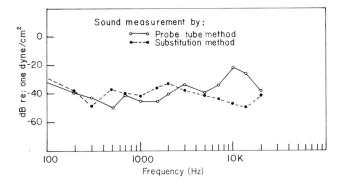

Fig. 1-18. Effects on the frequency function of the use of two different methods of sound measurement. The two curves shown were plotted using identical data from one guinea pig, but with two different calibration methods: the substitution method and the probe tube method. To obtain each point shown, the attenuation in decibels (of voltage sent to the speaker) required to elicit 1 μV of cochlear potential was first measured at each frequency. Then the sound pressure level corresponding to each of these readings was obtained using the probe tube method of calibration. Following this, the animal was carefully removed from the sound cannula (otherwise nothing was disturbed) and another microphone was sealed into the approximate position occupied by the eardrum. The sound pressure levels were again obtained, this time using the substituted microphone. The results show that the two methods do not yield identical results. The second method (the substitution method) is undoubtedly more vulnerable to error variations as it is not possible to duplicate exactly the conditions which prevailed during data collection. However, this method may be adequate for many purposes.

mary, we feel the probe tube method is superior for the reason that the stimulus conditions are more nearly identical during recording and calibration. Regardless of which procedure is used, the investigator must report in sufficient detail the way in which sound was measured.

b. GENERAL CONSIDERATIONS FOR CALIBRATION OF THE SOUND SYSTEM. When the output of the microphone is read out on an oscilloscope, it is important to keep in mind that the oscilloscope shows peak-to-peak voltage, whereas the microphone's calibration chart is based on rms output voltage. Figure 1-19 indicates the relation between these different voltage measurements and contains conversion formulas. The wave analyzer provides a more convenient readout of the microphone output for pure tones, as it automatically reads rms voltages, and its output is in numerical form.

This seems an appropriate place to explain why we require so strict a specification of stimulus frequency (± 1 Hz), as we mentioned earlier. We do this in order to ensure that when we calibrate the sound system we can return to the identical frequency used during stimulation of the animal. Because speakers are not "flat," and their efficiency may vary drastically with a small change in frequency, it is very important that the stimulating frequency be the same as that used when measuring the intensity of the stimulus.

1. ELECTROPHYSIOLOGY OF THE COCHLEA

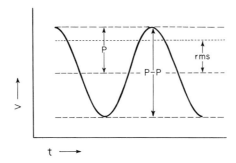

FIG. 1-19. Different ways of specifying signal amplitude. In the case of sound pressure or electrical energy which is varying sinusoidally, the amplitude is different at every instant in time. Although in neurophysiology (where sinusoidal signals are uncommon) it is usual to measure the peak-to-peak (P–P) amplitude of electrical signals, this is not the case in acoustics or in electrophysiology of the ear. In these latter contexts it is more useful to consider a sort of average amplitude, the root-mean-square (rms) amplitude. Wave analyzers are constructed so as to work in terms of rms values. If an oscilloscope is used to measure the cochlear potential, the observations of P–P voltages should be transformed according to the following conversion formulas:

$$\text{rms} = .707 \times \text{P} = .35 \times \text{P–P},$$
$$\text{P–P} = 2 \times \text{P} = 2.83 \times \text{rms}$$
$$\text{P} = 1.414 \times \text{rms}$$

Examples:
$$100 \text{ V P–P} = 100 \times 0.35 = 35 \text{ V rms},$$
$$100 \text{ V rms} = 100 \times 2.83 = 283 \text{ V P–P}.$$

where P is peak, P–P is peak to peak, and rms is root mean square.

Recalling that the operating characteristics of sound systems are susceptible to frequent, unpredictable changes, we wish to reiterate that the only safe course is to calibrate frequently. In our laboratory, calibration is done following every experiment. For the 14 or so frequencies we use most commonly, such a procedure takes about 10 min. Table 1-3 shows calibration runs from our laboratory for two successive days, obtained from one and the same sound system (all conditions identical). The variability, while not large, nevertheless would be sufficient to introduce significant experimental error into our measurements. Another reason why daily calibration is essential is that major changes in the sound system do occur, and only in this way can they be detected promptly.

G. Recording Apparatus and Methods

As we have already said, the cochlear potential is a high-fidelity analog of sound stimuli; thus it can reproduce a wide range of sound frequencies. In addition, it bears a linear relationship to stimulus intensity, and conse-

TABLE 1-3

COMPARISON OF SOUND MEASUREMENTS MADE ON CONSECUTIVE DAYS[a]

\multicolumn{3}{c}{Substitution method: dB of attenuation for 1 dyne/cm²}			Frequency (Hz)	\multicolumn{3}{c}{Probe tube method: dB of attenuation for 1 dyne/cm²}		
9/30/70	9/31/70	Diff.	Frequency (Hz)	2/25/71	2/26/71	Diff.
83	82	−1	100	83	74	−9
82	80	−2	200	81	71	−10
94	90	−4	310	83	76	−7
97	101	+4	500	85	82	−3
84	86	+2	700	84	84	0
79	75	−4	1000	71	69	−2
84	87	+3	1500	77	76	−1
87	91	+4	2000	77	75	−2
67	69	+2	3000	61	59	−2
65	58	−7	5000	68	71	+3
53	51	−2	7000	55	51	−4
60	58	−2	10,000	58	56	−2
46	34	−12	14,000	36	40	+4
28	40	+12	20,000	29	30	+1

[a] The point we are attempting to make here is that a single calibration of a sound producing system is not sufficient. The cases above were selected at random from our calibration records. In each set the two calibration runs were performed by the same investigator, on consecutive days, and with the same kind of animal.

In the case of the substitution method, done on cats, note that differences at various frequencies did occur, in fact exact agreement never occurred. This degree of variation is typical. Note also that the differences were not systematic, that is, one day was not consistently better or worse than its comparison day.

The comparison for the probe tube method, done on guinea pigs, reveals the same kind of day-to-day variability. The data confirm that a particular experimental setup on a given day has its own peculiarities. Therefore it is important to carry out measurements of the sound actually used in *each* setup and for *each* condition.

For a comparison of the two calibration methods when they were performed on the same animal, see Fig. 1-18.

quently at very low sound levels the amplitude of the cochlear potential is correspondingly small. This is the characteristic which has made it difficult to determine whether the cochlear potential possesses a true threshold: the electrical "noise" inherent in any recording situation obscures very small electrical signals from the cochlea. The problem of electrical noise has so far limited the measurement of cochlear potentials to levels no lower than approximately 0.1–0.03 μV, depending on frequency. It is to be hoped that the development of better, more noise-free recording techniques will extend future observations downward, for at present levels we are still 20–40 dB (depending on frequency) above the sound intensities required at hearing thresholds.

1. ELECTROPHYSIOLOGY OF THE COCHLEA

The ways in which the problem of electrical noise exerts a strong influence on the recording instrumentation used to record the cochlear potentials are discussed in detail below.

1. Shielding

The first line of defense against unwanted electrical signals or "noise," is to create an experimental environment which is as noise-free as possible. Most acoustically shielded chambers also provide some degree of electrical shielding, referred to as "RF" shielding. They attempt to make use of the fact that high-frequency signals transmitted through the air by radio and TV stations can be blocked from reaching the biological recording site by completely surrounding the preparation with a conductive metal sheet which is grounded.

A second type of noise results from the necessary presence of power lines which carry the alternating current which runs lights, equipment, etc. These lines are sources of very powerful electrical fields oscillating at 60 Hz. Because of their omnipresence and their strength, it is a constant fight to keep 60 Hz signals from being picked up by sensitive electronic circuitry such as that used in recording the cochlear potential. The electrically shielded room helps to cut down such 60 Hz interference.[17]

It is common practice to carry out bioelectric recording in a wire-screened enclosure, and for many purposes an adequate degree of shielding from RF and 60 Hz "noise" is obtained in this way. Unfortunately the screen room is usually not adequate for recording the cochlear potential at its lowest levels. To minimize noise contamination in the most sensitive cochlear records, an RF shield must be nearly perfect. It must have *no* holes or defects, and special care must be taken at metal joints to prevent RF "leaks." In practice, it is prohibitively difficult to achieve a perfect shield. Furthermore, it is usually necessary to bring in cables, electrode leads, sound tubes, etc. in order to conduct the experiment. Each such intrusion by a cable constitutes a break in the shielding, and the probability is high that even one such break will nullify the electrical noise-reducing effect of the shielding. During actual recording of the cochlear potential, all 60 Hz a.c. lines must be disconnected and removed from the chamber.

In summary, shielding of varying effectiveness is routinely used, and it may help to reduce noise contamination. However, it is not sufficient by by itself to solve the problem completely, and so additional noise-reducing

---

[17] Despite all attempts to eliminate them, stray 60 Hz signals may be strong enough to cause problems when recording cochlear potentials at frequencies equal to 60 Hz or to its harmonic multiples (at 120, 180, 240, or 300 Hz, for example). Since the choice of stimulus frequencies is quite arbitrary, it is easy to avoid exact multiples of 60 Hz. For this reason we often substitute 310 Hz for the 300 Hz stimulus we would otherwise use. With narrow-band filtering (provided by a wave analyzer as described in Section IV,H), electrical interference at 300 Hz is attenuated sufficiently that it does not interfere with the cochlear output at exactly 310 Hz.

features are incorporated into the recording instrumentation as we will show.

2. Biological Amplifier

Very special demands are placed upon the amplifier as a result of the requirement for very high fidelity with very small signals in the audio frequency range. First, the amplifier must be "flat," that is, capable of reproducing a wide range of signal frequencies with constant amplification. This is not difficult to achieve, and is in fact the simplest requirement placed upon the amplifier. The anticipated frequency range of the particular ear under study sets the required limits for the bioamplifier's performance. In studying cochlear potentials in animals whose frequency range is unknown, one must take care that the upper frequency limit of the recording system does not artificially limit the extent of the animal's response.

Second, the amplifier must minimize the ubiquitous electrical noise in favor of the desired signal from the cochlea. The requirement of being able to reproduce faithfully a 1 μV signal is unusually stringent from the point of view of most electronic instrumentation. Thus, the amplifier must be designed specifically to operate as a "low-noise" amplifier. In addition, in order to improve the signal-to-noise ratio at very low levels, it is customary to use a differential amplifier.

a. Differential Amplifiers. These are used frequently in biological recording because, in addition to the electrical noise present in the environment, living tissue is full of electrical events (such as heartbeats, brain activity, or muscle contractions) which may interfere with recording the signal of interest, in this case the cochlear electrical activity. Unwanted electrical signals, whether internal or external in origin, can be partially eliminated by differential amplification. This is accomplished by a double electrode system. The active electrode is placed as close as possible to the source of the signal of interest, in this case on the round window membrane or inside the cochlea. The second active[18] electrode is placed on tissue which is near the ear but from which no cochlear signal can be recorded. Usually the spot selected is in tissue of the head or neck. The differential amplifier subtracts one electrode input from the other, and amplifies only the difference between them. Since the cochlear signal is present at only one of the electrodes, it is amplified, while irrelevant signals such as heartbeats and 60-Hz "noise" are present at both electrodes and are thus canceled out.

[18] For clarity we prefer not to use the terms "active—indifferent—ground" which have sometimes been used to designate the three electrodes used with a differential amplifier. This usage is potentially confusing because "indifferent" may be used synonymously with "reference." As the common mode reference in many amplifiers is ground, calling a second active electrode an "indifferent" or "reference" may suggest erroneously that it is simply another "ground."

1. ELECTROPHYSIOLOGY OF THE COCHLEA   47

The ability of a differential amplifier to reject unwanted electrical signals common to both electrodes is referred to as its "common mode rejection ratio." A good amplifier can decrease noise by as much as 10,000 or 100,000 times (a decrease of 80–100 dB). Most differential amplifiers also use a third electrode as common mode reference, which is attached to the animal at some point such as a foot, and which serves to keep the amplifier and the biological preparation at the same level. Typically, the common mode reference is to ground; hence the third electrode is usually referred to as a ground electrode.

For accuracy it is very important to check the amplification (gain) factor of the amplifier. The nominal setting (for example, 1000×) cannot be assumed to be correct or invariant over time. It is always desirable to calibrate a differential amplifier with a "differential input," that is, a known input which mimics the biological recording situation in that the two active leads carry voltages which differ by some small amount. In Fig. 1-20 we present a schematic for an easily constructed circuit which will provide such a calibration input. Frequent, even daily, calibration of the biological amplifier is highly desirable, the more so because the cochlear signal is small relative to possible error variations in the amplifier.

Other tactics for reducing noise at the signal input include (1) the use of a battery-operated amplifier, thus eliminating the need for a power cable to the amplifier, with its attendant intrusion of 60 Hz and other noise; and (2)

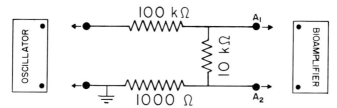

FIG. 1-20. Circuit to provide a calibration input for differential amplifier. The object of this circuit is to obtain voltages at the two active leads (A_1 and A_2) which differ only by a very small amount. The oscillator supplies some voltage V which is then divided by the circuit so that the voltage appearing at A_1 differs by 1/11 V from that at A_2. This small fraction of V is the voltage which will be amplified by the differential amplifier. To obtain a calibration input of known amplitude, first plug A_1 and A_2 into a differential oscilloscope amplifier (such as Tektronix 1A7A or 3A3) which has previously been calibrated with its own internally generated calibration signal. Be sure all amplifier filters are set appropriately for the frequencies used. Adjust the oscillator output to obtain exactly 1 mV of "differential" signal at some intermediate frequency, say, 1000 Hz. Without changing oscillator settings, connect A_1 and A_2 to the two inputs of the differential bioamplifier to be calibrated. If the bioamplifier gain setting is 1000×, for instance, then the output of the bioamplifier should be exactly 1000 × 1 mV = 1 V. If it is not, change the gain adjustment of the bioamplifier (usually an easily accessible screwdriver adjustment). The "flatness" of the differential bioamplifier should be checked by repeating the above procedure at each frequency used. (This circuit was provided by Jerald C. Jansen of Custom Systems Associates, Inc., Portland, Oregon.)

the use of electrical filters to limit the frequency range which is amplified. Many amplifiers incorporate adjustable filter settings, and these can be of great use in eliminating electrical noise which is higher or lower in frequency than the cochlear signal. It is extremely important to keep in mind that a filter may attenuate signals even well within the nominal frequency settings. A low cutoff at 100 Hz, for instance, means that the amplifier progressively attenuates frequencies starting somewhere above 100 Hz, with attenuation increasing as frequency decreases. The usual designation is for the nominal filter setting to describe the "3 dB down point," so that in this example a signal of exactly 100 Hz would be decreased exactly 3 dB, and signals lower than 100 Hz progressively more so. As filter specifications differ, the effect of using a particular filter setting should always be checked carefully over a representative frequency range.

b. SINGLE-ENDED AMPLIFIER. In those rare cases where extraneous electrical noise is not a problem, it is possible to use a single-ended amplifier rather than a differential amplifier. In this case, there is only one active electrode, and a ground electrode. There is no advantage to using a single-ended amplifier other than that it may be less expensive.

3. THE "DIFFERENTIAL ELECTRODE METHOD"

A specialized technique developed by Tasaki and his associates makes it possible to record highly localized cochlear potentials from particular regions of the cochlea (Tasaki & Fernandez, 1952; Tasaki, Davis, & Legouix, 1952). This method uses two intracochlear electrodes inserted on opposite sides of the cochlear partition, one in scala tympani and one in scala vestibuli. This placement of the electrode pair takes advantage of the fact that the cochlear potential measured in one scala at any instant is opposite in phase to the potential measured at the corresponding point in the other scala. The way in which this kind of electrode placement can be easily achieved in the guinea pig cochlea is shown in Fig. 1-21.

The essential aspect of the "differential electrode method" is that it makes use of differential amplification in the following way: The opposite-phase voltages are led to the two inputs of a differential amplifier, where only their difference is amplified. If the two voltages were exactly in phase, subtracting one from the other would exactly cancel them, thus resulting in zero voltage. However, since they are 180° *out* of phase, subtracting one from the other results in a doubling of the instantaneous signal voltage. At the same time that the cochlear potential receives this preferential treatment, electrical and physiological "noise" are rejected as usual during differential amplification. Hence, the cochlear signal-to-noise ratio is greatly improved by this recording technique.

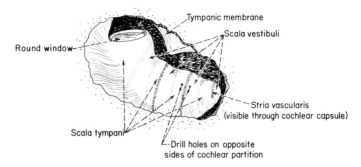

FIG. 1-21. The positioning of drill holes for the "differential electrode method." The cochlea is shown widely exposed through a large opening in the bulla. The stria vascularis is usually visible as a clearly defined dark band in each cochlear turn, except in the basal turn where the cochlear bone is usually too thick to see it. (The younger the animal, the easier it is to see through the cochlear bone.) The bony separation between each turn is indicated in this drawing as a dotted line. The actual position of the bony lamina is difficult to see unless a strong, highly focused light source is shone down through the external auditory meatus, thus providing translumination of much of the cochlea. The main consideration in placing drill holes for the differential electrode pair is to place them as nearly opposite each other as possible, with the stria vascularis (and other portions of the cochlear partition) centered between them. In each turn the scala vestibuli lies to the apical side of the stria vascularis. Hence the hole on the left in this picture lies in scala tympani, that on the right in scala vestibuli.

Another advantage of the technique is that if one wishes, the same electrode pair can be used to provide preferential recording of auditory nerve potentials while canceling out cochlear potentials. To accomplish this, a second amplifier is set up to receive signals from the two intracochlear electrodes, but in this amplifier the two signals are added together. Since the cochlear potentials in the two scalae are out of phase, adding them together results in their cancellation. However, the nerve action potentials as picked up in the two scalae are in phase with each other. Hence adding them doubles the resultant amplitude of the recorded nerve potentials.

There are many instances where it is desirable to be able to record both the cochlear potential and concurrent auditory nerve activity such as N_1. The differential electrode method was made to order for such situations, with only a slight increase in effort and risk (associated with the installation of the second intracochlear electrode). The reader who wishes a more detailed discussion of this ingenious method is referred to the work of Dallos (1969).

4. Electrodes

It is good recording practice to keep electrode leads, from animal to amplifier, as short as possible. This necessitates placing the animal very near the amplifier (6–12 inches is practicable).

a. ACTIVE ELECTRODE. The active electrode is placed on the round window membrane or inside the cochlea through a hole drilled for that purpose. Wire of small diameter (.001–.012 inch) is of a convenient size and flexibility for this purpose. Fine wire can be purchased insulated (with Teflon or other coatings) from sources listed in the Appendix. Silver or platinum wire is commonly used in acute experiments where tissue reactions are not a concern. For chronic preparations, stainless steel is strong, durable, and fairly inert in the body fluid environment. Platinum–iridium is nearly as good on these counts, and is much more malleable, especially after annealing, making placement easier.

The most important consideration for a round window electrode is that it should not inflict any damage on the delicate membrane it contacts. The simplest form of nontraumatic electrode is a balled ending obtained by melting the wire at its tip. Silver and platinum–iridium wires are easily balled, using a small heat source. Stainless steel is more difficult but it too can be balled, using a special electrical circuit described in the Appendix.

Another nontraumatic ending is a small loop. Still another is a small piece of platinum foil attached to a wire. Regardless of the type of electrode ending used, all electrode surfaces likely to come in contact with the round window membrane must be absolutely smooth and free from sharp points or edges. The cut edge of the insulating coating should likewise be carefully smoothed. Such precautions are especially important with small animals, such as guinea pigs, which have extremely delicate round window membranes.

The most important consideration for an intracochlear electrode (that is, one inserted through a hole drilled in the bony capsule of the cochlea) is that it be very small. Although microelectrodes may provide maximum assurance that damage to the inside of the cochlea is minimal, it is not necessary to go to such an extreme to obtain a long-lasting, stable preparation. We have found that wire diameters of .004 inch in the apical and middle turns, and .008 inch in the basal turn, are tolerated well. Much depends on the technique of drilling and insertion. The drill hole diameter should exactly equal the bare wire diameter, and the insulation should be cut back so as to expose no more than about 0.1–0.2 mm of the wire tip. The cut edge of the insulation then acts as a "stop" for the wire, preventing it from being inserted too deeply, thus eliminating the risk of damaging the cochlear partition. The use of a micromanipulator is desirable as it makes insertion of the wire tip into the drill hole controlled and secure. Tissue adhesive[19] is then applied around the insertion point, to seal the hole and

---

[19] Bucrylate Tissue Adhesive, Ethicon, Inc., Somerville, N.J. This quick-acting adhesive hardens only in a wet field.

prevent loss of intracochlear fluid. Greater mechanical stability can be achieved by leading the electrode through a small hole in the cut edge of the bulla (as in Fig. 1-9D) and immobilizing it at that point with dental acrylic.

b. SECOND ACTIVE ELECTRODE. The second active electrode can be either a fine or a larger wire. In a chronic preparation, it is convenient to use a fine wire cemented into a tight-fitting hole drilled for it in the skull. (The cement must not insulate the wire from its contact with the bone.) In acute preparations, a simple procedure is to use a mini-alligator clip, soldered to a lead, clipped to some accessible tissue which is moist, such as the cut edge of the incision.

c. GROUND ELECTRODE. For the ground electrode, a most convenient attachment is made by soldering a lead to a fine-gauge syringe needle. The chrome plating on the needle shoulder must be sanded off so as to expose the underlying brass, which will provide an area where solder will "take." The needle electrode made in this manner is then stuck through the skin of a leg and taped in place.

When recording d.c. potentials it is important to keep in mind the electrochemical junction potentials generated by the metal-to-fluid interfaces between electrodes and the body. In the interest of minimizing electrical effects due to such potentials, all electrodes should be of the same material.

d. CHRONIC ELECTRODES. For chronic preparations where skull screws are used to help immobilize a skull mounting (such as for a head holder or connecting pins for implanted electrodes), a skull screw can be used as second active electrode. During recording, connection can be made to the screw "electrode" using an alligator or other clip-on lead.

The durability of chronically implanted electrodes can be greatly improved by encasing the wire in fine polyethylene tubing. The tubing and wires should be soaked for at least 15 min in 70% ethyl alcohol as a sterile precaution. During the implanting procedure, bending and twisting the wire should be kept to a minimum.

H. Quantitative Measurements

In essence, the problem of measuring the a.c. cochlear output is simply a question of how best to measure the magnitude of the oscillating voltage it generates. Two methods of doing this will be described. With proper care, the two methods should yield identical results. However, the first method we will describe has a number of advantages, and therefore we recommend it strongly.

1. METHOD I: WAVE ANALYZER

The voltage generated by the cochlea is led, after amplification, to a wave analyzer, which is a continuously variable, tunable voltmeter. Its primary advantage is the use of highly restrictive filters so that the bandwidth of the energy measured is limited to a very narrow range surrounding the signal frequency being studied. As an illustration, suppose the acoustic stimulus is 1000 Hz; the output of the ear is then exactly 1000 Hz. In addition, despite precautions against electrical noise, there is still likely to be a noise contribution of 60 Hz, audio sections of radio signals, other physiological noise, etc. It is contaminations such as these which prevent many investigators from being able to record at the 1 μV level. If, on the other hand, this unwanted noise can be eliminated without degrading the signal in question, then the 1 μV level is possible. The wave analyzer does just that, because its filters tend to suppress all frequencies outside that narrow band which contains the frequency of interest.

There are several commercial wave analyzers available. It is our conviction that cochlear potential work requires a wave analyzer, and that it should have the following characteristics:

1. A "tracking mode," which provides an internal oscillator to start the signal-generating process which leads to acoustic stimulation of the animal (see Fig. 1-15): The advantage of the tracking mode is that the frequency analyzing function of the wave analyzer is automatically tuned to the signal frequency being generated by its oscillator. That signal is in turn identical to the signal generated by the ear. Thus one does not have to "hunt" laboriously for the frequency of the biological signal in question.

2. Narrow bandwidth of the input "window": For example, the General Radio 1900 wave analyzer that we use has three bandwidths from which to choose: 3 Hz, 10 Hz, and 50 Hz. The narrower the bandwidth, the greater the attenuation of unwanted frequencies above and below the signal frequency. The bandwidth is centered around the particular frequency to which the wave analyzer is tuned. For example, if one is using the 10 Hz bandwidth and the unit is tuned to 1000 Hz, it will then maximally pass only signals having frequencies from 995 to 1005 Hz.[20] Ordinarily we use a 3 Hz bandwidth, thus achieving maximal filtering out of unwanted input frequencies.

3. A dampened meter, which means that the action of the indicating meter is slowed sufficiently that erratic excursions due to transient noise are reduced.

[20] For the technically minded, the crystal filters provide the following attenuation of input frequencies:
      3 Hz bandwidth: 60 dB down at ± 15 Hz from center frequency.
    10 Hz bandwidth: 60 dB down at ± 45 Hz from center frequency.
    50 Hz bandwidth: 60 dB down at ± 250 Hz from center frequency.

1. ELECTROPHYSIOLOGY OF THE COCHLEA

4. The range of signal magnitudes the unit will accept should go down as low as a few microvolts.

5. Provision for calibration of the unit should be easily available.

6. The analyzer's frequency range should go from a low frequency of several hertz up to at least 50 kHz. These higher frequencies are needed for many animal preparations. It is also the case that the study of overtones (harmonic multiples of the stimulus frequency), even for relatively low frequency animals, quickly moves up into the higher frequency range. Our wave analyzer stops at 50 kHz but we often wish for 100 kHz capability.

7. We have one further refinement which, although somewhat expensive, is worthy of mention. During the recording of a frequency function, the tuning of each successive frequency is time consuming because we require signal accuracy for the acoustic stimulus to be ± 1 Hz. To speed this tuning, we have modified our wave analyzer so that it will accept an external push-button oscillator while still in the tracking mode. This modification allows us to provide signals at any desired frequency, at an accuracy of better than ± 1 Hz, as fast as we can push the button. Often such a concern for time is justified when for reasons of experimental manipulation the ear is undergoing rapid change.

Valuable as the wave analyzer is, nevertheless in our opinion it is not entirely sufficient for all kinds of inquiry. In many cases, in addition to the wave analyzer there should also be available an oscilloscope so that the waveform of the ear's output may be viewed. The magnitude of the ear's output signal should be determined by a wave analyzer; the shape of the output waveform can only be appreciated through visual display. Although the shape of the output signal can and should be monitored, a word of warning is necessary. It is easy to get an incorrect impression from an oscilloscope display. Overloading the ear is well known to result in a distorted waveform at high intensities of sound; however, such distortion may not be perceptible to the observer until overloading has proceeded dangerously far. Figure 1-22 shows some pictures we have taken of the output signal from a guinea pig's ear during measurement of an intensity function involving progressively increasing stimulus intensities. Note that the wave shape in part (3) of Fig. 1-22 is not bad—in fact to many eyes the results look perfectly sinusoidal; yet analysis by the wave analyzer revealed that there was significant distortion present. Part (6a) in this figure is perhaps the most dramatic. Again the waveform looks fairly good, yet the output signal was shown by the wave analyzer to be depressed by over 50 dB (remember a decrease of 6 dB means the output was cut in half).

The cochlear potentials shown in Fig. 1-22 were led to the oscilloscope in parallel with the input terminals of the wave analyzer. Many wave analyzers provide a special output terminal from which to monitor the highly filtered

FIG. 1-22. Changes in waveform of the cochlear potential with increasing stimulus intensity. The oscilloscope display of the cochlear response was photographed at successively higher intensity levels of a 5000 Hz tone. The graph plots the intensity function obtained at the same time. The linear portion of the curve has been extended upward (dotted line) in order to calculate the number of decibels by which each successive reading departed from linearity. The departure in decibels from the straight line predicted value is entered next to each photograph. The different oscilloscope gain settings were not recorded; instead, precise voltage magnitudes were measured by means of a wave analyzer in parallel with the oscilloscope. At (1) the sound intensity was not sufficient to produce any deviation from linearity. From (2) through (7) the deviation increased. Although deviation from linearity was substantial at (3), (down 9 dB), the waveform still appeared sinusoidal. Thus although significant distortion was occurring, this fact was not evident in the oscilloscope display. At (6) the waveform was photographed under two conditions: (a) unfiltered, which allowed accurate representation of all distortion present; and (b) filtered (the oscilloscope filters were set to pass the band from 1000 Hz to 10 kHz). It is evident from (b) that filtering can smooth out distortions, thus deceiving the observer into thinking no distortion is present.

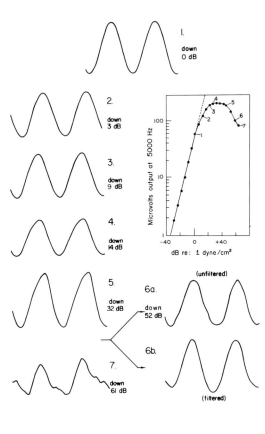

input signal whose rms value is registered on the meter. If one is interested in waveform of the input signal, the filtered output from the wave analyzer should not be used, for anything coming out here will be beautifully sinusoidal due to the "cleaning-up" action of the narrow bandwidth filters. We have deliberately put square waves as well as white noise into our wave analyzer and in each case the monitoring output yielded a perfect sine wave. This amounts to a good test of the effectiveness of the filters, as well as a demonstration of the dangers inherent in the use of filters whenever waveform is a parameter of interest.

In summary, for quantitative measurement of the cochlear potential a wave analyzer provides desirable features; the use of an oscilloscope as an additional monitor for waveform is also recommended. If, however, no wave analyzer is available, quantitative measures are still possible using only an oscilloscope, though such a procedure is more difficult.

2. METHOD II: OSCILLOSCOPE

The following precautions should be observed when using an oscilloscope to measure the cochlear potential:

1. Check the filter settings on the oscilloscope amplifier to ensure proper frequency range. Filter settings too close to the signal frequency distort both waveform and amplitude.

2. Low-frequency biological signals of small amplitude may not be discernible on an oscilloscope under usual noise conditions. The noise factor is a problem here because the filters of the oscilloscope do not have the rejection capabilities of the continuously tunable ones found in the wave analyzer. The addition of special external filtering may be helpful.

3. Remember the oscilloscope yields peak-to-peak readings which need to be transformed into rms, the traditional way of expressing cochlear voltage measurements(see Fig. 1-19 for conversion formulas).

I. Artifacts

The cochlear potential is unique in its vulnerability to certain kinds of artifacts. One of its most serious and least recognized problems is radiation artifact.

Transducers such as speakers which convert electrical energy into sound energy do so incompletely. Not only does sound issue from the speaker but also electrical emanations. These electrical waves have the same frequency and general characteristics as the biological signals produced in the ear, and the two types of electrical activity may be easily confused with each other. Consider for a moment a simplified picture of the process:

1. Electrical energy activates the speaker; it is, say, a 1000 Hz signal.
2. The speaker responds by producing a 1000 Hz tone plus some electrical radiations, also of 1000 Hz.
3. The tone activates the cochlea.

4. The cochlea responds by producing a bioelectric signal of 1000 Hz.

5. The relationship between the speaker and the animal preparation may be such that the radiated electrical signals can be impressed directly upon the recording electrodes.

Now, this electrical radiation (commonly called "jump-across") exactly resembles the cochlear potential. It is mandatory therefore that all data be demonstrably free of jump-across contamination.

There can be daily variations in the conditions leading to radiation artifact. As it must be repeated often, the procedure used for checking should produce as little disruption as possible. There are several ways to check for jump-across:

1. Move the electrode off the round window and into good contact with nearby tissues such as the bulla wall or exposed muscle. The new location should maintain about the same distance relative to the speaker. This action should effect a drastic reduction in the recorded potential. If it does not, jump-across must be suspected and conditions must be rearranged so as to eliminate it. Shielding the speaker is one recommended procedure here. Increasing distance between the ear and the speaker may also help. Because proximity of the transducer may so easily lead to jump-across, placing a transducer in or very near the ear canal should always be suspect. Similarly, the utilization of bone conductors during cochlear potential recording is particularly susceptible to jump-across.

2. Use a speaker which employs a polarizing voltage instead of a permanent magnet. In our laboratory, we use Western Electric 555 speakers, which require a 7 V d.c. bias. The oscillator signal is led to this speaker independently of the bias voltage, and the latter can be turned off without affecting the a.c. signal to the speaker in any way. When the bias voltage is turned off, the sound output of the speaker is reduced by about 25–30 dB, while the electrical emanations from the speaker are unaffected. Clearly, then, a convenient check for jump-across is to remove the polarizing voltage while observing a given level of cochlear potential. If the observed level does not drop at least 20 dB, then that level must be due primarily to jump-across.

3. Temporarily place a sound barrier between the speaker and the ear. Such an interruption will reduce the sound stimulus but will leave electrical radiation unaffected. Many experimental arrangements will easily allow for such treatment, particularly the closed sound tube system where the sound tube can simply be disconnected from the ear.

4. It will be recalled that, when doing an intensity function, eventually a level is reached where the output of the ear bends over. This fact can be used to identify a recording situation which is contaminated by jump-across. If

1. ELECTROPHYSIOLOGY OF THE COCHLEA

sound intensities sufficient to produce bend-over are used, and bend-over is not seen, then one must conclude that the signal being recorded is not biological in origin but is instead radiation artifact.

Jump-across checks should be made whenever high intensity sound is used, regardless of frequency. As procedure (4) is likely to damage the ear should jump-across actually be present, because traumatizing intensities are likely to be tried before one can be sure that no bendover is occurring, this procedure can be recommended only as a last resort.

Another common electrical artifact is the increase in wide-band electrical "noise" which results when the active electrode makes poor contact with the preparation. These increases in the noise are hard to identify as such if the signal is not being observed on an oscilloscope, but instead is being led "blind" into the wave analyzer for magnitude measurements. Because the noise is wide-band, a significant increase in the apparent signal may be seen at any frequency the wave analyzer may be tuned to, and this will appear to be an increase in cochlear potential output. This is one of the reasons it is desirable to monitor the cochlear signal with an oscilloscope in addition to the wave analyzer.

A biological artifact seen in recording from guinea pigs[21] is the change in signal level produced each time the middle ear muscles contract spontaneously. This brief twitch may produce a significant effect on the cochlear output which occurs at all frequencies. Because the twitches may occur as frequently as 20 to 30 times per minute, they can be an annoying impediment to accurate measurement of the cochlear potential. As we mentioned earlier, the spontaneous twitching can easily be eliminated by putting the animal on artificial respiration. This is another reason why artificial respiration helps to maintain a more stable, durable ear.

The story of hearing is far from being complete. Although there are many auditory theories, many facts have yet to be established. While electrophysiology of the cochlea may be only an initial aspect of the hearing process, nevertheless it has been and will continue to be an important tool for adding to our knowledge of the ear.

It is our hope that this chapter will provide useful information for those who wish to study the ear by the cochlear potential technique. Beyond this, we feel strongly that those who intend only to read about it nevertheless should understand it. Therefore, we hope that our present effort will provide the basis for a better understanding of this technique and of the information it provides.

[21] Other animals showing this activity are the agouti and the South American porcupine, the coendou (E. A. Peterson, personal communication).

Appendix

Instructions for making Dial®-urethane

(Adapted from information supplied by CIBA Pharmaceutical Co., Summit, New Jersey.)

CHEMICALS REQUIRED

Dial (diallybarbituric acid)	10.00 gm
Monoethylurea	40.00 gm
Urethane	40.00 gm
Distilled water to make	100.00 ml

MANUFACTURING DIRECTIONS

Place the urethane, monoethylurea, and Dial crystals into a suitable container. (Add 10 ml of distilled water.) Place on a water bath and stir until solution is complete. Cool to room temperature and add sufficient distilled water to bring to volume. (Note: Monoethylurea and urethane in equal parts are eutectic and form 90% of the liquid volume.) Filter the solution through a suitable filter. Keep refrigerated.

This results in a solution containing:

Dial	100 mg/ml
Monoethylurea	400 mg/ml
Urethane	400 mg/ml

Sources of Wire Suitable for Electrodes

1. Medwire Corp., 121 South Columbus Ave., Mount Vernon, New York 10553.
2. California Fine Wire Co., P.O. Box 446, Grover City, California 93433.
3. Thermal Wire & Electronics Corp., South Hero, Vermont.
4. Popper & Sons Inc., Bio-medical Instrument Division, 300 Park Ave. South, New York, New York 10010.
5. Cooner Wire Co., 9139 Lurline Ave., Chatsworth, California 91311.

Instructions for Balling Stainless Steel Wire

The following method is based upon a figure and accompanying instructions prepared by Dr. Blair Simmons, and received as a personal communication from him.

[22] See Riley (1949). This method has received further comment in Geddes and Baker (1968).

1. ELECTROPHYSIOLOGY OF THE COCHLEA

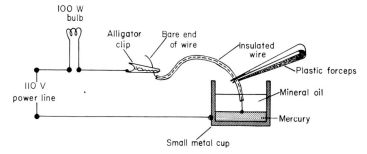

FIG. 1-23. Method for making balled ending on stainless steel wire.

Use 304-type stainless steel wire. The resistor shown in Fig. 1-23 (100 W light bulb) is appropriate for 7 mil wire. A different size of wire may require a different size of resistor.[22]

PROCEDURE

CAUTION: *Use extreme caution in handling apparatus connected to line voltage.*

1. Clean insulation from both ends of a length of electrode wire. Connect one end to the alligator clip. Hold the other end with plastic forceps or other nonconductor.
2. Connect the metal cup to one side of the power line.
3. Lower the free end of the wire through the oil to contact the mercury. The resulting current flow causes the wire end to melt and form a ball.

Do not leave any parts of the above apparatus connected to the power line without proper precautions against the hazard of severe shock.

Tail-End Attenuator

This attenuator is constructed so that all attenuation may be removed by simply switching to the "out" positions. Such an arrangement may be necessary where excessively intense sound levels are required. Attenuation can be added so that 20, 40, or 60 dB are available.

It is easy to demonstrate the need for this apparatus. If one listens to the output of a power amplifier without a signal applied it is very apparent that a great deal of noise is present. Introduction of the tail ender immediately and completely corrects this situation.

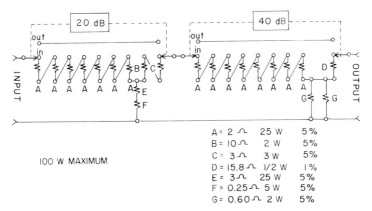

FIG. 1-24. Schematic diagram for a tail-end attenuator (with 16 ohm "T" pads).

In addition to the specifications given in Fig. 1-24, the resistors should be noninductive to ensure precise attenuation which is independent of frequency.

Acknowledgments

We wish to express appreciation for the careful reading of this manuscript to Drs. David Anderson, Robert Brummett, Catherine Smith and Richard Walloch, and to Curtin Mitchell. The mistakes which may remain are ours and not their responsibility. We also wish to thank Dr. Ruth Robbins who supplied material and helpful suggestions for the anatomical sections. Finally, our thanks are due to Herlene Benson who cheerfully typed the many drafts of this manuscript.

References

Adrian, E. D. The microphonic action of the cochlea: An interpretation of Wever and Bray's experiments. *Journal of Physiology (London)*, 1931, **71**, 28–29 P.

Butler, R. A. Some experimental observations on the dc resting potentials in the guinea-pig cochlea. *Journal of the Acoustical Society of America*, 1965, **37**, 429–433.

Crowley, D. E., Hepp-Raymond, M.-C., Tabowitz, D., & Palin, J. Cochlear potentials in the albino rat. *Journal of Auditory Research*, 1965, **5**, 307–316.

Dalland, J. I., Vernon, J. A., & Peterson, E. A. Hearing and cochlear microphonic potentials in the bat *Eptesicus fuscus*. *Journal of Neurophysiology*, 1967, **30**, 697–709.

Dallos, P. Comments on the differential-electrode technique. *Journal of the Acoustical Society of America*, 1969, **45**, 999–1007.

Davis, H. Some principles of sensory receptor action. *Physiological Reviews*, 1961, **41**, 391–416.

Davis, H., Deatherage, B. H., Eldredge, D. H., & Smith, C. A. Summating potentials of the cochlea. *The American Journal of Physiology*, 1958, **195**, 251–261.

Davis, H., Fernandez, C., & McAuliffe, D. R. The excitatory process in the cochlea. *Proceedings of the National Academy of Sciences*, 1950, **36**, 580–587.

Eyssen, H., de Somer, P., & Van Dijck, P. Further studies on antibiotic toxicity in guinea pigs. *Antibiotics and Chemotherapy*, 1957, **7**, 55–64.

Fernandez, C., Singh, H., & Perlman, H. Effect of short-term hypothermia on cochlear responses. *Acta Oto-Laryngologica*, 1958, **49**, 189–205.

Geddes, L. A., and Baker, L. E. *Principles of applied biomedical instrumentation.* New York: Wiley, 1968. Pp. 116–117.

Grinnell, A. D. Comparative physiology of hearing. *Annual Review of Physiology*, 1969, **31**, 545–580.

Guinan, J. J., & Peake, W. T. Middle-ear characteristics of anesthetized cats. *Journal of the Acoustical Society of America*, 1967, **41**, 1237–1261.

Gulick, W. L., & Cutt, R. A. The effects of abnormal body temperature upon the ear: Cooling. *Annals of Otology, Rhinology and Laryngology*, 1960, **69**, 35–50.

Gulick, W. L., & Cutt, R. A. Intracochlear temperature and the cochlear response. *Annals of Otology, Rhinology and Laryngology*, 1962, **71**, 331–340.

Harrison, J. B. Temperature effects on responses in the auditory system of the little brown bat *Myotis lucifugus*. *Physiological Zoology*, 1965, **38**, 34–48.

Hawkins, J. E., Jr. Hearing. *Annual Review of Physiology*, 1964, **26**, 453–480.

Kahana, L., Rosenblith, W. A., & Galambos, R. Effect of temperature change on roundwindow response in the hamster. *American Journal of Physiology*, 1950, **163**, 213–223.

Konishi, T., & Yasuno, T. Summating potential of the cochlea in the guinea pig. *Journal of the Acoustical Society of America*, 1963, **35**, 1448–1452.

Licklider, J. C. R. Basic correlates of the auditory stimulus. In S. S. Stevens (Ed.), *Handbook of experimental psychology*. New York: Wiley, 1951.

Meikle, M. B., Brummett, R. E., & Vernon, J. A. Spontaneous activity of the middle ear muscles in the guinea pig: An indicator of respiratory insufficiency, in preparation.

Misrahy, G. A., Shinabarger, E. W., & Arnold, J. E. Changes in cochlear endolymphatic oxygen availability, action potential, and microphonic during and following asphyxia, hypoxia, and exposure to loud sounds. *Journal of the Acoustical Society of America*, 1958, **30**, 701–704.

Petersen, H. Microphonics in vacuum tubes. *Brüel & Kjaer Technical Review*, 1958, **4**, 1–14.

Peterson, E. A., & Heaton, W. C. Peripheral auditory responses in representative edentates. *Journal of Auditory Research*, 1968, **8**, 171–184.

Peterson, E. A., Heaton, W. C., & Wruble, S. D. Levels of auditory response in fissiped carnivores. *Journal of Mammalogy*, 1969, **50**, 566–578.

Peterson, E. A., Wruble, S. D., & Ponzoli, V. I. Auditory responses in tree shrews and primates. *Journal of Auditory Research*, 1968, **8**, 345–355.

Rahm, W. E., Jr., Strother, W. F., & Gulick, W. L. The stability of the cochlear response through time. *Annals of Otology, Rhinology, and Laryngology*, 1958, **67**, 972–977.

Rice, E. A., & Shinabarger, E. W. Studies on the endolymphatic dc potential of the guinea pig's cochlea. *Journal of the Acoustical Society of America*, 1961, **33**, 922–925.

Riley, J. A. A simple method for welding thermocouples. *Science*, 1949, **109**, 281.

Ruben, R. J., Fisch, U., & Hudson, W. Properties of the eighth nerve action potential. *Journal of the Acoustical Society of America*, 1962, **34**, 99–102.

Schwartzkopff, J. Structure and function of the ear and of the auditory brain areas in birds. In A. V. S. de Reuck & J. Knight (Eds.), *Hearing mechanisms in vertebrates*. Ciba Foundation Symposium, Boston, Massachusetts: Little, Brown, 1968. Pp. 41–63.

Simmons, F. B. Permanent intracochlear electrodes in cats, tissue tolerance, and cochlear microphonics. *Laryngoscope*, 1967, **77**, 171–186.

Small, A. Audition. *Annual Review of Physiology*, 1963, **14**, 115–154.

Strother, W. F. Hearing in frogs. *Journal of Auditory Research*, 1962, **2**, 279–286.

Suga, F., Morimitsu, T., & Matsuo, K. Endocochlear DC potential: How is it maintained along the cochlear turns? *Annals of Otology, Rhinology, and Laryngology,* 1964, **73,** 924–933.

Suga, F., Nakashima, T., & Snow, J. B., Jr. Sodium and potassium ions in endolymph. *Archives of Otolaryngology,* 1970, **91,** 37–43.

Tasaki, I., Davis, H., & Eldredge, D. H. Exploration of cochlear potentials in guinea pig with a microelectrode. *Journal of the Acoustical Society of America,* 1954, **26,** 765–773.

Tasaki, I., Davis, H., & Legouix, J.-P. The space–time pattern of the cochlear microphonics (guinea pig), as recorded by differential electrodes. *Journal of the Acoustical Society of America,* 1952, **24,** 502–519.

Tasaki, I., & Fernandez, C. Modification of cochlear microphonics and action potentials by KCl solution and by direct currents. *Journal of Neurophysiology,* 1952, **15,** 497–512.

Vernon, J. A., Dalland, J., & Wever, E. G. Further studies of hearing in the bat, *Myotis lucifugus,* by means of cochlear potentials. *Journal of Auditory Research,* 1966, **6,** 153–163.

Vernon, J., Herman, P., & Peterson, E. Cochlear potentials in the kangaroo rat, *Dipodomys merriami. Physiological Zoology,* 1971, **44,** 112–118.

von Békésy, G. DC potentials and energy balance of the cochlear partition. *Journal of the Acoustical Society of America,* 1951, **23,** 576–582.

von Békésy, G. DC resting potentials inside the cochlear partition. *Journal of the Acoustical Society of America,* 1952, **24,** 72–76.

Wever, E. G. Electrical potentials of the cochlea. *Physiological Reviews,* 1966, **46,** 102–127.

Wever, E. G. Tonal differentiation in the lizard ear. *Laryngoscope,* 1967, **77,** 1962–1973.

Wever, E. G., & Bray, C. W. Action currents in the auditory nerve in response to acoustical stimulation. *Proceedings of the National Academy of Sciences,* 1930, **16,** 344–350.

Wever, E. G., & Vernon, J. A. The sensitivity of the turtle's ear as shown by its electrical potentials. *Proceedings of the National Academy of Sciences,* 1956, **42,** 213–220.

Wever, E. G., & Vernon, J. A. The problem of hearing in snakes. *Journal of Auditory Research,* 1960, **1,** 77–83.

Wever, E. G., & Vernon, J. A. Cochlear potentials in the marmoset. *Proceedings of the National Academy of Sciences,* 1961, **47,** 739–741. (a)

Wever, E. G., & Vernon, J. A. Hearing in the bat, *Myotis lucifugus,* as shown by the cochlear potentials. *Journal of Auditory Research,* 1961, **1,** 158–175. (b)

Wever, E. G., Vernon, J. A., & Peterson, E. A. The high-frequency sensitivity of the guinea pig ear. *Proceedings of the National Academy of Sciences,* 1963, **49,** 319–322.

Wever, E. G., Vernon, J. A., Rahm, W. E., & Strother, W. F. Cochlear potentials in the cat in response to high-frequency sounds. *Proceedings of the National Academy of Sciences,* 1958, **44,** 1087–1090.

Chapter 2

The Electroretinogram

Harold Koopowitz

Department of Developmental and Cell Biology
University of California
Irvine, California

I. Introduction	64
II. History of the Techniques	64
III. Recording Techniques	66
A. Recording Electrodes	66
B. Electrode Configurations	70
C. Electrode Position	70
D. Signal Amplification	71
E. Averaging Devices	72
IV. The Stimulus	73
A. Photostimulators	73
B. Direct Current Light Sources	74
C. Shutters	74
D. Filters	74
E. Light Guides	75
F. Monitoring of the Stimulus	75
V. Kinds of Preparations	76
A. Humans	76
B. Mammalian Preparations	78
C. Lower Vertebrates: Reptiles, Amphibians, and Fishes	80
D. Arthropods	81
E. Other Invertebrates	83
VI. Measurements and Interpretation	83
References	85

I. Introduction

The electroretinogram, or ERG, is a mass electrical response recorded from an eye that has been subjected to a light stimulus. These responses have been utilized for a considerable time to study various aspects of photoreception. Although these methods are now giving way to single unit recording situations, they are still valuable as experimental techniques, especially where intracellular recording is not feasible. Studies utilizing ERG recordings have ranged from human subjects to freshwater flatworms. The kinds of waveforms which can be recorded are very variable and depend not only on the species used to make the measurements but also the recording paradigm. Consequently, this chapter will not consider the interpretation and possible physiological bases of the various waveforms that can be measured; instead emphasis will be placed only on methodologies.

Basically the techniques needed to record an ERG are very simple and this may explain some of the appeal of this approach. Experimental situations tend to differ, depending on available resources and individual idiosyncrasies, and it is not the contention of this chapter to spell out exact experimental "recipes" but rather to indicate possible approaches and the various pitfalls that might be encountered. The ease with which ERGs can be measured, however, is countered by problems of interpretation. This is especially true in the case of the lower animals and these problems should make one hesitate before utilizing ERGs. Although the information which can be obtained from mass electrical responses measured outside the eye is limited, localized intraretinal recordings do tend to be more meaningful. In this chapter all extracellular recording situations will be considered to fall under the purview of the ERG although in a strict sense the term should be used for mass extraoptic records.

Following a short history on the evolution of techniques, experimental procedures will be discussed, followed by a description of a number of different kinds of preparations.

II. History of the Techniques

Animal electricity fascinated the early workers in the middle of the last century and they set about documenting its occurrence, even though its significance was not clearly understood. By 1849 Reymond du Bois had noted a resting potential between the front and back of the eye. Sixteen years later Frithiof Holmgren (1865) measured an oscillation in the electrical potential across a frog's eye produced in response to illumination. Dewar and McKendrick demonstrated for the first time (1873) the graded nature of the ERG and were able to show that the response amplitude could be correlated with the logarithm of stimulus intensity. Dewar (1877) was also the first person to measure a human ERG. He connected the recording electrode to a small clay cup which was filled with a salt solution and pressed so that the saline was in contact with his subject's eyeball. The fact that the ERG was actually produced by the retina was demonstrated by Kuhne and Steiner (1881) who had dissected the retina free from a frog eye and still recorded a recognizable ERG across it while the empty eyeball no longer responded to light.

Relatively little appears to have been done until the turn of the century when Gotch (1903) coined the term "off effect" to describe the potential oscillations at the cessation of the stimulus. He also noted that the latency of the "off effect" was shorter than that of the initial excitatory "on" response. Einthoven and Jolly (1908) labeled the various portions of the vertebrate ERG by letter and further tried to dissect the response into its constituent parts. It was soon evident that not all animals produced similar shaped ERGs. Some were quite simple depolarizations like those of the squid (Frölich, 1914, 1921) while others were quite complex such as those ERGs recorded from insect eyes (Hartline, 1928). Early workers on invertebrate eyes noticed that some of the more complex ERGs superficially resembled those of mammals except that the response appeared to be positive-going where it should have been negative and vice versa. In an effort to match the depolarizing invertebrate receptors with the hyperpolarizing eyes of vertebrates, it became conventional to invert invertebrate ERGs. It is only recently that invertebrate responses have begun to be portrayed correctly.

The first attempts to use ERGs as a clinical tool were made in 1924 (Kahn & Loewenstein, 1924). Unfortunately, the attempt was not successful and the project was abandoned. This first failure and the general lack of satisfactory equipment appear to have discouraged routine work on human subjects until the first contact lens electrodes were designed and used by Riggs (1941) and Karpe (1945).

The next real advance in methodology was the use of intraretinal electrodes by Tomita (1950). Prior to the advent of this approach, the ERG had been dissected by indirect means. Granit (1947; 1955, for reviews) had been able to divide the ERG into a number of basic components but it was rather difficult to assign specific cell types or layers to each component. Tomita (1950) advanced a microelectrode through a frog retina and noted changes in waveform at different levels.

Introduction of small electrodes which could be moved among the retinal cells gave a great impetus to vision research. The discovery that sodium aspartate appears to block synapses between the receptor cells and second-order neurons has allowed workers to isolate the receptor potentials (Sillman,

Ito, & Tomita, 1969). It is now possible to work on transduction processes with extracellular electrodes and some assurance that one has the isolated receptor potential and is not measuring the responses from other cell types. From this point it seems that advances in understanding integration mechanisms in the retina will come from intracellular recordings such as the brilliant work of Werblin and Dowling (1969), and Werblin (1971) on mudpuppy retinas.

Paralleling work on vertebrate retinas has been research on arthropod eyes, particularly insects. Many insects have ERGs with complex waveforms and major attempts at analysis have been made by Autrum (1958), Ruck (1958), and Mazokhin-Porshnyakov (1969).

III. Recording Techniques

A. Recording Electrodes

The kinds and configurations of electrodes are legion. Nearly any electrical conductor in contact with an eye can be used to measure an ERG, but certain electrodes, because of the ease with which they can be made or their recording characteristics, have been more favored than others. This section will deal with the most popular types. Basically, electrodes can be divided into two kinds, those utilizing liquid conductors such as salines and those which are solid metal conductors.

1. SALINE CONDUCTORS

Usually the saline is confined to glass tubing or contact lens electrodes, but wick electrodes in contact with the eye can also be used.

a. WICK ELECTRODES. Wick electrodes can be made of a number of different kinds of fibers impregnated with saline solution. Usual substances are cotton or synthetic fibers. The wick itself has to be connected to either an amplifier or preamplifier. Chlorided silver wire is probably the best kind of connection. The simplest way to chloride the wire is to pass current through the wire while it is immersed in an NaCl solution. Sufficient chloride will have been deposited when the wire is gray. Twist the chlorided wire to the wick so that it will be in good contact with the saline solution. In human subjects the wick can be placed at the outer corner of the eye. It is a good idea to fray the end of the wick so that it is somewhat spread out; this will ensure effective contact. The major use of a wick occurs where a certain amount of eye movement is desired and must be tolerated, but major difficulties can occur if the subject finds the wick irritating. A local anesthetic may be indicated. The greatest drawback to these electrodes lies in the fact that they tend to dry out very rapidly and hence cannot be used for extended periods of time without constant rewetting. To a certain extent this can be overcome by confining the wick inside a plastic sleeve. The following electrode type is more suitable for human and other mammalian eyes.

b. CONTACT LENS ELECTRODES. These are essentially contact lenses which have a small hole drilled into them and a metal electrode glued in the hole. The hole communicates to a small chamber next to the cornea which is filled with saline. A large number of different configurations have been designed (see Jacobson, 1961 for details). Some electrodes are held in position by suction and others have a number of apertures so that salines may be withdrawn or added during their use. In some variations wick electrodes have been used instead of metal and calcium alginate wool has been reported useful for this purpose as it is less irritating than other substances. One of the main problems caused by their use is irritation. When contact electrodes are used clinically, they are usually worn after a local anesthetic has been administered. Experimental subjects tend to be able to tolerate them for about 4 hr with some practice. Another disadvantage appears if the experimenter is concerned with accurate stimulus characterization. The electrodes may scatter or absorb light and this will necessitate additional correction factors in the calculations.

c. GLASS CAPILLARY ELECTRODES. One usually uses glass capillary electrodes to record from within cells but they can be modified for gross extracellular recordings. Electrodes meant for intracellular recording generally have very fine tips (less than 1 mμ) and consequently are rather fragile. Larger tips can be made relatively easily, by cutting the shank of these electrodes under a microscope. However, with a pore opening of about 35 μ, the saline filling the electrode tends either to run out of the tubing or is drawn into the preparation. This problem can be circumvented by filling the electrodes with an agar–salt solution. A 2% solution of agar should be made up in the saline and brought to a boil so that the agar is melted and stirred through the solution. By means of a Touhy adapter the glass electrode should be attached to a hypodermic syringe. The tip of the electrode is placed in the hot agar–salt solution and the plunger rapidly drawn out. Usually about 1 or 1½ inch of agar will flow into the tubing. The electrode can be withdrawn and allowed to cool until the agar has solidified. A chlorided silver wire may then be pushed into the agar to make contact with the electrolyte. Such electrodes can be reused many times and have comparatively low resistances (300 kΩ). These electrodes are also useful for measuring ERGs from eyes submerged in saline or sea water as the glass walls of the tubing provide ideal insulation.

d. SUCTION ELECTRODES. Often, especially with invertebrate preparations, it will be desirable to measure the ERG from an eye submerged in saline or even sea water. An external electrode which can be used for this purpose is the suction electrode. One way of making them follows: flexible electrodes can be made by drawing molten Tygon tubing, but the melting temperature for this is quite critical. An easy way to adjust this is by hand, holding a piece of tubing over a broad-tipped soldering pen or iron which is hot. Once the tubing starts to melt, it can be lifted out of the heat and pulled like a piece of capillary glass. The tubing can be held quite firmly in a Touhy adapter which itself will fit onto the male Luer fitting of a syringe. In some cases it may be necessary to interpose a three-way stopcock between the adapter and syringe, especially if difficulty occurs in maintaining suction. The saline of the bath itself can be used as the conducting fluid. A chlorided silver wire can be introduced into the electrode either through a hole in the wall of the tubing (this tends to be difficult to reseal) or through a small metal tube which mates snugly with the Tygon tubing. One of the main problems is ensuring that tiny air bubbles do not obstruct the conducting fluid in the electrode tip.

2. SOLID-CORE ELECTRODES

In this category one usually considers metal electrodes. The usual metals in common use for ERG are platinum, platinum–iridium, stainless steel, and tungsten. Silver wire tends to be too soft and malleable for very fine electrodes. The wire is usually etched electrolytically down to the desired tip diameter and configuration. This is best done under microscopic inspection with the microscope mounted on its side so that the electrode can be moved vertically into and out of the solution. The usual etching solutions for platinum and platinum–iridium are 50% sodium cyanide + 30% NaOH mixture (Wohlbarsht, MacNichol, & Wagner, 1960); for stainless steel, equal parts of concentrated HCl and 3 M KCl (Green, 1958) are used; the etching solution for tungsten is saturated aqueous potassium nitrite (Hubel, 1957). Electrodes are usually etched against a carbon rod and current flow adjusted with a variable transformer. For the final polishing very low voltages should be used. As most of the etching solutions are highly caustic or acidic, one can prevent or retard splashing from the electrolytic bubbling by applying a thin layer of xylene on the surface of the etching solution. If the electrodes are to be used for recording localized events, they must be insulated except for the tip. The usual insulators are glass or one of a variety of resins. Glass is probably the most satisfactory material but must be melted before it can be applied and one should be careful that one has glass with similar coefficients of expansion as the metal. A bead of glass is melted and usually the glass is placed on an Ω-shaped wire filament which can be heated by passing current through the filament.

2. THE ELECTRORETINOGRAM

With an appropriate variable resistance the amount of current and hence the heat of the glass can be carefully controlled. The electrode is usually passed vertically through the drop of molten glass a few times. Another way of applying glass is to place the etched electrode in a glass capillary tube and draw the tube as if making a glass electrode. The drawn glass will make a snug fit to the electrode if the correct tubing is used. In both these cases the glass will cover the electrode tip and has to be removed. The most satisfactory method for doing this is to put a drop of hydrofluoric acid on a sheet of acrylic about the size of a normal glass slide and place this on a microscope stage. If the glass-covered electrode is held horizontally in some sort of holder so that the tip can be seen under medium power of the microscope, then the small drop of acid can be advanced on the tip and the amount of glass digested by the acid can be carefully controlled. For obvious reasons, one should be careful not to get the acid on the microscope objectives. The other way of insulating the electrodes is by coating them with insulating varnishes or resins. Usually the electrode is dipped into the substance and then dried. Some varnishes require baking. Varnishes usually cover the tip of the electrode also, and these are removed by placing the tip in a conducting fluid and passing high voltage current through the electrode. If this is done with a microscope, it can be controlled to ensure that only the tip is uncovered. Bubbles will form wherever the insulator is not present or cracked. If too high a voltage is applied, holes will be blown along the shaft of the electrode as well as the tip. It is a good idea to examine these electrodes for signs of bubbles before using them. As with many of these techniques, the correct voltages and currents will have to be found by trial and error.

One of the simplest kinds of electrodes to make is fire-etched tungsten. We have used these consistently with good results. Tungsten wire 0.1 mm in diameter is cut into 3- or 4-cm lengths. The wire is bent into a right angle about 1 cm from the end and the bent portion inserted into the tip of a 24-gauge hypodermic needle. It is pushed in until only about 1 to 1½ cm is left protruding from the tip of the needle. The bend in the wire makes contact with walls of the needle and also tends to hold the wire firmly in place. The wire should be held steadily in a hot gas flame—a bunsen burner will do—until the tip starts to glow. After a while the tip will become white hot and burn. It is possible to adjust the length and diameter of the electrode shank by advancing or retreating the glowing tip through the flame. Tips 1 μ in diameter can be made in this way. The electrodes will not appear to be as highly polished as electrodes which are electrolytically etched but as the surface is covered with an insulator this hardly matters. We insulate tungsten electrodes by dipping them repeatedly into a thin Insulex solution and then air drying them. After each dunking, the electrode is held vertically with the tip uppermost. If one blows across the tip gently, the Insulex which is very thin there dries and as it does, shrinks a little, exposing the tip. We usually repeat the dunking and blowing three times. Electrodes can be examined for "electrical leaks" by passing current through them while in saline solution. These electrodes have fairly low resistances of between 10 and 15 kΩ. We find that if an electrode tip becomes bent it can be flame-etched again. The flame burns the varnish off quite rapidly. It is virtually impossible to differentiate between reetched and new electrodes.

The main disadvantage of metal microelectrodes is that they may polarize during slow potential changes. This will tend to produce a fair amount of drift. However, the amount of drift seems to be quite variable depending on the preparation.

B. Electrode Configurations

Recordings can usually be made single-ended against ground with the indifferent electrode placed somewhere on the animal's body. In a number of early experiments the indifferent electrode was placed on the cornea of the unstimulated eye but the possibility of interocular transfer has always made this procedure liable to criticism. Another popular configuration is to record differentially between two electrodes, one recording from the eye and the other elsewhere; in this case a separate ground electrode is necessary. Differential recording allows one to reject a certain amount of noise, provided that it is common and in-phase to both electrodes. A simple circuit to improve the amount of rejection is given by Offner (1967). Single-ended recording is most useful where the preparation is very small, for example, a tiny eye in a saline bath with recording done in a virtually noise-free Faraday cage.

More useful than simple differential recording is differential recording between two points within the retina. This technique can be used to localize the origins of potentials measured on the eye surface. Usually a double-barreled electrode is used. These are saline-filled microcapillary electrodes, and a number of configurations are possible. Electrodes can be made by fixing two capillaries together with the tips a set distance behind each other. This is not an easy task if very small tip diameters are desired. Another way is to produce one electrode within the barrel of the other. The electrodes are drawn separately; if one is of much narrower tubing than the other it can be slipped down the bore of the wider micropipette. Under a microscope one can ensure that the inner electrode does protrude beyond the outer and measure the distance between them. Getting good electrodes is quite difficult.

C. Electrode Position

As most visual organs are quite complex structures often having layers of different kinds of sense and nerve cells, the kind of response that one measures may reflect the position of the electrode. This is probably not so important if one is merely measuring the mass ERG from the surface of the eye but once the electrode is within the retina not only will one measure changes in amplitude but also one may find changes in waveform. Figure 2-1 represents a series of ERGs taken from the moth *Galleria mellonella* at different depths within the eye. Note the progression from simple to complex and back to simple waveforms. It is, therefore, mandatory that one should attempt to position an internally recording electrode at the same depth and position each time. Unless the tissue is very well known, it may be difficult to extrapolate from one recording situation to another. Volume conduction may become a problem if an electrode is moved in and out a number of times and one may find transients which normally do not occur in certain layers suddenly appearing. Presumably these flow down conducting channels made around the electrode shafts.

D. Signal Amplification

Resistance–capacitance coupled amplifiers (a.c.) appear to be used most commonly; however, they are prone to introduce a number of errors into the ERG waveform (Peckham, 1971). The components of the ERG tend to contain some rather slow waveforms which compounds problems introduced by a.c. amplifiers. Fidelity of the amplifier response depends on the ability to follow low frequency electrical signals. Generally the longest time constants will give the truest picture, but some amplifiers will be somewhat unstable at these frequencies. The effects of changing the time constant for the lower cutoff frequencies are displayed in Fig. 2-2a. Note that as the time constant decreases, the ability to hold the slow components disappears; also, the negative overshoot of the fast transient appears to increase. Even with a low frequency cutoff such as 2 Hz, the amplifier may be unable to handle slow ERG changes. In Fig. 2-2b the same response is displayed fol-

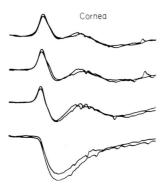

FIG. 2-1. Recordings made at successive depths within the retina of the bee moth, *Galleria*. Each pair of recordings was made after the tungsten electrode had been advanced 50 μ further into the retina. Vertical scale, 500 μV; horizontal, 5 msec.

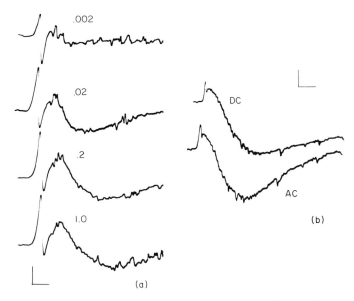

FIG. 2-2. (a) Recordings using an a.c. preamplifier. The numbers above each trace refer to the time constant of the preamplifier in seconds. Vertical scale, 200 μV; horizontal, 10 msec. (b) The same response after d.c. preamplification but with d.c. or a.c. coupling at the input stage to a Tektronix 3A72 vertical amplifier. Vertical scale, 500 μV; horizontal, 20 msec.

lowing both (low frequency cutoff 2 Hz) a.c. and d.c. amplification. Any studies on the time course of the response would be quite inaccurate if a.c. measurements were made. It follows then, that wherever possible, measurements should utilize d.c. recordings. There are, however, some disadvantages to using d.c. amplifiers. They tend to be unstable and one usually experiences some drift problems. However, many of the modern solid-state d.c. preamplifiers are more stable and these, together with d.c. balance adjust controls available on some oscilloscope amplifiers (e.g., Tektronix 3A9), will eliminate most drift problems.

E. *Averaging Devices*

An externally recorded ERG is generally of quite small amplitude, often in the order of microvolts, and it may be submerged in a great deal of background electrical noise. This may be particularly apparent if the subject is not confined to a shielded room or Faraday cage. The simplest but perhaps least desirable method of averaging merely consists of repeated superimposed recordings either on film or the screen of a storage oscilloscope. These can be used to determine if a small transient potential change is real or merely noise. However, because it is almost impossible to superimpose the traces exactly and because of the amount of noise, it is very difficult to make accurate measurements on this kind of record. Signal-averaging computers provide the best method for pulling an average signal out of a high noise background. There are a number of different kinds available. It may be useful to record data on an FM tape recorder so that it can be reexamined in different ways. One should be aware that in certain computers changing the analysis window may result in a change in amplitude scale. This will make direct comparisons between different responses difficult once the analysis sweep time itself has been changed. Useful information may be obtained from Perry and Childers (1969) and Pearlman (1966) on how these devices work and how to use them.

IV. The Stimulus

Both the amplitude and the latency of the response are dependent on parameters of the stimulus; it is, therefore, essential that the stimulus be well defined. The response of the receptor cells will be dependent not only on the total number of quanta but also on the light flux intensity; dynamic components of the receptor potential will reflect the latter parameter rather than the former. It is useful, therefore, to know both the intensity and duration of the stimulus, as well as its quantal content.

Two main kinds of stimuli are (1) those produced by discharge tubes such as the commercial photostimulators which are available, and (2) steady light sources which need shutters to regulate the stimulus duration. Fluorescent tubes should not be used because the light intensity fluctuates with the line current and many animals are able to follow the flicker fluctuations. For the same reason the line voltage is not suitable for incandescent lamps either. The best light sources should have low-ripple d.c. voltage power supplies, and one should be able to adjust and set the voltage and amperage to predetermined values.

A. Photostimulators

These tend to put out a blue-white light and the flashes are generally of the order of microseconds. Because of the short duration, the shape of the stimulus is not very important. It is usually not possible to measure the light intensity with a thermopile as these seem unable to follow very high frequencies. The quantal content of the flash would have to be calculated from foot-candle measurements supplied by the manufacturer. One should be careful of using flash frequencies of greater than ten per second as the intensity of the individual flashes tends to decrease with increasing frequency. One should also be careful that the photostimulator does not introduce a rapid transient stimulus artifact into the records.

B. Direct Current Light Sources

Perhaps the best light is monochromatic light from a high intensity monochromator. These sources, however, tend to be expensive and it may be more reasonable to use a light source with filters to cut out the unneeded wavelengths. (See Section IV,D.) Halogen lamps which have a halogen in the walls of the glass are better than normal incandescent globes. They are brighter and longer lived with more blue and less red light. The spectral characteristics of lamps are usually available from the manufacturers.

C. Shutters

The ideal stimulus is a well-defined square stimulus with the light abruptly reaching its greatest intensity well before the receptors have had time to respond. A number of shutters are available. Iris diaphragm shutters are commonly available but do not give well-defined square stimuli. Focal plane shutters present a straight edge which moves across the aperture; these are difficult to obtain as a separate instrument but can be obtained from dismantled cameras. Window shutters also present a straight edge. They consist of an opaque disk which interrupts the light source and has one or more windows cut in it. As the disk revolves, it spins, and the window will allow the stimulus to pass through. The duration depends on the size and speed of the window. Mechanically operated shutters will have to be recalibrated every so often. Modern camera film allows a wide latitude for exposure error and camera shutters may be up to 20% out in their designated shutter speeds.

D. Filters

Neutral density filters are usually used to cut down light intensity. The spectral emission of a lamp depends on the temperature of the filament and hence on the current flowing through it. If light intensity is adjusted by changing the voltage of the source, then the spectral composition will also change. Neutral density (n.d.) filters and wedges interposed between the source and the preparation will cut down all parts of the visible spectrum equally. Usually the filters cut down the intensity by logarithmic values and hence a log 3.0 filter will let much less light through than a log 0.5 n.d. filter. Filters are made of gelatin film or glass sandwiches with silver deposited between the plates.

Narrow band-pass filters for isolating specific colors are available. It will be necessary to get the transmission specifics for each filter. Interference filters will usually have a second band pass on another region of the spectrum and this will need to be masked. Gelatin type filters usually have very wide band passes and should only be used for masking. The band pass is usually measured by the peak wavelength (λ_{max}) the half-band pass width (the wavelengths on each side of the λ_{max} which are transmitted at half of the peak transmission value), and the amount of light at the λ_{max} wavelength allowed through the filter.

Nearly all incandescent light sources put out considerable heat. It is, therefore, necessary to use infrared filters to screen this from the preparation. Specific infrared filters are obtainable or they can be made from glass-sided containers filled with water. Cobalt chloride solutions are efficient at absorbing infrared but will also cut down light at the red end of the spectrum. The efficiency of solution filters should be monitored from time to time and one should ensure that the solution is well filtered and clear.

E. Light Guides

Localized and confined stimuli can be applied by means of light guides. These are most useful if one wishes to stimulate small parts of a retina or individual facets of a compound eye. There are a few precautions which should be taken. The light intensity emerging at the tip of the guide is dependent on the amount of reflection from the inner surface of the guide. If low light intensity is used, then there may be a significant change in the stimulus intensity if the light guide is twisted into a new position. Another problem concerns light scatter. One can expect considerable scatter as the light emerges from the guide and this may stimulate adjacent areas unless the guide is pressed tightly against the organ. One way of reducing this is to extend an opaque sleeve beyond the tip of the light guide.

F. Monitoring of the Stimulus

One often needs to record the onset and offset of a stimulus as these changes can be used to trigger oscilloscopes, computers, and other devices. One may also want a measure of the stimulus as an illustration in a photograph. A rapidly responding circuit which is useful not only for triggering but also calibrating shutter speeds can be made from a photoconductive cell, resistor, and current source in series. The photoconductive cell acts as a diode and only allows current to flow when it is illuminated. This can be picked up across the resistor and used to trigger the oscilloscope. Useful guides to measuring light stimuli can be found in the volumes by Seliger and McElroy (1965) and Crawford, Granger, and Weale (1968) and the reader is referred to those books.

V. Kinds of Preparations

A. Humans

Because of man's egocentric nature he has a tendency to render his science in terms of himself. If the ultimate goal is to understand man and his place in the universe, then surely for those interested in vision the major goal is understanding all aspects of human photoreception. It is quite clear from trends in recent years that the major accomplishments in retinal research will involve single unit studies. It appears that only in this way can one approach the complexities involved with interactions between succeeding strata in the retina and also between adjacent cells sharing the same levels. For obvious reasons this kind of approach is not amenable to human studies. Here external recording techniques and hence mass electrical events must be utilized. ERGs are, however, quite useful for studying psychophysical phenomena, especially the retinal contributions towards them (see Riggs & Wooten, 1972 for a recent review). Other studies which might profitably utilize ERGs are studies on sensitivity changes and adaptational states where experimental procedures may require recordings for time periods longer than usually available with intracellular techniques.

One of the problems posed by human preparations is the dual nature of the photoreceptor layer, and some means of separating the rod and cone contributions to the ERG is often desirable. The easiest way to achieve a separate rod retinogram is by stimulating the eye with violet light. The rods are stimulated by this light (± 400–410 nm) and cone responses will not be obtained unless the stimulus is intense. Violet light may be obtained by by either of two means. The simplest way is to use interference filters in front of the light source, but one should endeavor to have filters which have only a very narrow band pass. Gelatin or cellophane type filters of broad band pass usually allow light of other wavelengths through. Most normal incandescent light sources only have a relatively small violet component, and high intensity lamps must be used. Grating monochromators are another source of violet light which will have a narrow band pass.

Gouras (1970) has suggested a comparatively simple method for isolating a pure cone ERG. This is based on the fact that the cones are more sensitive to red light than the rods. The subject is dark adapted and given a flash of violet light, of just above threshold value. Subjectively the flash should appear to be whitish or colorless. The subject then matches a red flash to coincide with the violet. Neutral density filters can be used to dim the lights until they match. ERGs recorded from these flashes should be nearly identical in amplitude. Now a white background illumination is introduced. As this adapting light is gradually brightened, the response to the violet light will start to disappear as the rods bleach out or fail to respond. A background intensity will be reached where the violet light will not produce an effect. The response to the red light at the same background illumination will be due to the cone components since the rods will no longer be contributing to the ERG for that particular intensity light flash.

One of the problems often faced when working with ERGs is producing a homogeneous stimulus over the whole retina. This kind of stimulus is known as ganzfeld. The simplest way of producing a ganzfeld is to cover the eye with half a ping-pong ball. This will tend to diffuse the stimulus fairly evenly. Other devices with the same end result are fairly easy to construct.

The kinds of electrodes used on human subjects are wick or more commonly contact lens electrodes. Jacobson (1961) is one of the best sources of information if one wishes to conduct experiments with human subjects. A recent review of ERG literature which may prove helpful is that of Riggs (1965).

Clinical Uses

At a first glance it is surprising, in view of the long history of ERG usage for research, that relatively little use has been made of the phenomenon for medical diagnosis. A considerable body of information has been built up correlating a variety of pathogenic states with aberrant ERG waveforms. Why then has the ERG not evolved into the common clinical tool that the ECG and EEG have for cardiovascular and brain dysfunction problems respectively? Actually there are a number of clinics where ERGs are used, but it can hardly be considered a routine procedure for the majority of eye specialists. There are a number of reasons why there has been an apparent reluctance to resort to ERGs for clinical use. Among the more important reasons is the fact that the eye is one of the very few organs which is available for internal visual examination from outside the body. It is a simple matter to look inside an eye and see if something is not normal, hence this sort of examination is popular and more preferred. Another problem posed by ERGs is that the waveform is usually small and buried in noise. In order to examine small nuances in ERG shape, an averaging computer is required, a piece of equipment which is relatively expensive for a clinician. Another difficulty is posed by the methodology. The electrodes are usually placed under local anesthesia and the patient should be dark adapted—processes which do not lend themselves to rapid patient turnover. The final stumbling block concerns interpretation of the measured response. Except for cases where an entire component of the waveform may be missing, it is difficult to be sure of the real causal factors for slight changes. Despite these problems and shortcomings, the ERG still remains the best way of examining the physiology of the retina, especially where obvious morphological changes have not occurred. It can be expected that most use will be derived from retinal degenerative pathologies not only for diagnosis but an understanding of the pathology itself. More information can be obtained from Jacobson's classic (1961) and the various proceedings of the ISCERG (International Symposia on Clinical ERG).

B. Mammalian Preparations

A number of mammalian preparations are more popular than others, some because they are common laboratory animals and others for their special properties. At the present time, the trend appears to be toward working with excised eyes or retinas; however, it is not clear if these preparations are worth the extra time and trouble required to maintain them. Localized microelectric recording is probably easier in the excised organ but for gross ERG studies intact animals are probably simpler to use.

There are some general points which can be applied to nearly all mammalian preparations. The eyelids and the nictitating membrane can affect the response in a number of ways. They may mechanically move the electrode, which will cause drift with d.c. amplifiers. A conducting electrode in contact with the lids will pick up electrical activity such as muscle spikes from those organs and these will interfer with the ERG records. In acute preparations the eyelids and nictitating membrane can be cut away. However, if the animal is not to be sacrificed afterward, the lids may be held open with a speculum. The nictitating membrane is a problem and must be held retracted. The practice in some laboratories of cutting away the nictitating membrane in nonacute preparations merely to facilitate experimental procedure may fall outside animal cruelty and vivisection laws in some areas.

A variable that has to be controlled in human as well as other vertebrate eyes is pupil diameter. Usually the eye is dilated by application of a few drops of an atropine solution, before the animal has been dark adapted. A local anesthetic may also be required to reduce sensitivity of the cornea and decrease eye movements. Animals will have to be held quite firmly in a harness or other device. It is usual to set up and even perform experiments while the animal is under a general anesthetic. Under these conditions careful controls must be undertaken to ensure that the effects of the anesthetic on both ERG shape and performance are understood.

Contact lens electrodes are probably the best kind for external recording but one might consider substances other than glass for making or getting them made.

Some comments on specific mammalian preparations are as follows:

1. CATS

These have been used for a considerable time and Granit's ground-breaking experiments were performed on these animals. It might be mentioned here that many nocturnal mammals have been considered to possess all rod retinas, yet very few "pure-rod" retinas seem to have borne up on close inspection.

2. RATS

Albino rats have usually been considered to have all rod retinas; however, there is a growing body of evidence that either cones or rodlike cones may be present in the eye (Massof & Jones, 1972).

3. GROUND SQUIRRELS

The ground squirrels, *Cittelus* spp., are thought to have all cone retinas and are adapted to vision in bright light (Crescitelli, 1972). Another animal thought to have an all cone retina is the tree shrew, *Tupaia glis*.

4. RABBITS

As with rats these are relatively cheap and common laboratory animals, on which numerous studies have been performed. A useful source of information on rabbit eyes can be found in the book by Prince (1964), but there is not overly much on actual ERG techniques in this reference.

5. MONKEYS

Among the monkeys, *Aotes trivirgatus* has been claimed to have an all-rod retina. Other monkeys have cones although the nocturnal prosimians probably have only or mainly rods too. Rhesus monkeys are the ones more commonly used.

6. BIRDS

Although not mammals, they are included here as they are homeotherms and require similar approaches. Pigeons have been used. Most of the comments which have been made about techniques in mammals can be applied here. In birds the retina appears to be dominated by cones and there are relatively few rods. There are a number of morphologically different cone types (Morris & Shorey, 1967). Contact lens electrodes are usually used. Birds are usually more sensitive to anesthetics than mammals and care must be exercised.

C. Lower Vertebrates: Reptiles, Amphibians, and Fishes

Frog eyes are probably one of the most popular preparations but any of these three classes could be approached the same way. They are in a way easier to work with than mammals or birds as one does not have to worry about keeping the eye at a warm temperature. In fact the preparation will last better at cold temperatures (about 15°C). Although whole animal preparations are feasible, the tendency is to use enucleated (excised) eyes. Methods have been worked out for using detached, perfused retina and these promise to be the most valuable kind of preparation.

1. Whole Animal Preparations

It is probably best to keep the animal paralyzed under curare. Frogs will last a long time but care should be taken to keep their skins moist to prevent dehydration. With fish it may be necessary to pump water through the gills. If a tube is introduced into the mouth, water can be pumped through it and after passing over the gills will emerge automatically. The major advantage for using whole animals is that the eye remains functional as long as the animal is alive, and long periods might be needed for dark adaptation studies. Wick electrodes are often used with these kinds of preparations. If one were to study marine fish, in which the entire animal is submerged in a conducting medium, suction electrodes would be useful.

2. Excised Eyes

Excised eyes only last for a relatively short period of time (2–3 hr) and must be considered to be deteriorating preparations. Nevertheless, they are convenient to use as they lack much of the background activity and noise that may be superimposed on intact animal ERGs. The animal should be completely dark adapted before enucleation, and the operation should be carried out rapidly and under very dim red illumination. This will require some practice beforehand to familiarize oneself with the placement of oculomotor muscles and the optic nerve. The muscles should be trimmed off as close to the eyeball as possible. Previously a soft agar–saline solution should be set in a dish. The excised eye then can be positioned half in the agar with the cornea uppermost. A recording wick electrode can be placed on the cornea with the indifferent electrode in the dish. The iris should be kept dilated with a few drops of atropine. If one wishes to use microelectrodes or bathe the receptors in certain solutions, then one will have to remove the cornea, iris, and lens. One might either slice through the eye at a point behind the lens or carefully cut away each structure separately. The vitreous humor can be left behind and the microelectrode advanced through this. One should be careful if the vitreous humor is removed as the retina is held in position by the vitreous humor and can become crumpled and folded if handled carelessly. Saline should not be introduced in the eye cup as it may change the characteristics of the ERG. A thin layer of vitreous humor will help to keep the retina moist.

3. Perfused Retina

After the vitreous humor has been removed, it is possible to cut free large pieces of retina. They can be held firmly between two rings and supported on a porous disk. The tissue can be oriented with either the receptors or ganglion cell layers uppermost. Details of the preparation procedures may be found in the articles by Sillman *et al.* (1969) and Tomita (1950). If asparate is added to the perfusate, it is possible to isolate the receptor potentials and hence it is possible to work on transduction processes with extracellular electrodes in a relatively clean situation.

D. Arthropods

Electroretinograms have been measured from a number of different invertebrate types; however, they tend to be used routinely only in the arthropods, where most of the studies appear to be centered. This section will deal mainly with insects.

1. Whole-Animal Preparations

When entire animals are used, a number of difficulties may be encountered. The major one concerns immobilizing the preparation. Small insects can be partially embedded in soft wax (surgical periphery wax). Larger insects can be immobilized in a low-melting-point hard wax. One should place a large drop of molten wax on a platform; as the wax starts to congeal, the insect thorax should be pressed into it. This will effectively entangle the legs. Insects may either be anesthetized with CO_2 gas or by placing them in a cold refrigerator (temperature about 4°C) for about 1 hr. Carbon dioxide is fast and the animals also recover rapidly from it. We have found that the ERG under anesthesia will be quite different from unanesthetized animals but recovery is complete. It is also necessary to wax the head to the thorax to avoid movement and one may wish to wax the abdomen down too. There are some precautions which should be taken. If the insect is scaly, such as a moth, the scales will stick to the wax but the animal may wiggle free. One should be careful not to wax over all of the spiracles so that the animals can still breathe. Once the insect has been mounted and is in position, a small hole can be made in the cornea. This should be done under a dissecting microscope. Minuten pins mounted on a handle are good for making small holes. Care should be taken that the hole merely breaks the cornea and does not damage the underlying cells. It usually takes some practice to make a small hole. If the cornea is very hard, then it may be impossible to punch a hole without distorting the eye. In this case, it may be necessary to cut away part of the exoskeleton around the eye and penetrate the retina from that side. The electrode is positioned and advanced into the hole with a micromanipulator. The indifferent electrode is often placed on the cornea of the other eye, but if this is the case, one should be careful to ensure that the light stimulus is confined to the recording eye. We have found it convenient to place the indifferent electrode in the abdomen of the animal.

Whole animal preparations offer a number of advantages. The eye remains functional as long as the animal is alive and hence one can make a preparation which will often function for days if desired. The animal efficiently oxygenates and nourishes the organ, functions which must be performed by the experimenter if only the eyes or heads are used.

2. Head and Eye Preparations

For some experimental setups it may be more convenient to use a detached insect head or even an excised eye. In this case it is usually necessary to submerge the organ in an oxygenated Ringer's solution. If the eye or head is very small with respect to the size of the bath, oxygenation may not be necessary, especially if the preparation is kept at a cool temperature. The cornea should be raised above the level of the solution, so that the solution cannot enter through the hole for the electrode. The ionic environment of the retinula cells appears to be quite different from the general hemolymph (Swihart, 1972), and saline solutions may have odd effects on the recorded response. There are a number of insect Ringer's solutions (Hale, 1958; Pringle, 1938) but it is usually difficult to find one ideally suited to the animal unless one is working on the more common laboratory species. With unusual species the best that one can do is either search the older literature to see if a saline has been developed or use one of the general ones. It should be obvious that it is very important that the basement membrane of the retina must remain intact. If this is broken the ERG will change its characteristics and may not represent the normal condition. With some preparations it may be necessary to slice off the surface of the cornea and this would allow the preparation to become dessicated. One way to avoid drying without using a saline is to position the electrode in the retina and then cover the cut surface with petroleum jelly. Gently heating the Vaseline will melt it and it can be applied round the open surface (Järvilehto & Zettler, 1971). Once the jelly has congealed it is still possible to position the electrode, which may be advanced or retracted as long as the tip is not brought in contact with the jelly.

E. Other Invertebrates

One of the main problems involved with working on many invertebrates is holding the animal steady while measurements are being taken. There are a number of ways of attacking the problem. A terrestrial arthropod can be waxed down on a platform as described for insects. Carbon dioxide gas will act as an anesthetic on most terrestrial arthropods. The immobilized animal then can be waxed down or operated upon. If the eyes are to be cut out, then the comments from the previous section can be applied. Small aquatic arthropods can often be immobilized by pressing them into a large glob of Vaseline. A number of animals especially some of the worms, are difficult to keep stationary even if only pieces of the animal are being used. Isotonic $MgCl_2$ can be used as an anesthetic for many marine animals but is not effective on all kinds. An ingenious way of immobilizing freshwater planarians was developed by Brown and Ogden (1968). They held the animal in place with an electron microscope specimen grid, and the electrode was introduced into the eye through one of the holes in the grid.

As with arthropods, salines have not been worked out for many marine invertebrates but sea water makes a fair substitute and pieces of most invertebrates will survive for many hours in cold oxygenated sea water. Suction electrodes will be necessary if an external ERG is measured from an eye in sea water but because it is such a good conductor the indifferent electrode may merely be placed in the bath.

VI. Measurements and Interpretation

The amplitude of the various waveforms which go to make up the ERG is the easiest parameter of the response to measure. This is, however, one of the most variable of parameters dependent on the state of light adaptation, stimulus, and general physiological state of the preparation. If amplitude measurements are to be used, the test stimuli will need to be repeated a number of times. Usually there is considerable variation between response amplitudes, even to the same stimulus. It is also difficult to extrapolate conditions from one preparation to another. Normalizing the data may be of some help and one will soon learn how reliable these kinds of data are. There are some insects which do lend themselves to amplitude measurements, but most animals do not. A more reliable measure is latency, that is, the time from the onset of the stimulus to the onset of the response. Both latency and amplitudes tend to have maximum and minimum possible values and if one is working toward the ends of the range, data may be difficult to interpret. Perhaps the most reliable data involve threshold measurements. Often, however, workers may use a stimulus which is slightly above threshold and then relate this to changes in threshold; such measurements are considered to be "relative threshold" data. If one is investigating problems concerned with processing information, then threshold or near threshold stimuli may not reveal ways that the retina has for coping with sensory modulation and moderating intensity stimuli.

There is a considerable body of criticism against the use of ERGs in the study of vision. Perhaps the most valid criticism which can be levied is that the retina comprises an outpocket of the brain and as such the ERG represents a simplistic view of the tremendously complex interactions of its individual parts. Vision often entails discriminations involving a mere handful of these cells and it seems obvious that one will be unable to deduce the interactions of these receptors and neurons from a mass electrical recording. Further, there has been a tendency to study ERGs in an effort to understand the phenomenon rather than how it might relate or reflect visual processes. Most of the other criticism should be directed at the slippiness of the workers and the tendency to interpret their data casually and without stringent controls. Problems in interpretation of insect compound eye ERGs have been summarized by Horridge (Bullock & Horridge, 1965) but these also apply to many other kinds of eyes as well. Present emphasis is on single unit intracellular techniques but these tend to represent a biased sample of the retina. The larger cells get the most attention and in many eyes the cells are simply too small to allow these techniques. Furthermore, it is difficult to localize the site of recording accurately without dye injecting the impaled cells, a process which does not lend itself readily to every impaled cell. Future techniques may simplify or even make these techniques routine. Until then, however, localized extracellular recordings will be necessary for building up a picture of the physiology of the majority of animal eyes. At the present time, the ERG still remains the easiest means for recording activity from the eyes of most animals.

Acknowledgments

Much of the lore of electrophysiology is passed by word of mouth, hence it is often difficult to assign credit where it is due. I would like to apologize to those who feel that they may have been slighted in this fashion. I would like to thank Professors F. Crescitelli and R. Josephson for their time and discussion and F. E. Dudek who helped consolidate some of my ideas. The responsibility for both comments and any possible mistakes is mine.

References

Autrum, H. Electrophysiological analysis of the visual systems in insects. *Experimental Cell Research,* Supplement, 1958, **5,** 426–439.

Brown, H. M., & Ogden, T. E. The electrical response of the planarian ocellus. *Journal of General Physiology,* 1968, **51,** 237–253.

Bullock, T. H., & Horridge, G. A. *Structure and function in the nervous systems of invertebrates.* Vol. II. San Francisco: Freeman, 1965.

Crawford, B. H., Granger, G. W., & Weale, R. A. (Eds.) *Techniques of photostimulation in biology.* Amsterdam: North-Holland Publ., 1968.

Crescitelli, F. The visual cells and visual pigments of the vertebrate eye. In H. J. A. Dartnall (Ed.), *Photochemistry of vision. Handbook of sensory physiology.* Berlin and New York: Springer-Verlag, 1972. Pp. 245–363.

Dewar, J. The physiological action of light. *Nature,* 1877, **15,** 433.

Dewar, J., & McKendrick, J. G. On the physiological action of light. *Journal of Anatomical Physiology,* 1873, **7,** 275–282.

du Bois, R. E. Untersuchunger über thierische Electricitet. Berlin: Reimer, 1849.

Einthoven, W., & Jolly, W. A. The form and magnitude of the electrical response of the eye to stimulation by light at various intensities. *Quarterly Journal of Experimental Physiology,* 1908, **1,** 373–416.

Fröhlich, F. W. Beiträge zur allgemeinen Physiologie der Sinnesorgane. *Zeitschrift Sinnesphysiologie,* 1914, **48,** 28–164.

Fröhlich, F. W. *Grundzüge einer Lehre vom Light- und Farbensinnen. Ein Betrag zur allgemeinen Physiologie der Sinne.* Jena: Fischer, 1921.

Gotch, G. The time relations of the photoelectric changes in the eyeball of the frog. *Journal of Physiology,* 1903, **29,** 388–410.

Gouras, P. Electroretinography: some basic principles. *Investigative Ophthalmology,* 1970, **9,** 557–569.

Granit, R. *Sensory mechanisms of the retina.* London and New York: Oxford Univ. Press, 1947.

Granit, R. *Receptors and sensory perception.* New Haven: Yale Univ. Press, 1955.

Green, J. D. A simple microelectrode for recording from the central nervous system. *Nature (London),* 1958, **182,** 962.

Hale, L. J. *Biological laboratory data.* London: Methuen, 1958.

Hartline, H. K. A quantitative and descriptive study of the electrical response to illumination of the arthropod eye. *American Journal of Physiology,* 1928, **83,** 466–483.

Holmgren, F., Method att objectivera effecten av ljusintryck pa retina. *Upsala läkareförening Förhandlingar,* 1865, **1,** 177–191.

Hubel, D. H. Tungsten microelectrode for recording from single units. *Science,* 1957, **125,** 549–550.

Jacobson, J. H. *Clinical electroretinography.* Springfield: Thomas, 1961.

Järvilehto, M., & Zettler, F. Localized intracellular potentials from pre- and postsynaptic components in the external plexiform layer of an insect retina. *Zeitschrift für vergleichende Physiologie,* 1971, **75,** 422–440.

Kahn, R., & Loewenstein, A. Das Electroretinogram. *Graefe's Archiv für Ophthalmologie,* 1924, **114,** 304.

Karpe, G. Basis of clinical electroretinography. *Acta Opthalmologica Supplement,* 1945, **24.**

Kuhne, W., & Steiner, J. Elektrische Vorgange im Sehorgane. *Untersuchungen aus dem Physiologischen Institut der Universität Heidelberg,* 1881, **4,** 64.

Massof, R. W., & Jones, A. E. Electroretinographic evidence for a photopic system in the rat. *Vision Research,* 1972, **12,** 1231–1239.

Mazokhin-Porshnyakov, G. A. *Insect vision.* New York: Plenum, 1969.

Morris, V. B., & Shorey, C. D. An electron microscope study of types of receptor in the chick retina. *Journal of Comparative Neurology,* 1967, **129,** 313–339.

Offner, F. F. *Electronics for Biologists.* New York: McGraw-Hill, 1967.

Pearlman, J. T. Computer averaging technique for routine ERG studies: A normative series in rabbits. In H. M. Burian and J. H. Jacobson (Eds.), *Clinical electroretinography.* Oxford: Pergamon, 1966.

Peckham, R. H. The AC amplifier artifact in electroretinography. *American Journal of Optometry and Archives of the American Academy of Optometry,* 1971, **48,** 932–935.

Perry, N. W., & Childers, D. G. *The human visual evoked response: Method and theory.* Springfield: Thomas, 1969.

Prince, J. H. (Ed.), *The rabbit in eye research.* Springfield: Thomas, 1964.

Pringle, J. W. S. Proprioception in insects. 1. A new type of mechanical receptor from the palps of the cockroach. *Journal of Experimental Biology,* 1938, **15,** 101–113.

Riggs, L. A. Continuous and reproducible records of the electrical activity of the human retina. *Proceedings of the Society of Experimental Biological Medicine,* 1941, **48,** 204–207.

Riggs, L. A. Electrophysiology of vision. In C. H. Graham (Ed.), *Vision and visual perception.* New York: Wiley, 1965.

Riggs, L. A., & Wooten, B. R. Electrical measures and psychophysical data on human vision. In D. Jameson and L. M. Hurvich (Eds.), *Visual psychophysics. Handbook of sensory physiology.* Berlin and New York: Springer-Verlag, 1972. Pp. 690–731.

Ruck, P. A comparison of the electrical responses of compound eyes and dorsal ocelli in four insect species. *Journal of Insect Physiology,* 1958, **2,** 261–274.

Seliger, H. H., & McElroy, W. D. *Light: Physical and Biological action.* New York: Academic Press, 1965.

Sillman, A. J., Ito, H., & Tomita, T. Studies on the mass receptor potential of the isolated frog retina. I. General properties of the response. *Vision Research,* 1969, **9,** 1435–1442.

Swihart, S. L. Variability and the nature of the insect electroretinogram. *Journal of Insect Physiology,* 1972, **18,** 1221–1240.

Tomita, T. Studies on the intraretinal action potential. Part I. Relation between the localization of micropipette in the retina and the shape of the intraretinal action potential. *Japanese Journal of Physiology,* 1950, **1,** 110–117.

Werblin, F. S. Adaptation in a vertebrate retina: Intracellular recording in *Necturus. Journal of Neurophysiology,* 1971, **34,** 228–241.

Werblin, F. S., & Dowling, J. E. Organization of the retina of the mudpuppy, *Necturus maculosus.* II. Intracellular recording. *Journal of Neurophysiology,* 1969, **32,** 339–355.

Wohlbarsht, M. L., MacNichol, E. F., & Wagner, H. G. Glass insulated platinum electrode. *Science,* 1960, **132,** 1309–1310.

Chapter 3

Recording of Bioelectric Activity: The Electro-Olfactogram

David S. Phillips

> Department of Medical Psychology
> University of Oregon Medical School
> Portland, Oregon

I. Introduction	87
II. Technique	92
A. Electrodes	92
B. Recording Equipment	92
C. Animals	93
D. Stimuli	93
References	94

I. Introduction

In 1956 Ottoson published an article describing a number of experiments conducted in his laboratory on the slow potential changes found in the olfactory epithelium. He called these slow potentials electro-olfactograms or EOGs. The existence of these potentials had been known for 20 years prior to Ottoson's work, but they had received little attention (Hosoya & Yashida, 1937). Ottoson's contribution was to systematically study some of the parameters influencing these potentials and to propose that they were generator potentials.

Ottoson studied the most common type of EOG, the slow electronegative monophasic potential sometimes referred to as the negative "on" EOG. This potential shows an initial rapid rise with stimulus onset followed by a decline to a static level which lasts as long as the stimulus is present. At the cessation of stimulation, the response gradually returns to base line. Ottoson demonstrated that within certain limits this response is proportional to the logarithm of the stimulus intensity (Fig. 3-1) and that a similar relationship exists between response amplitude and the volume of stimulating air for a given stimulus strength. An increase in stimulus intensity is reflected in a shorter latency, a more rapid rise time, a broader crest, and a lengthened decay time of the potential (Fig. 3-2).

The argument that the EOG is a generator potential is based upon several facts.

1. Like the generator potential of the Pacinian corpuscle (Gray & Sato, 1953), the EOG shows graded activity and is not abolished by cocaine applied in concentrations sufficient to block the response in the olfactory nerve.
2. As indicated above, it exhibits a definite relationship to the intensity of the stimulus.
3. Microelectrode recordings show that the response is maximal close to the surface of the olfactory epithelium but decreases rapidly with further penetration to the point that the response recorded from a depth of 150–170 μ is 80% of that recorded from the surface.
4. The EOG is not affected by antidromic stimulation of the olfactory nerve.
5. The EOG decreases in amplitude and eventually disappears following sectioning of the olfactory nerve. These changes in amplitude parallel retrograde degeneration of the olfactory epithelium (Takagi & Yajima, 1965).

Since Ottoson's original work, four other types of EOG responses have been found. In 1959 Takagi and Shibuya reported a negative "off" wave (Fig. 3-3). Takagi and his co-workers have studied this potential in great detail, including its underlying ionic mechanisms (Takagi, Kitamura, Imai, & Takeuchi, 1969b).

Positive EOG potentials have also been discovered. Takagi, Shibuya, Higashino and Arai (1960) first recorded the positive "on" potential in frogs, and Shibuya (1960) recorded a positive "off" potential in fish (Fig. 3-4). MacLeod (1959) has reported a positive afterpotential type of EOG in rabbits.

At first glance, one may be inclined to think that the variety of EOGs reported might be due to the different techniques, electrodes, and animals

3. THE ELECTRO-OLFACTOGRAM

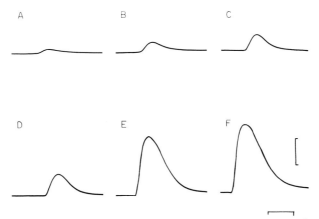

FIG. 3-1. Increase of amplitude of potential at increase of stimulus strength. Records of responses evoked by stimulation: (A) With purified air; B–F, with butanol. (B) 0.001 M. (C) 0.005 M. (D) 0.01 M. (E) 0.05 M. (F) 0.1 M. Volume of air 0.5 cm^3. Vertical line in F, 1 mV. Time bar 2 sec. (From Ottoson, 1956.)

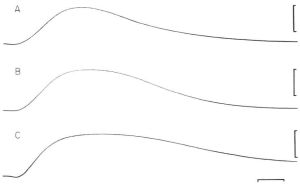

FIG. 2. Changes in shape of potential at increase of stimulus strength. Records of responses evoked by stimulation: (A) With 0.001 M. (B) 0.01 M. (C) 0.1 M butanol. Volume of air 1 cm^3. Vertical line in A 1 mV, in B 1.5 mV, in C 4 mV. Time bar .5 sec. (From Ottoson, 1956.)

employed in the various studies; however, more recent work indicates that some of the differences are due to the stimuli used. Takagi, Aoki, Iino, and Yajima (1969a) recorded EOGs in the frog to 122 odorants, each presented in three different concentrations. They found that 106 of these stimuli produced only negative EOGs which increased in amplitude with increases in concentration. Six of the stimuli produced only positive EOGs, again showing a positive correlation between response amplitude and stimulus concentration. Three stimuli produced only negative EOGs at the lowest concentration and only positive EOGs at the highest concentration. The remaining seven stimuli yielded a mixed EOG, a negative followed by a positive.

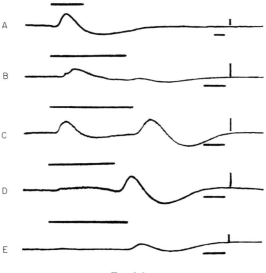

Fig. 3-3.

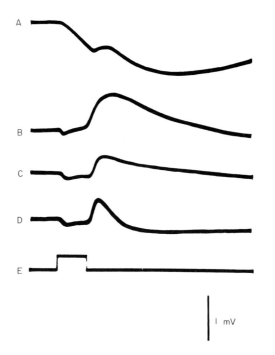

1 mV

Fig. 3-4.

3. THE ELECTRO-OLFACTOGRAM

The question of whether one or all of the different types of EOGs are true generator potentials has yet to be completely settled. Takagi has published a number of papers on this subject, and in his 1967 paper he noted a number of facts that raise problems for the generator potential interpretation. He charges that changes in the amplitude of the negative "on" response are not paralleled by changes in the induced wave in the olfactory bulb and reports an inverse relationship between the concentration of ethyl ether and the amplitude of the negative "on" EOG. Kimura (1961), recording from strands of the olfactory nerve, has shown that activity in this structure does follow the negative "on" EOG, and Doving (1964) has shown this also to be the case in the second-order neurons. These reports, together with other observations to be cited shortly, raise questions as to whether ether or chloroform should be used as a stimulus for EOG studies.

Takagi (1967) has also criticized the generator potential interpretation on the basis that (1) induced waves may be recorded from the olfactory bulb to both negative "on" and positive "on" EOGs and, further, (2) off-induced waves in the olfactory bulb may precede the "off" EOG. In response to these criticisms, it should be noted that Ottoson made the generator potential claim only for the negative "on" EOG. Ottoson reported the occurrence of polyphasic responses but regarded them as abnormal since they were seen only in animals exposed to strong stimulus concentrations or in deteriorating preparations. Reports by Takagi and Shibuya (1959) that the magnitude and shape of their "on–off" potential varied from time to time, and by Higashino and Takagi (1964) that some of the responses could only be obtained with saturated vapors of chloroform or ether, appear to bear out this point and to raise some doubt as to whether polyphasic responses "normally" occur in the olfactory mucosa unless, perhaps, they are due to interactions between activity in different areas of the mucosa. It is well known that "off" responses may be recorded from second- or third-order neurons in other sensory systems but reports of "off" responses from receptors are rare.

FIG. 3-3. On- and off-responses. (A) On-response. (B), (C), and (D) Various types of on-off-responses. The on-response is bigger than the off-response in (B) and vice versa in (D). (E) Off-response. The horizontal lines above the potentials on the left indicate the durations of olfactory stimulation. The vertical and horizontal bars on the right show 50 μV and 1 sec, respectively. (From Takagi & Shibuya, 1959.)

FIG. 3-4. Positive EOGs of various types. (A) Typical EOG elicited by chloroform vapor. This shape is always observed in the ceiling olfactory epithelium of the nasal cavity. (B) Small on-potential is followed by a large negative off-potential. This is often followed by a small positive afterpotential. (C) and (D) Intermediate shapes between A and B. The EOGs in B, C, and D are most often observed in the olfactory eminentia. Time of stimulation is indicated at bottom right. (From Takagi, Aoki, Iino, & Yajima, 1969a.)

Shibuya (1960) reported that the EOG may be abolished by placing a small piece of absorbant paper on the olfactory mucosa and then removing the paper prior to stimulation. While this does destroy the EOG, responses may still be recorded from olfactory nerve fibers. Shibuya was careful to identify the area of the mucosa to be "blotted" by antidromic stimulation of the nerve fibers; however, since all of the mucosa was not treated, the question may be raised as to whether fibers arising outside of the treated area might not have contributed to the fiber response.

At the Third International Symposium on Olfaction and Taste, Takagi (1969) stated his belief that the negative EOG is usually a composite response composed of true generator potentials which are negative and two positive potentials which are probably of a secretory nature. At the same symposium, Shibuya (1969) stated that the dc-positive shift may be the true generator potential. Obviously, much more work remains to be done before a satisfactory conclusion can be reached on this matter.

II. Technique

A. Electrodes

Since it has been shown that certain types of electrodes will record sustained nonbiological potentials, the selection of the proper type of electrode for EOG recording is essential. Mozell (1962) investigated platinum, tungsten, nichrome, 3 M KCl, and Ringer–agar silver–silver chloride electrodes and found that the latter were the best for recording EOGs. This type of electrode either does not record nonbiological potentials or, if it does, they are of such a magnitude as not to present problems.

The manufacture of Ringer–silver–silver chloride electrodes is a relatively simple process. A 2-inch section of small diameter silver wire is immersed in a saturated NaCl solution and connected to the anode of a 1.5 V flashlight battery. After a few hours, the immersed section of wire will have a light gray coating. The wire is then removed from the solution and inserted into a glass pipette which has been filled with Ringer's solution. The tip diameter of the pipette should be 0.1–0.2 mm. These electrodes may be stored in Ringer's solution until they are needed.

B. Recording Equipment

The EOGs may be recorded in two different ways depending upon the preference of the investigator and the equipment available. One method is to connect the electrodes to a low-level d.c. preamplifier such as a Grass

P1 or P5, then into an amplifier and write the signal out on polygraph paper. The second method is to connect the electrodes to a symmetrical direct-coupled amplifier (Haapanen, 1953) and display the signal on an oscilloscope. This signal can then be photographed from the oscilloscope face using a Grass camera to provide a permanent record.

C. Animals

The variety of animals used in EOG recording has been relatively restricted owing to the problems involved in exposing the olfactory mucosa. Ottoson (1954, 1959a) has recorded these potentials from rabbits, Shibuya (1960, 1964) from fish, Hosoya and Yashida (1937) from dogs, and Osterhammel, Terkildsen, and Zilstorff (1969) from man. An electroantennogram (EAG) has been recorded from insect antennae (Schneider, 1957). The most popular animals for this type of research have been frogs and toads (Doving, 1964; Gesteland, Lettvin & Pitts, 1965; Ottoson, 1956, 1959b; Takagi et al., 1969b). The problem of exposing the mucosa in these animals is relatively minor and, once exposed, the receptor sheet is relatively flat. Furthermore, these animals may be decapitated, the head placed in a chamber of Ringer saturated cotton, and the EOG recorded for up to 6 hr before the preparation begins to deteriorate. This procedure eliminates the possible effects of anesthetics.

It has been observed in all animals in which EOG activity has been investigated that these responses occur only in the pigmented part of the olfactory epithelium. Within the pigmented epithelium, response amplitude is usually greatest near the midline and anteriorly.

D. Stimuli

Any odorous compound may be used as a stimulus to elicit the EOG. As noted previously (Takagi et al., 1969b), the odorants used may influence the type of EOG reported. Certain compounds such as ether or chloroform have been found to destroy the olfactory epithelium and, hence, permanently abolish the EOG when presented in strong concentrations and/or for prolonged time periods. The procedure preferred by most investigators is to use stimulus presentations of 1 sec duration with 1–15 min between stimulations. The length of the interstimulus interval should be adjusted depending upon the odorant used and the concentration at which it is presented. This is to allow the epithelium to recover completely. Generally, strong concentrations are to be avoided. In between stimulus presentations pure air should be passed over the olfactory epithelium to eliminate any lingering traces of the odorant.

References

Döving, K. B. Studies of the relation between the frog's electro-olfactogram (EOG) and single unit activity in the olfactory bulb. *Acta Physiologica Scandinavia*, 1964, **60,** 150–163.

Gray, J. A. B., & Sato, M. Properties of the receptor potential in Pacinian corpuscle. *Journal of Physiology*, 1953, **122,** 610–636.

Gesteland, R. C., Lettvin, J. Y., & Pitts, W. H. Chemical transmission in the nose of the frog. *Journal of Physiology*, 1965, **181,** 529–559.

Haapanen, L. A direct coupled amplifier for electrophysiological investigations. *Acta Physiologica Scandinavia*, 1953, **29** (Suppl. 106), 157–160.

Higashino, S., & Takagi, S. F. The effect of electrotonus on the olfactory epithelium. *Journal of General Physiology*, 1964, **48,** 323–335.

Hosoya, Y., & Yashida, H. Ueber die bioelektrische Ersheinungen an der Riechschleimhaut. *Japanese Journal of Medical Sciences III Biophysics*, 1937, **5,** 22–23.

Kimura, K. Olfactory nerve response of the frog. *Kumamoto Medical Journal*, 1961, **14,** 37–46.

MacLeod, P. Première données sur l'électro-olfactogramme du lapin. *Journal of Physiology (Paris)*, 1959, **51,** 85–92.

Mozell, M. M. Olfactory mucosal and neural responses in the frog. *American Journal of Physiology*, 1962, **203,** 353–358.

Osterhammel, P., Terkildsen, K., & Zilstorff, K. Electro-olfactograms in man. *Journal of Laryngology and Otolaryngology*, 1969, **83,** 731–733.

Ottoson, D. Sustained potentials evoked by olfactory stimulation. *Acta Physiologica Scandinavia*, 1954, **32,** 384–386.

Ottoson, D. Analysis of the electrical activity of the olfactory epithelium. *Acta Physiologica Scandinavia*, 1956, **35** (Suppl. 122), 1–83.

Ottoson, D. Studies on the slow potentials in the rabbit's olfactory bulb and nasal mucosa. *Acta Physiologica Scandinavia*, 1959, **47,** 136–148. (a)

Ottoson, D. Comparison of slow potentials evoked in the frog's nasal mucosa and olfactory bulb by natural stimulation. *Acta Physiologica Scandinavia*, 1959, **47,** 149–459. (b)

Schneider, D. Elektrophysiologische Untersuchungen von Chemo-und Mechanorezeptoren der Antenne des Seidenspinners *Bombyx mori* L. *Zeitschrift für Vergleich. Physiologie*, 1957, **40,** 8–41.

Shibuya, T. The electrical responses of the olfactory epithelium of some fishes. *Japanese Journal of Physiology*, 1960, **10,** 317–326.

Shibuya, T., Dissociation of olfactory neural responses and mucosal potential. *Science*, 1964, **143,** 1338–1340.

Shibuya, T. Activities of single olfactory receptor cells. In C. Pfaffmann (Ed.), *Olfaction and taste III*. New York: Rockefeller Univ. Press, 1969. Pp. 109–116.

Takagi, S. F. Are EOG's generator potentials? In T. Hayashi (Ed.), *Olfaction and taste II*. Oxford: Pergamon, 1967. Pp. 167–179.

Takagi, S. F. EOG problems. In C. Pfaffmann (Ed.), *Olfaction and taste III*. New York: Rockefeller Univ. Press, 1969. Pp. 71–91.

Takagi, S. F., & Shibuya, T. "On"- and "off"-responses of the olfactory epithelium. *Nature*, 1959, **184,** 60.

Takagi, S. F., & Yajima, T. Electrical activity and histological change in degenerating olfactory epithelium. *Journal of General Physiology*, 1965, **48,** 559–569.

Takagi, S. F., Shibuya, T., Higashino, S., & Arai, T. The stimulative and anaesthetic actions of ether on the olfactory epithelium of the frog and toad. *Japanese Journal of Physiology*, 1960, **10,** 571–584.

Takagi, S. F., Shibuya, T., Higashino, S., & Arai, T. The stimulative and anaesthetic actions of ether on the olfactory epithelium of the frog and toad. *Japanese Journal of Physiology*, 1960, **10,** 571–584.

Takagi, S. F., Aoki, K., Iino, M., & Yajima, T. The electropositive potential in the normal and degenerating olfactory epithelium. In C. Pfaffmann (Ed.), *Olfaction and taste III.* New York: Rockefeller Univ. Press, 1969. Pp. 92–108. (a)

Takagi, S. F., Kitamura, H., Imai, K., & Takeuchi, H. Further studies on the roles of sodium and potassium in the generation of the electro-olfactogram: Effects of mon-, di-, and trivalent cations. *Journal of General Physiology*, 1969, **53,** 115–130. (b)

Effector Processes

Chapter 4

Recording of Human Eye Movements

Bernard Tursky

Department of Political Science
State University of New York
Stony Brook, New York

I. Introduction	100
II. Neuromuscular Control Mechanisms	100
III. Eye Movement Recording Methods	104
A. Direct Observation	104
B. Mechanical Coupling	104
C. Photography and Corneal Reflection	105
D. Photoelectric Recording	110
E. Recent Developments	110
F. Electro-Oculography	111
IV. History of Electro-Oculography	113
V. Electro-Ocolography Recording Problems	114
A. Instrumentation	114
B. Stability of the Electrode Skin Circuit	120
C. Head Movement	123
VI. Applications	128
VII. Conclusions	130
References	131

I. Introduction

The accurate recording of the position and movement of the eyes has been of scientific and clinical interest for many years, and the methods used to record the various types of eye movements have been thoroughly reviewed (Carmichael & Dearborn, 1947; Shackel, 1960a, 1967; Kris, 1960; Young, 1963). This chapter briefly describes some of the early methods used to record eye movements and the physiological mechanisms that control these movements. The major interest of this chapter is focused on the development and use of the three most often utilized techniques for recording and measuring human eye movements: photography, photoelectric recording, and electro-oculography (EOG). Though the utility of these methods may overlap, it is important to report on all of them because the choice of method is frequently dictated by the range and accuracy of the movements to be measured, the amount of constraint that the clinical or experimental procedure imposes on the subject, and the availability of recording equipment in the laboratory.

II. Neuromuscular Control Mechanisms

The structure of the eye may be classified under three major functional headings, one dealing directly with the function of sight, the second with the elements that protect the eyes (eyelids and eyebrows) and the third with the neuromuscular functions that control their movement. Volumes have been written on the optical properties of the eyes. Polyak (1948) thoroughly reviewed the history of the investigation of the optical system. Granit (1947) investigated the electrical potentials related to excitation and inhibition in the retina and the optic nerve, dark and light adaptation, and color reception. The function of the movement of the eyes is also important in the visual process. Recent work by several investigators (Stark, Vossius, & Young, 1962; Young & Stark, 1963; Fender, 1964; Steinback & Held, 1968) has utilized dynamic eye tracking methods to demonstrate that tracking movements of the eyes are divided into smooth pursuit and saccadic jumps that serve separate functions in the viewing process.

The eyes and their protective appendages (eyebrows and eyelids) are controlled by a complex musculature (Fig. 4-1) inervated by several cranial nerves. The eyelids are movable folds which protect the eye from injury caused by exposure. Three muscles control the eyelid. The obicularis muscle acts as a sphincter of the lids. It lies directly beneath the skin, surrounding the opening between the lids (palpebral fissure) and is inervated by the seventh cranial nerve. The lid is opened and closed by the levator palpebrae

4. RECORDING OF HUMAN EYE MOVEMENTS

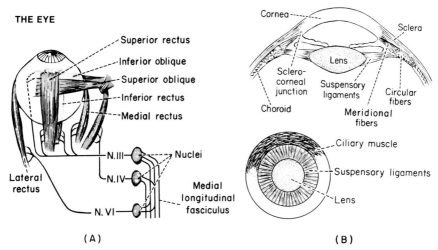

FIG. 4-1. Neuromuscular mechanisms that control the movement and focusing of the eyes. (A) The extra ocular muscles. (B) The focusing (accommodation) mechanisms. (From Guyton, 1969.)

superioris muscle which is activated by the oculomotor or third cranial nerve. Mueller's muscle is a layer of unstriped muscle tissue that is sympathetically inervated and aids in widening the palpebral fissure. One of the functions of these muscles is to initiate eye blinks, which in turn sweep tears generated in the lachrimal apparatus across the eye to help keep the eyeball moist, thus maintaining the transparency of the cornea and helping to keep it free of dust particles.

Gross movements of the eyes are controlled by six extrinsic muscles (Fig. 4-1). Four recti muscles aid in the control of vertical and horizontal movements of the eyes. These muscles originate at the apex of the eye and extend forward around the globe in divergent directions connecting with the sclera near the limbus. The principal function of the recti muscles is to direct the line of vision. Vertical movements are controlled by the superior and inferior rectus, horizontal movements by the medial and lateral rectus. The oculomotor nerve inervates the superior, inferior, and medial muscles, but the lateral rectus is controlled by the abducens (sixth cranial) nerve. Rotary or oblique movements of the eye are aided by the inferior and superior oblique muscles which rotate the eyeball around the antero–posterior axis. The inferior muscle is inervated by the oculomotor nerve and the superior oblique muscle is inervated by the trochlear (fourth cranial) nerve. The eyes are focused on the visual field by changing the curvature and thickness of the lens. This is accomplished by the ciliary muscles which are inervated by the oculomotor (third cranial) nerve. Finally, pupil size is controlled by two intrinsic muscles that are antagonistic in action. Constriction of the pupil

TABLE 4-1
Eye Movements

Type	Description	Size (degrees of arc)	Latency (msec)	Speed (degrees/sec)	Possible recording methods
1. Saccadic movements	Conjugate fast eye movements that carry the eyes from one fixation point to another	.5–50	100–500	100–500	Photography Corneal reflection Photoelectric Electro-oculography (EOG)
2. Smooth pursuit movements	Conjugate involuntary slow eye movements to follow slowly moving targets	1–60	200	1–30	EOG Photoelectric Photography
3. Compensatory movements	Smooth conjugate involuntary movements used to compensate for passive or active movements of the head or body	1–30	10–100	1–30	Photography Photoelectric EOG
4. Vergent movements	Nonconjugate movements of the eye to maintain binocular vision. Vergence movements are smoother and slower than conjugate pursuit and compensatory movements	1–15	—	6–15	Corneal reflection Photography Photoelectric EOG
5. Torsional or rolling eye movements	Involuntary movements around the line of gaze that compensate for the displacement of the visual vertical	—	—	—	EOG Photographic Photoelectric

4. RECORDING OF HUMAN EYE MOVEMENTS

		Less than 1°			Contact lens	Corneal reflection			Photography
6. Miniature eye movements	Tiny involuntary movements that occur during periods of fixation. These movements have been classified into these categories		—	—	—	—			
a. Flicks	Sharp saccadic movements								
b. Drifts	Slow movements between flicks								
c. Tremor	Rapid oscillating movements								
7. Nystagmoid	Eye movements of an oscillating or unstable nature classified into three categories	—	—	—	EOG	Photoelectric	—	—	
Occular or optikinetic	Movement of the eyes trying to follow a nonhomogeneous field that is continuously moving past the observer	—	—	—					
Vestibular	Compensating movements to overcome problems due to impairment of vestibular nerve.	—	—	—					
Spontaneous or central nystagmus	Occurs when the gaze is directed peripherally and is usually a sign of impairment of the central visual and vestibular pathways	—	—	—	—	—	—	—	
8. Intraocular movements	Pupillary reflex contraction to change in illumination	—	—	—	—	—	—	—	Photography

is controlled by the sphincter pupillae which surrounds the pupil and is inervated by the parasympathetic nerves to the eye. Dilation of the pupil is controlled by the sympathetically inervated dilator pupillae which consists of radiating muscle fibers.

The eyes can be moved to fix on objects within a circular area having a diameter equal to approximately 100° of visual angle. Horizontal movements to left and right are equal but there is a greater ability for downward vertical movement than for upward. The control ability of the ocular system is indeed remarkable when you consider the complex feedback system that operates to ensure the extreme coordination that must exist in the eye muscles to achieve and maintain useful binocular vision. Such simple tasks as reading require simultaneous conjugate movements and a slight convergence, while a change in fixation from a near to a distant point requires a simultaneous three-way change: dilation of the pupil, accommodation of the lens, and divergence of the two eyes. Eye movements have been carefully classified by size, rate, and function into several categories. Table 4-1 lists and describes the various types of eye movement, their physical properties, and some of the possible methods of recording each type.

III. Eye Movement Recording Methods

A. Direct Observation

The simplest and least interfering of the methods used to systematically study human eye movements is direct observation. This method is still the primary technique used by clinicians to detect gross disorders associated with saccadic or pursuit movements. Techniques have been devised to make this type of observation as unobtrusive as possible and instruments have been designed to make it possible precisely to observe eye movements from a distance (Newhall, 1928). Early investigators (Lamansky, 1869; Bruckner, 1902) who were interested in more quantitative measures of eye movement devised ingenious methods of using the afterimage of a bright light focused on the pupil to obtain quantifiable data on the movements of the eyes. The obvious disadvantage of these methods for experimental and clinical purposes was the lack of a permanent record of the phenomenon being observed, making it difficult to compare data from subject to subject and across time on the same subject.

B. Mechanical Coupling

Crude but imaginative permanent recordings of eye movements were first achieved in the last decade of the nineteenth century by the use of a

4. RECORDING OF HUMAN EYE MOVEMENTS

direct mechanical linkage between the eye and a smoked drum kymograph recorder. Delabarre (1898) made a plaster cast of an artificial eye and obtained a smooth concave surface that would fit over the anesthetized eye of his subject. A hole about the size of the pupil was drilled through the cast and a wire ring was embedded in the cast surrounding the hole. A thread connected the ring to a recording lever that recorded eye movements on the smoked surface of a kymograph drum. Huey (1898) independently developed a similar technique (Fig. 4-2) and also introduced the use of a bite board coated with soft wax to ensure the permanent fixation of the subject's head with reference to the recording apparatus. These interesting devices set an early standard for the quantitative recording and measurement of eye movement. The drawbacks to this type of recording were obvious. The insertion of a large foreign objects in the eye was painful and constricting, small saccadic movements were lost in the inertia of the system, and the use of kymograph drums restricted the recording time.

C. Photography and Corneal Reflection

The cumbersome mechanical transducers were replaced in the early twentieth century by the direct photographic recording of the displacements of the light–dark boundary between the sclera and the iris which produced a continuous record of horizontal eye movements. Dodge developed a falling plate photographic process in 1899 to produce the first recording of eye movements on film (Dodge & Kline, 1901). The development of continuous film drives (Dearborn, 1906) and a kinetoscope camera (Weiss, 1911) that recorded eye movements on individual film frames improved the resolution of the eye movement data and increased the recording time. Further improvement in the photographic method of recording eye movements was

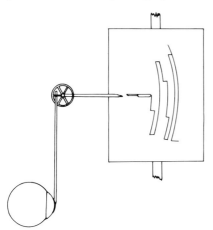

FIG. 4-2. Recording of eye movements by direct coupling. Apparatus consists of a smooth cup capping the cornea coupled by means of a light rod and wheel to a glass pointer that writes a kymograph record proportional to eye movements. (From Huey, 1899.)

also achieved by improving the reflective qualities of the eye. Bright foreign objects such as a flake of Chinese white (Judd, McAllister, & Steele, 1905) or a silver bead (Dodge, 1907a, b) were attached to the part of the eye that was being photographed. Though these photographic methods were a great improvement over the direct linkage systems they still suffered from the drawbacks of not being sensitive enough to record small saccadic movements, and they were especially prone to artifacts produced by head movements.

Further improvement in the photographic methods of recording eye movements was achieved by capturing on film the reflection of light from the cornea (Dearborn, 1906). A beam of light can be partially reflected from the smooth spherical surface of the cornea, forming a bright spot or highlight on the cornea. The position of this highlight moves in the same direction as the cornea and the recorded distance of movement is directly proportional to the actual eye movement. This reflected beam can be easily photographed either on moving film or by successive frame photography for future examination.

The most sensitive of eye movement recording methods employing light reflection is achieved by the use of a contact lens which can be fitted directly over the cornea (Riggs, Ratliff, Cornsweet, & Cornsweet, 1953). A plane mirror attached to the lens reflects light from a fixed source which may be observed directly or photographed for future analysis. There are several advantages in the use of contact lenses over direct corneal reflection. Artifacts of reflection caused by irregularities in the surface of the cornea are eliminated and head movement artifacts are reduced. Byford and Stuart (1961) attached a small lamp to the contact lens and used the direct light from the lamp to improve the sensitivity of the recording. As the lamp moves with the eye, the amount of light detected by a photomultiplier through a narrow aperture varies in proportion to the eye position. Fender (1964) used a reflective contact lens mirror system to record small tracking movements and then improved his technique by placing a tiny medical lamp in the stem of the mirror so that it shone into a photomultiplier tube mounted perpendicular to the subject's line of sight. The output current of the tube varied as the light fluctuated with movement, making it possible to detect minute eye tremors as well as large saccades. The use of contact lenses has of course the obvious problems and disadvantages of having to insert a foreign body in the eye. A second problem that restricts the usefulness of this method is the possibility of the lens slipping and causing errors in measurement. This problem limits the accuracy of measurement to $5.0°$ of arc and therefore makes this method suitable for recording only small eye movements.

4. RECORDING OF HUMAN EYE MOVEMENTS

In recent years several important innovations have been introduced that combine corneal reflection recording with real time photography of the subject's visual field. Wendt (1952) synchronized a stimulus film with the eye movement recording camera at 24 frames per second and then projected both films on the same screen, thus superimposing the eye fixation dot on the field of vision picture. Mackworth and Mackworth (1958) developed a television camera technique that superimposed the corneal reflection on the scene being viewed by the subject. In this instance two TV cameras were used, one trained on the eye, the other on the scene and the output of these cameras was electronically mixed and displayed on one TV monitor which was photographed by a 16-mm motion picture camera running at six frames per second. This technique provided simple, directly interpretable permanent records for the study of the eye movement patterns, and the information yielded by this apparatus permitted studies of search patterns and fixation sequences as well as defining the specific areas that are looked at or avoided by the subject. This technique produced excellent records of eye movements and fixation but did not give any information about the natural interaction between head and eyes in the viewing process. Shackel (1960b) devised a system that utilized a TV camera mounted on a crash helmet that could be worn by the subject. The camera, though bulky, was always pointed directly forward and moved with the head, ensuring the recording of the visual scene. Horizontal and vertical eye movements were recorded by electro-oculographic means and the EOG potentials were amplified and used to displace a spot on a calibrated cathode ray tube to show the two dimensions of eye movement. The visual scene from the head camera was displayed on a TV monitor and the eye fixation spot was picked up from the oscilloscope by a second camera and superimposed on the visual scene to show exactly where the subject was looking. Shackel criticized his own apparatus as being too heavy and unwieldy and suggested the possibility of lighter equipment in the form of a system that employed an 8-mm motion picture camera and utilized superimposed corneal reflection to map eye position.

Mackworth and Thomas (1962) reported a head-mounted apparatus that used a lightweight 8-mm motion picture camera. The camera was rigidly mounted and turned with the subject's head to record the field of vision through one lens. A second lens provided an image of the corneal reflection of a built-in light source, and a periscope and a beam-splitting prism were used to combine the visual scene and the corneal reflection into a permanent film record of the area of regard. This portable system permitted the subject freedom of movement under a wide range of conditions. A recent improvement in this method produced and marketed by the Polymetric Co. of Hoboken N.J. reduced the bulkiness of the system by separating it into two component parts (Fig. 4-3). The motion picture camera is carried by a shoulder strap and the optical unit, weighing less than 2 lb, is secured to the subject's head by a head band and bite board. The optical unit transmits the image of the area viewed by the subject; the corneal reflection from the subject's eye is superimposed on this field, and the images are transmitted to a 16-mm camera by a pair of high-resolution fiber optic cables. These techniques have been used extensively in the study of driver and consumer behaviors and other viewing behavior that requires freedom of head and body movement.

One important area of human ocular behavior that cannot be easily investigated by the previously described methods is the viewing behavior of the human infant. Unique problems are introduced when dealing with the nonverbal or understanding child. The normal photographic, photoelectric, or corneal reflection methods are usually not feasible because they require the cooperation of the subject in both the calibration and recording procedure. Kessen and Hirshenson (1963) tried to photograph the corneal reflection of the newborn and found that eye closing produced by normal photographic lighting conditions made it impossible to obtain a useful record. Their introduction of infrared lighting techniques reduced this problem and permitted them to record reflections from the infant's eyes. Haith (1969) reasoned that direct photographic procedures using infrared lighting and a closed circuit TV system (Fig. 4-4) could be used to accurately record and measure the occular behavior of the human infant. The system developed by Haith used a filtered window in the light spectrum between 900 and 1200 mμ to illuminate the infant's eyes. The radiation produced by this type of lighting is approximately 1/200 of that produced by scattered light on a sunny day and therefore is extremely safe. A TV camera records the baby's eye on videotape and any segment of tape can be stored on a video disk to improve the quality of measurement. The specification of eye position, pupil dilation, or extent of lid opening is found by specifying the Cartesian coordinate position of two points on the recorded image. (The pupil center and the infrared reflection.) Since the TV monitor image is generated by a single flying spot moving from top to bottom of the screen in 525 horizontal sweeps, it becomes reasonable to determine the Y component of the measurement point by counting the number of sweeps. A similar process is used to determine the X coordinate by counting at a 10 MHz rate on each horizontal line. A photosensitive probe placed on the spot to be measured activates the two counters, one determining the horizontal position, the other the vertical. These numbers are stored on magnetic tape for further processing. This unique system, though complex, permits the accurate unconstrained recording of occular behavior in the infant.

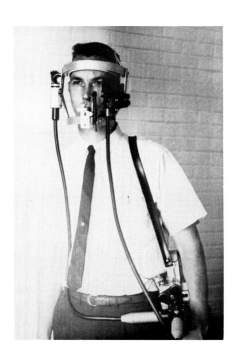

FIG. 4-3. Mobile photographic eye movement recorder (produced by Polymetric Co., series V-0165). Optical unit is worn on head: weight 2 lb; 16-mm motion picture camera is carried by means of shoulder strap.

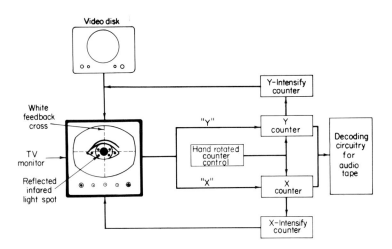

FIG. 4-4. Schematic diagram of TV system used to record and measure ocular behavior in the human infant. (From M. M. Haith, Infrared television recording and measurement of ocular behavior in the human infant, *American Psychologist,* 1969, **24,** 3, 279–283. Copyright 1969 by the American Psychological Association, and reproduced by permission.)

D. Photoelectric Recording

In recent years photoelectric methods (Torok, Guilleman, & Barnothy, 1951; Cornsweet, 1958; Smith & Wartis, 1960) have been developed to overcome some of the problems encountered in the use of corneal reflection and photographic recording methods. These techniques operate on the principle of optically detecting the position of the boundary between the sclera and the iris. By using photomultipliers, this optical information was converted into a voltage that could be recorded directly on an oscilloscope or pen recorder. These measurements have high resolution limited only by the noise of the electronic system and do not interfere in any way with normal eye movements. The early photoelectric devices were still subject to the problems caused by head movement during recording periods. Stark and Sandberg (1961) developed a photoelectric eye movement recording system designed to overcome the experimental problems caused by restricting the subject's head movements. This system incorporates the light source and light-sensitive transducer into a pair of eyeglass frames (Fig. 4-5A) worn by the subject. The lights are mounted to illuminate large circles of light on one eye (Fig. 4-5B). Photoresistors or phototubes were mounted on the frames to record the amount of diffuse reflected light from the exposed sclera on each side of the eye, thus indicating the horizontal eye positions. Figure 4-5C is a record of sinusoidal tracking. This method permitted free head movement and ensured accurate recording of larger eye movements than simple corneal reflection. The use of a mirror galvanometer to project a controllable tracking target in conjunction with the photoelectric recording system was used by Stark *et al.* (1962) and Young and Stark (1963) to demonstrate the importance of studying the response of biological control systems to predictable and unpredictable input signals.

E. Recent Developments

The photoelectric method of eye movement recording has recently been incorporated into an instrument designed to be used in evaluating oculomotor activity in the laboratory, classroom, and clinical situation (Newman, 1970). This system, called Eye-Trac (Fig. 4-6), is described by its designer as, "a new easy-to-use eye movement monitor." The instrument, produced and marketed by Biometrics Inc. of Cambridge, Massachusetts, includes a complete optical system, solid state electronics, and a built-in strip chart recorder. Operation is simple and does not require the attachment of any devices to the subject or any special skill on the part of the operator. Figure 4-7 is a diagram of the system. The subject's eyes are illuminated by two incandescent lamps contained within the unit; both eyes are imaged on a ground glass screen at the rear of the instrument and the

4. RECORDING OF HUMAN EYE MOVEMENTS

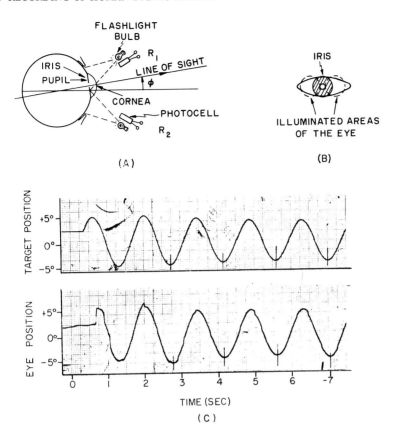

FIG. 4-5. Mobile photoelectric eye movement recording system. (A) Light bulb and photocell arrangement. (B) Front view of illuminated areas of the eye. (C) Sinusoidal tracking record showing saccadic jumps and smooth movements.

operator positions two pairs of photocells to monitor the movements of both eyes. The photocell signals are amplified and recorded on a calibrated strip chart record. The apparatus is equipped with a number of attachments and stimulators for evaluation of reading patterns, visual acuity, optikinetic nystagmus, and the detection of phorias and tropias.

F. Electro-Oculography

While the previously described oculographic recording methods are useful in many experimental and clinical situations, they do not readily permit the recording of eye movements and positions from the closed eyes or from a subject confined to a darkened room. The recent expansion of interest in

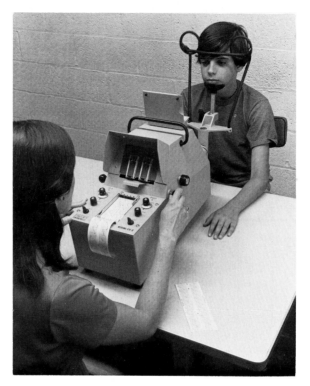

Fig. 4-6. "Eye-Trac" photoelectric eye movement recording system in operation (produced by Biometrics, Inc.).

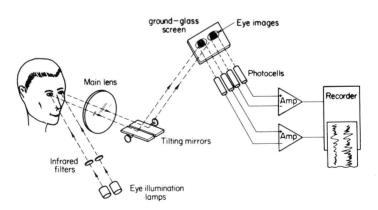

Fig. 4-7. Functional diagram of Eye-Trac photoelectric eye movement recording system.

the psychophysiological investigation of sleep and dreaming (Kleitman, 1939; Oswald, 1962; Luce, 1968) produced a need for adequate procedures for recording movement from the closed eyes. Only two recording methods have been reported that allow the accurate recording of eye movements under conditions where the eyes are closed. These are the use of strain gauges taped to the eyelids (Baldridge, Whitman, & Kramer, 1963) and electro-oculography (Mowrer, Ruch, & Miller, 1936). EOG, which is the direct measurement of the variation in the standing potential between the cornea and retina as a function of the position and movement of the eye, has become the method of choice in recording eye movements in sleep and dream research. The remainder of this chapter will be devoted to the background, techniques, problems, and advantages of EOG recording.

IV. History of Electro-Oculography

Measurement and recording of potentials generated in the eye has a long history. DuBois Reymond (1849) is credited with first establishing the existence of the standing corneoretinal potential. He recorded the voltage from the eye of a fish and discovered that when one electrode is placed on the cornea in the dark and a second electrode placed anywhere on the body, a potential in the order of 5 mV could be measured across the electrodes. This potential is reasonably steady and has been called the steady status or standing potential. If an illuminating stimulus is introduced to the eye, a multiphasic change is produced which is called the photoelectric or action potential. The recording of this potential change was named the electro-retinogram. Several investigators observed the variations in the standing potential caused by eye movements but attributed these fluctuations to artifacts or extra ocular muscle potentials.

DeWar (1877) experimented with the electro-retinogram of the human eye and showed concern for the artifacts introduced by eye movements. Schott (1922) attempted to record nystagmus using a string galvanometer and thought the observed potential changes were a function of the mechanical displacement of the eye. Meyers (1929) was the first to record eye potentials from the closed eyes but attributed the observed potential changes to the action potentials generated by the extra ocular muscles. Jacobson (1930) recorded eye movement potentials by using a vacuum tube amplifier but he too thought he was recording muscle action potentials. Mowrer *et al.* (1936) repeated the work of Meyers and Jacobson but reasoned that the galvanometer effects associated with eye movement were associated with a cornea retinal potential difference. They proved this point by demonstrating that these potentials could be recorded from passively moved eyes and from freshly exised turtle eyes and that the galvanometric effect disappeared when the retina was destroyed by chemical means. The establishment of the relationship between the steady potential and eye movements created a furor of activity in the investigation of eye movement potentials; the work in this area is superbly reviewed (Marg, 1951; Kris, 1960; Shackel, 1967).

V. Electro-Oculography Recording Problems

Three major problems exist in the accurate recording of eye position and eye movement potentials: (1) the stability and sensitivity of the recording instrument, (2) the stability of the electrode skin circuit, and (3) the effect of head movements on the recording of eye position and movement. We will attempt to discuss each of these in detail and present some possible solutions.

A. Instrumentation

The recording of eye position and eye movement potentials requires the use of a d.c. amplification and recording system. The amplifiers should have an input impedance of about 1 MΩ, an internal noise level of less than 5 μV, and a drift rate of less than 100 μV per hour with shorted input. Since a degree of horizontal eye movement will produce a voltage change of approximately 20 μV (Shackel, 1960a), the system should be sensitive enough to detect a change of 5 μV (approx. 15 min of arc), and the full-scale deflection of the recording system should cover a range of 500 μV ($\pm 25°$ of arc). The frequency response of these amplifiers must be suitable to record the fastest possible eye movement as well as long periods of eye fixation (d.c. to 30 Hz).

Recording systems designed specifically for eye movement recording have been described (Kris, 1958; Shackel, 1958), but the state of the electronic art in physiological recording is now at a stage that permits the accurate recording of eye position and movement with standard off-the-shelf recording equipment. Multichannel d.c. physiological amplifiers and recorders are now commercially available that exceed the specifications required to record EOG signals. These systems are designed to accurately record continuous records of physiological activity, and controls are provided to adjust the sensitivity, scale, and frequency response of the instrument to suit the recording need. Output jacks are provided in many of these instruments to enable the experimenter to view higher frequency information on oscilloscopes or to store this information on FM tape for future computer analyses.

Time Constants

The problem of distortion in EOG recording as a function of the time constant used has been considered by several investigators (Jung, 1939; Francois, Verriest, & DeRouk, 1957). Kris (1958) recorded d.c. and a.c. eye movements simultaneously and demonstrated some of the distortions introduced by a.c. recording. The purpose of this segment of the chapter is to stress the role of the time constant chosen in EOG recording. Unfortunately this is often expediently determined by the recording equipment that is familiar and available to the investigator, such as the EEG recorder in the sleep laboratory, rather than the time characteristics of the eye movements to be recorded. Aserinsky and Kleitman (1953, 1955) differentiated two types of eye movement in sleeping subjects: rapid eye movements (REMs), associated with dream report, and slow shifts in eye position. They found the slower shifts impossible to record with their instrumentation and had to revert to direct observation of subject's closed eyes to establish their presence. These studies, and the great majority of sleep studies following them, have relied on standard EEG machines for recording eye movements. These instruments impose serious restrictions on the type of signal that can accurately be recorded. The short EEG time constants (0.1–1.0 sec) produce differentiated and attenuated records of all slow activity.

With the development in recent years of stable d.c. recording systems, it has become readily possible to produce undistorted records of slow eye movements. Although the accuracy of such measures has been demonstrated (Kris, 1958), this type of recording has not become widespread.

Tursky and O'Connell (1966) conducted a detailed study on the effects of a range of time constants on several types of eye movements. The results of this study are presented here in detail in the hope that they will encourage investigators to evaluate the effect of their recording methods on their eye movement data.

In this study the subject was seated in a comfortable lounge chair in an air-conditioned, soundproof room adjacent to the instrument room. Subjects were instructed to move their head as little as possible during recording periods, but no physical provision was made to immobilize the head.

It is important in recording d.c. potentials from the body surface to use an electrode system that is free of drift caused by polarization. Nonpolarizable, low-bias silver–silver chloride electrodes originally developed for skin potential recording (O'Connell & Tursky, 1960) were modified for application to electro-oculography (O'Connell & Tursky, 1962). Two of these electrodes were taped securely to the skin near the outer canthi of the eyes to detect horizontal eye movements. Contact with the skin was made with Sanborn Redux electrode paste. The leads from these electrodes were connected to a Beckman Type R dynagraph in the instrument room through a lead selector, which made it possible to make simultaneous recordings on five separate channels from the same pair of electrodes. Time constants for these channels were: d.c., 3, 1, 0.3, and 0.1 sec.

Each recording session consisted of the following six test periods:

1. Spontaneous eye movements, eyes open.
2. Spontaneous eye movements, eyes closed.
3. Moving eyes slowly from side to side, eyes open.
4. Moving eyes slowly from side to side, eyes closed.
5. Slow pursuit movements, eyes open (following the tip of a pointer).
6. Rapid rhythmic eye movements, eyes open (following a metronome).

Each open-eyed test period was preceded and followed by 15 sec of straight-ahead fixation. During the equivalent periods of closed eye tests, the subject was instructed to hold his eyes as still as possible. Test periods were separated by rest periods of at least 30 sec.

The results of this study clearly demonstrate that simultaneous recording of d.c. and a.c. eye movement potentials was quite feasible. Baseline drift in the d.c. record was not apparent, the pen tracings returning to the same positions during successive fixation periods between eye movements. All subjects showed similar records, typical examples of which are given in Fig. 4-8.

In Fig. 4-8A both rapid and slow eye movements appear as the subject's gaze moves about the room. The differentiation between slow and rapid movements, however, is lost as the time constant used becomes shorter. With a.c. recordings, of course, steady eye positions cannot be determined. It is apparent that a small deflection on the 1 sec time constant channel could result from either a rapid eye movement or a large slow movement attenuated by the time characteristics of the a.c. channel.

In Fig. 4-8B the subjects show spontaneous movements of the closed eyes that look very much like open eye movements. They reported that they were not aware of these spontaneous movements while their eyes were closed, but then neither were they aware that they were making similar movements during normal periods of inattentive open-eyed vision. Their ability to keep their eyes steady when instructed to do so at the beginning and end of this period appears to be comparable to their ability to do so during open-eyed fixation.

The saccadic nature of normal voluntary slow eye movements is clearly apparent with d.c. recording in Fig. 4-9A. The a.c. channels show progressively less of the slow components of these movements as the time constants become shorter. Rapid eye movements and saccadic components of slow

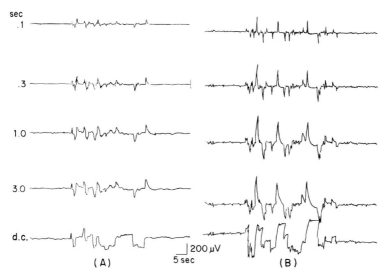

FIG. 4-8. EOG recording of spontaneous eye movements. (A) Eyes open. (B) Eyes closed. (From B. Tursky and D. N. O'Connell, A comparison of AC and DC eye movements, *Psychophysiology,* 1966, **3**(2), 157–163. © The Williams & Wilkins Co., Baltimore, Maryland.)

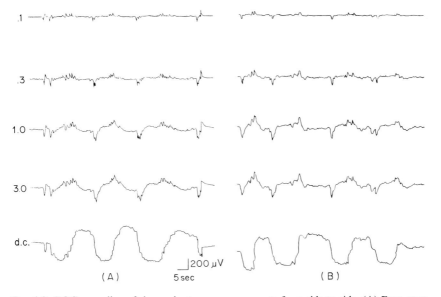

FIG. 4-9. EOG recording of slow voluntary eye movements from side to side. (A) Eyes open. (B) Eyes closed. (From B. Tursky and D. N. O'Connell, A comparison of AC and DC eye movements, *Psychophysiology,* 1966, **3**(2), 157–163. © The Williams & Wilkins Co., Baltimore, Maryland.)

eye movements become increasingly difficult to distinguish. Direct current recording allows the determination of the relative position of the eyes in the head during slow movements. With the gain adjusted to give maximal deflection (up to 5 cm or more) with maximal eye movement, this type of recording not only allows the determination of eye position with considerable accuracy, but also facilitates the differentiation of eye movements from EEG artifacts on the basis of relative amplitude.

Eye movements under closed conditions (Fig. 4-9) appear similar to those under corresponding open eye conditions, showing again the saccadic nature of slow voluntary eye movements.

In Fig. 4-10 subjects were instructed to track the tip of a pointer as it was slowly moved back and forth across their visual field. They were instructed to keep their heads still and follow the moving pointer with their eyes only. These smooth pursuit movements produced records strikingly different from those of slow voluntary movements. The saccadic components drop out, and the resulting smooth movements soon become much less detectable as the time constant becomes shorter, until at .1 sec they have disappeared from the record altogether. Note the marked attenuation even with as long a time constant as 3.0 sec.

In Fig. 4-11 quite rapid rhythmic eye movements in response to the metronome (two swings per second) are followed well by all channels, although there is a marked attenuation as the time constant is decreased. There is even here, however, a distortion that appears as a sharpening of the periodic signal on the a.c. channels as a result of their decay characteristics. This type of distortion is shown even more clearly on records of the somewhat less rapid rhythmic movements. Figures 4-11A and B clearly demonstrate the regular return of the tracing to baseline during fixation periods, which again reflects the stability of the electrodes used.

These records demonstrate a fact that becomes obvious upon reflection: a.c. recording of eye movements with short time constants cannot differentiate between fast eye movements and the saccadic components of slow eye movements, cannot follow slow smooth eye movements, and cannot show the position of the eyes when not in motion. Direct current recording provides an accurate means of registering these classes of events. For some applications, for example, the detection of rapid conjugate eye movements and the recording of nystagmus, these limitations of a.c. recording may not be a serious handicap. In other situations, however, they may be quite inappropriate.

Indeed, one wonders to what extent the emphasis in current investigations of sleep on the significance of rapid eye movements has come about by default through the infrequent use of apparatus capable of recording d.c. bioelectric signals.

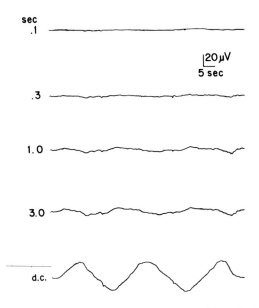

FIG. 4-10. EOG recording of slow pursuit eye movements. (From B. Tursky and D. N. O'Connell, A comparison of AC and DC eye movements, *Psychophysiology*, 1966, **3**(2), 157–163. © The Williams & Wilkins Co., Baltimore, Maryland.)

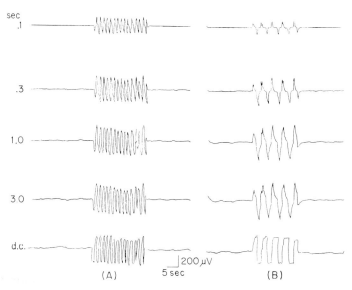

FIG. 4-11. EOG recording of oscillating eye movements. (A) Rapid. (B) Slow. (From B. Tursky and D. N. O'Connell, A comparison of AC and DC eye movements. *Psychophysiology*, 1966, **3**(2), 157–163. © The Williams & Wilkins Co., Baltimore, Maryland.)

B. Stability of the Electrode Skin Circuit

A major problem encountered in recording physiological information from electrodes placed on the skin surface is the electrochemical interaction between the skin and the electrode. These problems are not too significant in the case of some physiological measures recorded from surface electrodes (muscle, EMG, and cardiac, ECG, potentials), since contact is made with the skin purely for the purpose of transferring comparatively high frequency signals generated by an underlying source. In recording d.c. potentials such as eye position and slow eye movement, four interfaces are involved in the electrode skin circuit, two between the skin and the electrolyte and two between the electrolyte and the electrodes. The interaction between these interfaces forms an active electrochemical system that can produce drift and noise that have the same general characteristics as the signal to be observed.

These interface problems are further complicated in EOG recording by the presence of a varying skin potential sometimes of equal or greater amplitude than the EOG. The problems introduced by these factors can result in a record only remotely related to the position and movement of the eyes. To reduce these problems to a tolerable level, precautions must be taken in the choice of recording electrodes, the preparation of the skin at the recording site, and the control of head movements.

1. Problems Related to Recording Electrodes

Much effort has been devoted to the development of electrodes for the recording of low frequency physiological signals such as electrodermal activity. Ideally, electrodes used for this type of recording should be nonpolarizable. That is, no difference in potential should be generated between the electrode and the skin due to a chemical change in the electrode by elements in the electrolyte. The bias potential (difference in potential between the electrodes when they are shorted through the electrolyte) should be minimal and negligible in comparison to the magnitude of the effects being studied; the electrodes should be maximally stable over time.

A number of elements, silver, zinc, and lead have been used in the fabrication of electrodes used in recording electrodermal activity. O'Connell & Tursky (1960) developed a silver–silver chloride (Ag/AgCl) sponge electrode using the technique originated by Janz and Tamiguchi (1953) which showed excellent bias and drift characteristics. O'Connell, Tursky, and Orne (1960) conducted an evaluation of the types of electrodes most often used for skin potential recording. They found that Ag/AgCl sponge electrodes showed lower bias potentials: under 50 μV for matched pairs and a drift rate of less than 20 μV/hr. These results were independently confirmed by Feder (1963). All other electrodes tested demonstrated higher bias potentials and drift

4. RECORDING OF HUMAN EYE MOVEMENTS

rates; however, Venables and Sayer (1963) were able to demonstrate that carefully prepared and selected Ag/AgCl disk electrodes produced a bias potential of 100 μV and a drift rate of 100 μV/hr. These studies indicate that Ag/AgCl electrodes are the electrodes of choice for recording d.c. potentials from the skin.

The need for stable electrodes in recording eye movement potentials has been recognized by other investigators. (Ford & Leonard, 1958; Lykken, 1959). Shackel (1958) designed a Ag/AgCl suction cup electrode with a drift rate of 50 μV/min, which he considered to be adequate; however, he does not report the bias potential across these electrodes.

A special modification of the Ag/AgCl sponge electrodes (O'Connell *et al.*, 1960) was developed specifically for the recording of eye movements. Exact details of the preparation of the electrode core are described by O'Connell and Tursky (1960) and O'Connell *et al.* (1960). Figure 4-12 shows the details of the construction of the electrode; the hat-shaped casing is made of black Lucite and the wide brim permits easy attachment to the skin by means of

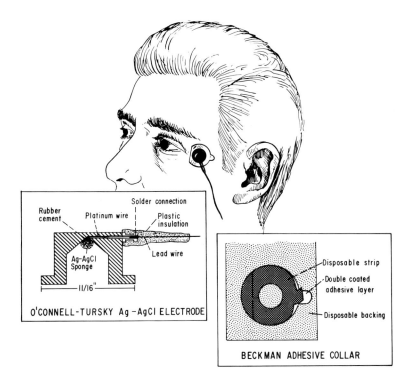

FIG. 4-12. Silver–silver chloride sponge electrode, and adhesive collar. Electrode attached to subject for horizontal eye movement recording.

commercially available adhesive collars. These electrodes have been made in several sizes, the smallest, ¼ inch in diameter, can easily be used for monocular recording.

2. Problems Related to Varying Skin Potentials at the Recording Site

The discovery by Féré (1888) and Tarchanoff (1890) that the electrical resistance as well as the potential difference between two electrodes placed on the skin vary with stimulation has become the basis for a great body of psychological and physiological research (Wang, 1957). The methodology involved in recording both electrodermal measures has been carefully and thoroughly documented (Venables & Martin, 1967; Edelberg, 1967; Wilcott, 1958; Grings, 1953). The electrical activity of the skin has been attributed partly to the distribution and activity of the accrine sweat glands (Kuno, 1934) and partly to neural processes (Richter, 1929; Edelberg & Wright, 1964). Rickles and Day (1968) and Tursky and O'Connell (1969) demonstrated that electrodermal activity could be recorded from any body site, and that responses from nonpalmar skin areas could be elicited by exercise or strong stimuli. The presence of this naturally varying skin potential can create a major problem in EOG recording, since it can be similar in frequency and of greater magnitude than the observed measure. Methods have been devised to treat the skin at any recording site to reduce the impedance and neutralize the natural electrical activity. These methods are commonly used in electrodermal recording to provide an inactive skin area to act as a reference for the active site (usually the palm). Similar methods can be used in EOG recording to neutralize all recording sites and thereby eliminate much of the problem created by ongoing electrodermal activity.

3. Preparation of the Skin at the Recording Site

To achieve electrical neutralization at each recording site, the skin is carefully cleaned with alcohol or acetone and then abraded to the point where the epidermis is ruptured so that ions from the electrolyte can pass freely in either direction without producing a change in potential. Shackel (1959) used a rapidly rotating dental burr to erode the skin. In using this method great care must be taken not to drill too deep since an abrasion of the skin will react painfully with the sodium chloride electrolyte. Rubbing the skin with fine sandpaper can achieve a similar result. O'Connell and Tursky (1960) and Tursky and Watson (1964) found that vigorous rubbing of the skin at the recording site with Sanborn Redux paste which contains a quartz abrasive and no metal ions reduced skin impedance and potential to a low constant level; this stabilization could be maintained for a long time.

C. Head Movement

There is little difference in EOG recordings obtained by moving the head with eyes fixed or moving the eyes with head fixed. In both cases the difference in standing potential is affected by the relationship of the cornea to the iris. If the position of the eyes is to be related to points in the visual field, some method must be provided to control or track head movements. The usual solution to this problem in the laboratory and in the use of most commercial eye tracking instruments has been to prevent the head from moving during recording periods so that all observed movement signals can be attributed to a change in eye position relative to the head. Devices for controlling head movement range from simple head clamps and wooden bite boards (Shackel, 1960a) to complicated head rests that provide an adjustable chin rest and head holder (Fig. 4-6). Special bite boards have been devised that use dental impression compound to make a locating bite for the subject's teeth to ensure that he will not move during the recording session (Ford & Leonard, 1958; Shackel, 1967). Though these methods are useful in recording eye movements from cooperative subjects in situations that permit constraint, in situations where constraint is not reasonable or when the interaction between head and eye movements is one of the points of experimental interest, a method must be devised to accurately track both head and eye movements relative to the visual field. Trevarthen and Tursky (1969) devised a method to simultaneously and independently record the movement of the head, the eyes, and the target. The method described here in detail has proved to be extremely comfortable for the subject who senses only that he is wearing a light hat and has three small electrodes attached to his face. He becomes quickly adapted to these attachments. The apparatus has been found highly suitable for studies with adult subjects as well as infants. It is easy to employ, accurate, and reliable for studies of natural shifts in the horizontal direction of gaze.

1. Measurement of Head Rotation

Various optical and electromagnetic methods were considered as possible means of inertialess recording of head movement, but these were rejected because of their complexity or inaccuracy. Finally a counterweighted mechanical coupling was designed which provided an elegant and accurate solution (Fig. 4-13).

In this mechanism a low torque potentiometer is linked to a hatlike harness designed for comfort and easy attachment. A band of vinyl plastic material provides a nonskid grip on the skin of the forehead. It is continued by an elastic band under the back of the head and the hat is held firm by an adjustable adhesive connection of Velcro tape. An elastic band over the top

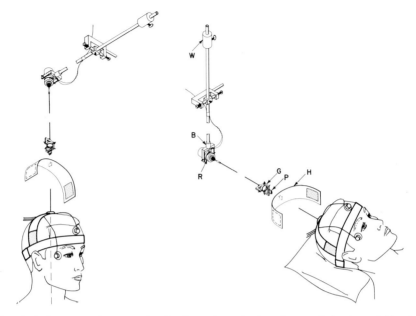

Fig. 4-13. Apparatus for measuring head rotations. A pivoted arm of aluminum tubing is mounted horizontally or vertically. The weight of the potentiometer (R), which is suspended within a gimbal attached to the arm by a precision ball bearing (B), is balanced by the adjustable weight (W). A 0.05 inch diameter, hard stainless-steel tubing, no less than 18 inches long, connects the potentiometer through a ring coupling (G) made of aluminum to a moulded hatband (H) of $\frac{1}{16}$ inch thick plastic. The connection is by a plug (P). The subject wears a light hat, described in the text, and EOG electrodes.

of the head carries patches of Velcro tape for the attachment of the coupling to the potentiometer. The microtorque potentiometer is connected as one arm of a Wheatstone bridge, the output of which is recorded on one d.c. channel of a polygraph.

An adult in an upright position maintains his head in balance between gravitation tending to pull the head forward, and a muscular pull in the opposite direction from the back of the skull. The balanced head is rotated around an axis that is behind the center of gravity of the head. This axis remains fixed relative to the head while the inclination of the head on the shoulders is changed by tilting or bending the head forward or backward.

When the potentiometer is attached as shown in Fig. 4-13, the rotations of the head to turn it to the left or right are measured accurately even when the head is tilted somewhat out of the normal balanced position. With a seated subject, the normal head inclinations do not introduce an error of more than 1–2° in the measure of horizontal head rotation. The largest inclinations are commonly in a forward direction to lower the gaze for inspection in planes nearby and below eye level. The eyes usually remain within $\pm 20°$ of the position which they assume when the gaze is on the horizon and the head is held comfortably upright.

When there is an angle between the axis of rotation of the head and the shaft of the potentiometer, the transmission through the coupling G is non-linear. For an angle of 20° between these axes, errors varying between $\pm 5\%$ will be introduced which depend upon the rotary position of the coupling. It is a simple matter to remove these errors by presenting the subject with a visual field in which the information he is interested in lies close to the horizontal plane through the eyes.

Normal head rotations, even the most rapid, are highly damped; therefore the lightweight mechanism attached to the head does not produce an inertial force big enough to be sensed by the subject. Even the erratic movements of a 3-month-old infant do not appear to be modified by the apparatus. As shown in Fig. 4-13, the apparatus may be positioned to record from a prone subject.

2. Measurement of Horizontal Eye Movement

Modified silver–silver chloride electrodes (O'Connell & Tursky, 1960) are employed to record horizontal eye movements. These highly stable, non-polarizable electrodes are mounted in small tight plastic holders to fit the size requirements of experimentation with infants. Two electrodes are directly connected to a d.c. channel of the recorder (Fig. 4-14A).

The following steps are taken in preparation for recording: the electrodes are stabilized for 1 hr in contact with a volume of Sanborn Redux paste. Stabilization is carried out with the electrodes filled with paste and in close contact, face to face, to ensure against evaporation of the enclosed paste. The electrodes are attached to the subject after the hatband for the head rotation measurement is in place. The skin is first cleaned with alcohol or simply wiped with a damp cloth, and then dried. The spot of skin which will be under the electrode should be moistened with a drop of paste which is rubbed in gently for a second or two. The electrode must be attached immediately after this, and the paste inside it must be sealed from the air and be without bubbles. The electrode resistance must be below 5000 Ω.

With these electrodes it is possible to obtain drift-free d.c. recordings of eye rotations. A noise compounded of effects from EEG, EMG, and the recording equipment, and equivalent to approximately 1 degree of eye movement, is inevitable. High frequency components (above 60 Hz) may be filtered out and appropriate RC filter circuits are included (Fig. 4-14A) following the amplifiers. Blink artifacts and galvanic skin potential changes are also recorded, but both may be recognized and are not troublesome with most subjects. With infants under 1 month of age and with drowsy or

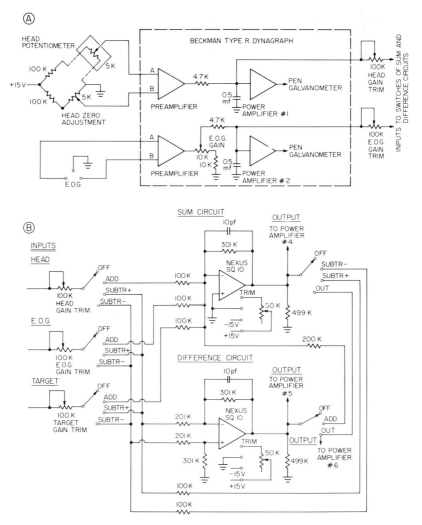

FIG. 4-14. (A) Recording circuits for head rotation and EOG. A bridge circuit identical with that for the head rotation is used to record rotation of a target about an axis in line with the vertical axis of the head. The sum and difference amplifiers, and switching circuits.

frightened subjects, less controllable artifacts may appear. Crying, talking, chewing and the like produce myographic and electrode movement artifacts.

In visual tracking tests the target is mounted on an arm that has the same axis of rotation as the head potentiometer. Rotation of the target is recorded from a 5000 Ω potentiometer which is connected as one arm of a Wheatstone bridge similar to that described for recording head movements. The output of this bridge is displayed on a third d.c. channel of the recorder.

3. DERIVATION OF THE DIRECTION OF GAZE AND TRACKING ERROR

The amplitude of the direct eye-movement potentials is adjusted by proper setting of the Beckman preamplifier sensitivity controls and the supplementary continuous gain controls shown in Fig. 4-13A. The outputs of the head movement and target bridges are similarly adjusted to give equal amplitude signals for equal arcs of rotation.

The head channel is first accurately calibrated to produce a known signal for 10° of head rotation with respect to the room in which the apparatus is suspended. With both infants and adults the eye rotation signals may then be standardized against the head rotations with the aid of automatic compensatory reactions of the subject while he is making spontaneous reorientations with head and eyes. Alternatively, head rotations are imposed while the subject is fixating on a point in the room. This latter is a more convenient method with adult subjects who may also be asked to fixate points of known location in the room while keeping the head immobile.

Because of the early maturation of accurate compensatory adjustments, a record of 1 or 2 min of spontaneous visual exploration of a stationary array occupying at least 90° of the field is generally sufficient for accurate matching of head and eye signals with an attentive subject over the age of 2 months. When this adjustment is correctly made, the summed record of gaze displacement shows almost complete compensation for head rotations in both directions and the portions of the trace between eye saccades are flat (Fig. 4-15A). Consistent deviations indicate that the gain for the EOG is too high or too low.

The outputs of the amplifiers for the EOG and both bridges are connected through an appropriate switching circuit to an electronic sum and difference circuit. The switching at the input to these analog devices is arranged so that any of the signals can be added to or subtracted from any other signal or any combination of signals (Fig. 4-14B). The output of these networks is displayed on separate channels of the polygraph. Thus the angle of the direction of gaze in the horizontal plane is the sum of the head and eye voltage, and the difference between this sum and the target voltage is the tracking error (Fig. 4-15B).

4. DIRECTION-OF-GAZE SWITCH FOR THE CONTROL OF VISUAL EXPLORATION

The d.c. "line-of-sight" signal has been employed to regulate a stimulus. An angle of visual space is designated, and the stimulus may be made to

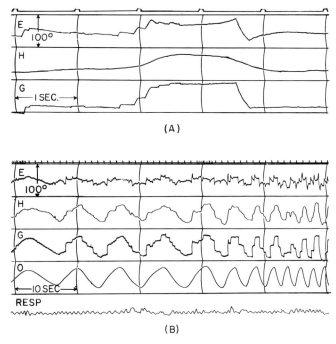

FIG. 4-15. (A) Sample recordings: adult. Large saccades coupling head and eyes in visual exploration of an extensive array. (B) Sample recordings: nine-week-old infant. Tracking a $1\frac{1}{2}°$ target rotating at 36 inches from the head axis. E, EOG; H, head rotation; G, angular direction of gaze $= E + H$; O, angular direction of target object; RESP, respiration. The time marker records seconds.

appear only when the subject is looking outside it. Electronic level detectors have been used to delineate the width of the field, the boundaries of which may be adjusted independently to obtain an "on" or "off" zone of any width in any part of the field. In other words, it is possible to arrange for a stimulus to appear only at a designated locus in the visual field, or in a large or small part of the field.

This method has been used to verify infant fixation while testing peripheral vision.

VI. Applications

There has been growing evidence in recent years of the importance of eye movements in the visual and perceptual process. Early eye movement research and application dealt primarily with evaluation of reading ability and visual fatigue (Carmichael & Dearborn, 1947; Tinker, 1958). While these are still valid applications, the recent realization that eye movements are physiologically necessary and that the eye must move to provide information about objects that are to be examined has led to research into the processes involved in the control of eye movements. The eye movement control system has been characterized as a biological servomechanism that functions by utilizing feedback from the retinal image. Models have been evolved (Stark et al., 1962; Young & Stark, 1963) to describe the relationship between eye movements and the visual process.

Eye movements have also become an important tool in developmental studies. These measures can serve as both a physiological and behavioral indication of development in the nonverbal child. Dayton and Jones (1964) used EOG recordings to analyze the characteristics of the fixation reflex in children ranging in age from 8 hr to 6 months. These studies indicated a much more rapid development of visual acuity in these age groups than was expected. Visual behavior in infants related to the perception of brightness, color, and contrast have been investigated by recording eye movements related to these stimuli (Salapatek & Kessen, 1966). Kessen, Salapatek, and Haith (in press) used closed-circuit TV photography to investigate the visual responses of newborns to edges and contours, and Bergman, Haith and Mann (1971) extended this work to record infant responses to human faces. They were interested in observing the difference in scanning behavior between the child looking at its mother compared to looking at a stranger.

The use of eye movements as a behavioral measure is not limited to studies of the nonverbal infant. Faw and Nunnally (1968), working with elementary school children, used eye movements to test the complexity, novelty, and affective value of visual stimuli and found a direct connection between these variables and amount of fixation time. Mackworth and Morandi (1967), using scanning camera techniques, tested the eye patterns of college students on photographic material previously rated for informativity, and found that students concentrated their gaze on areas of the photographs that provided the most information.

Eye movement recording has played a major role in the investigation of states of consciousness and the association of these and clinical states to the electrical activity of the brain, adding information in some cases and contributing confounding artifact in others.

The investigation of sleep and dreaming was dramatically accelerated by the discovery (Aserinsky & Kleitman, 1955; Dement & Kleitman, 1957) of the relationship between rapid eye movements and EEG patterns related to stages of sleep and dreaming. The investigation of the cyclic nature of the sleep and dream phenomenon and its relationship to the physical well-being of the human organism is a prime example of the importance of eye movement recording.

Eye movements have been used extensively in evaluating the effect of suggestions on physiological functions in hypnosis. Schwarz, Bickford, and Rasmussen (1955) found that hypnotized subjects produced no recordable eye movements when urged to look at an object after being given suggestions of blindness. Barber (1958) reported that his subjects given similar suggestions typically focused on all parts of the room except where the object was situated. Aschan, Finer, and Hagbarth (1962) studied the effect of hypnotic suggestion on vestibular nystagmus.

Eye movements have been demonstrated to be a major source of artifact in clinical EEG recording and have confounded the use of EEG in evaluating other states of consciousness. The recent interest in alpha training (Kamiya, 1962; Mulholland, 1968) and its association with meditative states (Wallace, 1970; Brown, 1970) is often complicated by the role of eye movements in the production of alpha (Dewan, 1967; Mulholland & Evans, 1966). Recently a similar association was made between eye movements and the recording of the steady potential shift recorded from the scalp. The conjugate negative variation (CNV) has been experimentally associated with attention and anticipation. The extent of CNV–EOG interaction has been carefully investigated (Wasman, Morehead, Lee, & Rowland, 1970), the relationship between the measures clearly established, and a reduction in contamination of the CNV effected by using a fixation procedure.

VII. Conclusions

The importance of the development of sophisticated methods of recording eye movements is easily justified, and these measures have been used as clinical and experimental tools in a wide range of applications. This chapter has attempted to review briefly the historical methods and to report in some detail on the latest developments in the three most commonly used eye movement recording methods: corneal reflection photography, photoelectric recording, and electro-oculography. We have tried to avoid direct comparison between these recording methods, since we feel that their uses, though overlapping, are not redundant and that each method has its own applications and limitations. One such comparison was recently conducted by Metz, Scott, O'Meara, and Stewart (1970). In this study a direct comparison was made between EOG and photoelectric recordings on the effect of lateral rectus palsy on occular saccades. The pathological condition was easily detected by direct observation but documentation of severity and progress required a permanent eye movement record. Both measures in this instance produced accurate records of the impairment to the system, and the authors concluded that in this instance EOG was more sensitive but the photoelectric system was easier to use and interpret.

There are many possibilities for new methods of tracking eye movements as well as for clinical and experimental use of these measures. New developments in highly controllable laser techniques raise the possibility of using these techniques in developing remote accurate eye-movement tracking devices. On-line computers which analyze the data produced by existing or new recording methods can be used to provide feedback to patients suffering from visual problems related to their eye movements. It has been demonstrated that direct biofeedback techniques can alter physiological functioning and successful cases have been reported of the clinical value of alterations of heart rate (Engel & Melmon, 1968), blood pressure (Benson, Shapiro, Tursky, & Schwartz, 1971), and other physiological functions. It seems reasonable that similar procedures can be used to alter abnormal eye movement patterns.

Acknowledgments

This work was supported by Grant No. M.H.-0417 (11) from U.S. Public Health Service and much of it was done at the Massachusetts Mental Health Center, Boston, Massachusetts, and at the Department of Social Relations, Harvard University. I wish to thank Milton Lodge for his editorial help.

References

Aschan, G., Finer, B. L., & Hagbarth, K. E. The influence of hypnotic suggestion on vestibular nystagmus. *Acta Oto-Laryngology*, 1962, **55**, 97–110.

Aserinsky, E., & Kleitman, N. Regularly occurring periods of eye mobility and concomitant phenomena during sleep. *Science*, 1953, **118**, 273–274.

Aserinsky, E., & Kleitman, N. Two types of ocular motility occurring during sleep. *Journal of Applied Physiology*, 1955, **8**, 1–10.

Baldridge, B. J., Whitman, R., & Kramer, M. A simplified method for detecting eye movement during dreaming. *Psychosomatic Medicine*, 1963, **25**, 78–82.

Barber, T. X. Hypnosis as a perceptual cognitive restructuring IV. Negative hallucinations. *Journal of Psychology*, 1958, **46**, 187–201.

Benson, H., Shapiro, D., Tursky, B., & Schwartz, G. E. Decreased systolic blood pressure through operant techniques in patients with essential hypertension. *Science*, 1971, **173**, 740–742.

Bergman, T., Haith, M. M., & Mann, L. Development of eye contact and facial scanning in infants. Paper read at meeting of Society for Research in Child Development, 1971.

Brown, B. B. Recognition of aspects of consciousness through association with EEG alpha activity represented by a light signal. *Psychophysiology*, 1970, **6**, 442–452.

Bruckner, A. Ueber die anfangsgeshwindikeit der augenbewegungen *Plügers Arch.*, 1902, **90**, 73–79.

Byford, G. H., & Stuart, H. G. An apparatus for measuring small eye movements. *Journal of Physiology*, 1961, **159**, 2–10.

Carmichael, L., & Dearborn, W. F. *Reading and visual fatigue*. Boston, Massachusetts: Houghton, 1947.

Cornsweet, T. N. New techniques for the measurement of small eye movements. *Journal of the Optical Society of America*, 1958, **48**, 808–811.

Dayton, G. O., & Jones, M. H. Analyses of characteristics of fixation reflex in infants by use of direct current occulography. *Neurology*, 1964, **14**, 1152–1156.

Dearborn, W. F. The psychology of reading. *Archives of the Phil. Psychol. Sci. Methods*, 1906, **No. 4**.

Delebarre, E. B. A method of recording eye movements. *American Journal of Psychology*, 1898, **9**, 572–574.

Dement, W. C., & Kleitman, N. The relation of eye movements during sleep to dream activity: An objective method for the study of dreaming. *Journal of Experimental Psychology*, 1957, **53**, 339–346.

Dewan, E. M. Occipital alpha rhythm, eye position and lens accommodation. *Nature*, 1967, **214**, 975–977.

DeWar, J. The physiological action of light. *Nature*, 1877, **15**, 433–5, 453–4.

Dodge, R. An experimental study of visual fixation. *Psychological Monographs*, 1907, **2**, 193–199. (a)

Dodge, R. An experimental study of visual fixation. *Psychological Monographs*, 1907, **8**, 1–95. (b)

Dodge, R., & Kline, T. S. The angle velocity of eye movements. *Psychological Reviews*, 1901, **8**, 145–157.

DuBois, R. E. Untersuchungen über thierische Electricitat. *Berlin*, 1849, **2/1**, 256–7.

Edelberg, R. Electrical properties of the skin. *Methods in psychophysiology*. Baltimore, Maryland: Williams & Wilkins, 1967. Pp. 1–53.

Edelberg, R., & Wright, D. J. Two galvanic skin response effector organs and their stimulus specificity. *Psychophysiology*, 1964, **1**, 39–47.

Engel, B. T., & Melmon, K. L. Operant conditioning of heart rate in patients with cardiac arrhythmias. *Conditional Reflexes*, 1968, **3**, 130–140.

Faw, T. T., & Nunnally, J. C. The influence of stimulus complexity, novelty, and affective value on children's visual fixations. *Journal of Experimental Child Psychology*, 1968, **6**, 141–153.

Feder, W. Silver–silver chloride electrode as a nonpolorizable bioelectrode. *Journal of Applied physiology*, 1963, **18**, 397–401.

Fender, D. H. Control mechanisms of the eye. *Scientific American*, 1964, 24–33.

Féré, C. Note sur des modifecations de la résistance électrique sous l'influence des éxcitations sensorielles et des émotions. *Compte Rendu de la Société Biologie (Paris)*, 1888, **8**, (5), 217–219.

Ford, A., & Leonard, J. L. Techniques for recording surface bioelectric direct currents, *U.S. Navy Electronic Laboratory Report*, 1958, **839**.

Francois, J., Verriest, G., & DeRouk, A. L'électro-occulographie en tant qu'examen fonctionnel de la rétina. *Progress in Opthalmology*, 1957, **7**, 1–67.

Granit, R. *Sensory mechanisms of the retina*. With an appendix on electroretinography. London and New York: Oxford Univ. Press, 1947.

Grings, W. W. Methodological considerations underlying electrodermal measurements. *Journal of Psychology*, 1953, **35**, 271–282.

Guyton, A. C. The eye. *Function of the human body*, Chapter 24. 3rd ed. Philadelphia: Saunders, 1969.

Haith, M. M. Infrared television recording and measurement of ocular behavior in the human infant. *American Psychologist*, 1969, **24**, 3, 279–283.

Huey, E. B. Preliminary experiments in the physiology and psychology of reading. *American Journal of Psychology*, 1898, **9**, 575–586.

Jacobson, E. Electrical measurement of neuromuscular states during mental activities: I. Imagination of Movement involving skeletal muscle. *American Journal of Physiology*, 1930, **91**, 567–680; II. Visual imagination and recollection. *American Journal of Physiology*, 1930, **95**, 694–702.

Janz, G. J., & Tamiguchi, H. The silver-silver halide electrodes. *Chemical Review*, 1953, **53**, 397–437.

Judd, C. H., McAllister, C. H., & Steele, W. M. General introduction to a series of studies of eye movements by means of kinetoscope photographs. *Psychological Monographs*, 1905, **7**, 1–16.

Jung, R. Eine electrische. Methode zur mehrfachen registierung von augenwegungen und nystagmus. *Klin. Wshr.*, 1939, **18**, 21–24.

Kamiya, J. Conditioned discrimination of the EEG alpha rhythm in humans. Paper presented at the meeting of the Western Psychological Association in San Francisco, 1962.

Kessen, W., & Hirshenson, M. Occular orientations in the human newborn infant. Paper presented at the American Psychological Association Meetings, Philadelphia, 1963.

Kessen, W., Salapatek, P., & Haith, M. M. The visual response of the human newborn to linear contour. *Journal of Experimental Child Psychology*, in press.

Kleitman, N. *Sleep and wakefulness*. Chicago: Univ. of Chicago Press, 1939.

Kris, C. E. A technique for electrically recording eye position. USAF WADC Technical Report, 1958, No. 58-660 (ASTIA Document No. AD 209385).

Kris, C. E. Vision: *Electro-oculography, medical physics*, Vol. III. Chicago, Illinois: Yearbook Publ., 1960. Pp. 691–700.

Kuno, Y. *The physiology of human perspiration*. London: Churchill, 1934.

Lamansky, S. Bestimming der Winkelgishwindigkest der bleck bewegung respect we ougenbewegung. *Pflügers Archives*, 1869, **2**, 418–422.

Luce, G. G. Current research on sleep and dreams. Public Health Service Publication No. 1389, 1968.

Lykken, D. T. Properties of electrodes used in electrodermal measurements. *Journal of Comparative Physiological Psychology*, 1959, **52**, 629–634.

Mackworth, J. F., & Mackworth, N. H. Eye fixations recorded on changing visual scenes by the television eye marker. *Journal of the Optical Society of America*, 1958, **48**, 439–445.

Mackworth, N. H., & Morandi, A. J. The gaze selects informative details within pictures. *Perception and Psychophysics*, 1967, **2**, 547–552.

Mackworth, N. H., & Thomas, L. E. Head mounted eye movement camera. *Journal of the Optical Society of America*, 1962, **52**, 713–716.

Marg, E. Development of electro-oculography: Standing potential of the eye in registration of eye movement. *Archives of Ophth.*, 1951, **45**, 169–185.

Metz, H. S., Scott, A. B., Omeara, D., & Stewart, H. L. Ocular saccades in lateral rectus palsy. *Archives of Opthalmology*, 1970, **84**, 453–460.

Meyers, I. L. Electronystagmography: A graphic study of the action currents in nystagmus. *Archives of Neurology and Psychiatry*, 1929, **21**, 901–918.

Mowrer, O. H., Ruch, R. L., & Miller, N. E. The corneo-retinal potential difference as the basis of the galvanometric method of recording eye movements. *American Journal of Physiology*, 1936, **114**, 423–428.

Mulholland, T. Feedback electroencephalography. *Activ. Nerv. Supplement*, 1968, **10**, 410–438.

Mulholland, T., & Evans, C. R. Occulomotor function and the alpha activation cycle. *Nature*, 1966, **211**, 1278–1279.

Newhall, S. M. Instrument for observing ocular movements. *American Journal of Psychology*, 1928, **40**, 628–629.

Newman, J. S. Eye movement measurements. *Medical electronics and Data*, 1970, **1** (3), 82–84.

O'Connell, D. N., & Tursky, B. Silver–silver chloride sponge electrodes for skin potential recording. *American Journal of Psychology*, 1960, **73**, 302–304.

O'Connell, D. N., & Tursky, B. Special modifications of the silver–silver chloride sponge electrodes for skin recording. *Psychophysiology Newsletter*, 1962, **8**, 31–37.

O'Connell, D. N., Tursky, B., & Orne, M. T. Electrodes for the recording of skin potentials. An evaluation. *American Medical Association Archives of General Psychiatry*, 1960, **3**, 252–258.

Oswald, I. *Sleeping and waking*. Amsterdam: Elsevier, 1962.

Polyak, S. L. *The retina*. Chicago, Illinois: Univ. of Chicago Press, 1948.

Richter, C. P. Physiological factors involved in the electrical resistance of the skin. *American Journal of Physiology*, 1929, **88**, 596–615.

Rickles, W. H., & Day, J. L. Electrodermal activity in non-palmar skin sites. *Psychophysiology*, 1968, **4**, 421–435.

Riggs, L. A., Ratliff, R., Cornsweet, J. C., & Cornsweet, T. N. The disappearance of steadily fixated test objects. *Journal of the Optical Society of America*, 1953, **43**, 495.

Salapatek, P., & Kessen, N. Visual scanning of triangles by the human newborn. *Journal of Experimental Child Psychology*, 1966, **3**, 115–167.

Schott, E. Über die Registrierung des Nystagnus und anderer augenbewegungen Vermittels des Saitengalvanometers. *Geutsches Arch. für Klin. Med.*, 1922, **140**, 79–90.

Schwarz, B. E., Bickford, R. G., & Rasmussen, W. C. Hypnotic phenomena including hypnotically activated seizures studied with the electroencephalogram. *Journal of Nervous and Mental Disease*, 1955, **122**, 564–574.

Shackel, B. A rubber suction cup surface electrode with high electrical stability. *Journal of Applied Physiology*, 1958, **13**, 153–158.

Shackel, B. Skin drilling: A method of diminishing galvanic skin potentials. *American Journal of Psychology*, 1959, **72**, 114–121.

Shackel, B. Review of the past and present in oculography. Proc. 2nd International Conference Medical Electronics, Paris 1959, Iliffe, London, 1960. (a)

Shackel, B. A note on mobile eye viewpoint recording. *Journal of the Optical Society of America*, 1960, **50**, 763–768. (b)

Shackel, B. Eye movement recording by electro-oculography. *Manual of psychophysiological methods*. New York: Wiley, 1967. Pp. 298–334.

Smith, W. M., & Wartis, P. J. Eye movement and stimulus movement; new photoelectric electromechanical system for recording and measuring tracking motions of the eye. *Journal of the Optical Society of America*, 1960, **50**, 3, 245–250.

Stark, L., & Sandberg, H. A simple instrument for measuring eye movements. *Quarterly Progress Report No. 62 Research Laboratory of Electronics, M.I.T.*, 1961, 268–271.

Stark, L., Vossius, G., & Young, L. R. Predictive control of eye tracking movements. *IRE Transactions on Human Factors in Electronics*, 1962, **HFE-3**, 52.

Steinback, M. J., & Held, R. Eye tracking of observer-generated target movements. *Science*, 1968, **161**, 187–188.

Tarchanoff, J. Uber die Galvaneschen erscheinungen in der Hout des Menschen bei Reigungen der sines organe und bei vesheidinen Formen der psyshischen Tatikeit. *Pflüger's Arch. Ges. Physiol.*, 1890, **46**, 46–55.

Tinker, M. A. Recent studies of eye movements in reading. *Psychological Bulletin*, 1958, **55**, 215–231.

Torok, N., Guillemin, V., & Barnothy, J. M. Photoelectric nystagmography. *Ann. Otol. Phin. and Laryngology*, 1951, **60**, 917–923.

Trevarthan, C., & Tursky, B. Recording horozontal rotations of the head and eyes in spontaneous shifts of the gaze. *Behav. Res. Meth. & Instru.* 1969, **8**, 291–293.

Tursky, B., & O'Connell, D. N. A comparison of palmer and nonpalmer electrodermal responses to electrocutaneous stimulation. Paper presented at 9th Annual meeting of Society for Psychophysiological Research 1969, Monterey, California.

Tursky, B., & O'Connell, D. N. A comparison of AC and DC eye movements. *Psychophysiology,* 1966, **3** (2), 157–163.

Tursky, B., & Watson, P. D. Controlled physical and subjective intensities of electric shock. *Psychophysiology,* 1964, **1,** 151–162.

Venables, P. H., & Martin, L. Skin resistance and skin potential. *Manual of psychophysiological Methods.* New York: Wiley, 1967.

Venables, P. H., & Sayer, E. On the measurement of the level of skin potential. *British Journal of Psychology,* 1963, **54,** 251–260.

Wallace, R. K. Physiological effects of transcendental meditation. *Science,* 1970. **167,** 1751–1754.

Wang, G. H. The galvanic skin reflex. A review of old and recent works from a physiologic point of view. *American Journal of Physical Medicine,* 1957. **36,** 295–320.

Wasman, M., Morehead, S. D., Lee, H. Y., & Rowland, V. Interaction of electro-ocular potentials with the contingent negative variation. *Psychophysiology,* 1970, **7,** 103–111.

Weiss, O. Die Zeitlicke dour des Lidshlages. *Zeitscher für Sinnesphysiol.,* 1911, **45,** 307–312.

Wendt, P. R. *Psychological Monograms,* 1952, **66**.

Wilcott, R. C. Correlation of skin resistance and potential. *Journal of Comparative Physiology and Psychology,* 1958, **51,** 691–696.

Young, L. Measuring eye movements. *American Journal of Medical Electronics,* 1963, **2,** 300–307.

Young, L., & Stark, L. A discrete model for eye tracking movements. *IEEE Trans. on Mil.* 1963, **7,** 113–115.

Chapter 5

Electromyography: Single Motor Unit Training

John V. Basmajian

> *Emory University Regional Rehabilitation Research and Training Center*
> *and*
> *Georgia Mental Health Institute*
> *Atlanta, Georgia*

I. Introduction	137
A. History	138
B. Basic EMG	138
II. General EMG Techniques	140
A. Equipment	140
B. Electrodes	142
C. Surface Electrodes as Necessary Substitutes	143
III. Basic Motor Unit Training	143
A. Preliminaries	143
B. Sample Routines for Single Motor Unit Training	145
IV. Further Psychophysiologic Techniques Employing SMUT	147
A. Reaction-Time Experiments	147
B. Studies of Covert Operant Conditioning	151
C. Studies of Personality Factors Affecting SMU Control	152
References	153

I. Introduction

Single motor unit training (SMUT) in electromyography (EMG) depends upon the fortuitous discovery that human subjects can isolate and activate tiny twitches in striated muscles with no other apparent feedback than auditory or visual cues from EMG. This technique offers a novel tool in the search for ultramodels of the learning process and a methodology for investigating psychophysiological control systems. While it may be argued that it is not the molecular basis of learning (if such exists), it certainly provides a microcosmic model for neuromuscular training, operant conditioning, and conscious control over spinal motoneurons.

A. History

This new approach to an old problem owes its origins to the improvements in electromyography during the past two decades. Smith (1934), Lindsley (1935), Gilson and Mills, (1940, 1941), and Harrison and Mortensen (1962) gave accounts of man's ability to discharge single motor units (SMUs). However, electrophysiologists took this phenomenon for granted and neither they nor psychologists performed systematic studies of it. Knowledge of the work of Harrison and Mortensen who would report that subjects of kinesiologic studies on the tibialis anterior muscle could recruit a number of isolated motor units serially, prompted my series of systematic studies of motor-unit isolation and control (Basmajian, 1963; Basmajian, Baeza, & Fabrigar, 1965). Our coincidental development of a special intramuscular electrode allowed rapid progress (Basmajian & Stecko, 1962). Other laboratories have been developing SMU research both for neurophysiological studies and for psychophysiological and behavioral studies. Perhaps the most dramatic application has been the use of SMU controls to operate myoelectric prostheses and orthoses. This, in turn, has demanded greater depth in our understanding of the physiological and psychological phenomena.

B. Basic EMG

1. Motor Units

A motor unit includes only one motor neuron (motoneuron) in the spinal cord—or in the case of cranial nerves, in the brainstem. Thus, the activity of the motor unit is a reflection of the activity of the motoneuron. From the cell body of a motoneuron a single axon runs in a nerve to a small group of striated muscle fibers. Collectively, the motoneuron (cell body and axon) plus the muscle fibers it supplies, constitute the motor unit (Fig. 5-1). A twitchlike contraction of the muscle fibers of the unit is detected by a recordable myoelectric potential and indicates an activation of the neuron.

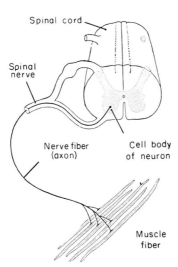

Fig. 5-1. Diagram of a motor unit.

Obviously, it is much easier to place an electrode in a human being among the muscle fibers rather than in his spinal cord. Thus, refined electromyography provides us with a simple way to record the activity of the motoneurons, one neural pulse leading to one twitch under normal conditions. In other words, the electromyogram can be an accurate reflection of the electroneurogram with none of the overwhelming technical problems (Gasser & Newcomer, 1921; Adrian & Bronk, 1929).

A contraction of a motor unit is a twitch lasting a few milliseconds. Immediately after each twitch the fibers of the motor unit relax completely and they only twitch again when an impulse arrives along the nerve fiber. Individual muscles of the body consist of many hundreds of such motor units and it is their summated activity that develops the tension in the whole muscle.

2. Motor Unit Potentials

The motor unit potential has a brief duration (with a median of 9 msec) and a total amplitude measured in microvolts or millivolts. Most SMU potentials recorded by conventional techniques (surface or needle electrodes) are sharp triphasic or biphasic spikes; but when fine-wire electrodes are used, the potentials are much more complex and so are recognizable by their shapes (Fig. 5-2). Generally, larger motor unit potentials come from larger SMUs. However, distance from the electrode, the type of electrodes, and equipment used (and many other factors) influence the final size and shape.

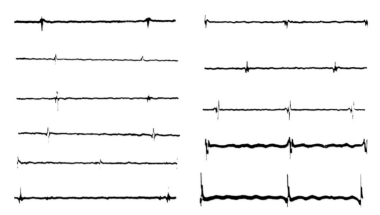

FIG. 5-2. Complex waveforms of eleven different SMUs under conscious control from the same set of fine-wire electrodes.

3. BIOMECHANICS

The motor units twitch repetitively and asynchronously. It is this very randomized activity that causes the smooth tension of the whole muscle rather than a jerky one. The amount of external work or force produced by SMUs is quite small. In a living human being it is usually insufficient to cause any external movement of a joint spanned by the whole muscle of which it is a part. Even in the case of small joints, such as those of the thumb, at least two or three motor units are needed to give a visible movement.

Under normal conditions small motor units are recruited early and, as the force is automatically or consciously increased, larger and larger motor units are recruited (Henneman, Somjen, & Carpenter, 1965) while all the motor units also increase their frequency of twitching. The upper limit in man is about 50 per second, but rates much slower than this are the rule. There is no single, set frequency; individual motor units can fire very slowly and will increase their frequency of response on demand.

II. General EMG Techniques

A. *Equipment*

Electromyographs are high-gain amplifiers with a preference for frequencies from about ten to several thousand cycles per sec. An upper limit of 1000 Hz (cycles per second) is acceptable for SMUT, but 2000 Hz or more are preferable. Ideally, the recording device either should be photographic, or it should employ electromagnetic tape recording. In recent years, multitrack tape recorders have provided a relatively cheap method of storing EMG signals, especially for SMUT studies.

A minimal configuration would include inexpensive commercial equipment designed for other general uses. A cheap single-channel assembly consisting of, for example, a Tektronix oscilloscope (Type 502), a Tektronix low level preamplifier (Type 122) and a Heathkit audioamplifier (Type A-9C) with an ordinary loudspeaker is the minimal equipment requirement. Equivalent units of other manufacturers can be, and are, substituted. For example, the preamplifier might be an Argonaut type LRA-042 or a Sanborn type 350, etc. Many inexpensive general-purpose oscilloscopes could be substituted for the Tektronix 502 for visual display. One should aim for a minimum sensitivity of about 25 μV per inch.

Small audio or high-fidelity amplifiers (using either transistors or tubes) are readily available. They can be satisfactory if they have a power output of about 3 W (this is not critical) and provision for magnetic phono input. They can be connected to the output of the preamplifier for audio monitoring in conjunction with a standard loudspeaker. A start can be made in SMUT electromyography without spending huge sums of money. Further, the simple assemblies will still be useful as adjuncts when more elaborate and comprehensive apparatus has been obtained.

An early secondary acquisition for SMUT would include a storage oscilloscope (for example, Tektronix 564B or a Hewlitt-Packard 141B or 1201A), a four-channel FM tape recorder (for example, HP 3960A, or Thermionic T3000), standard photographic equipment for recording from the storage oscilloscope, and perhaps a direct-writing galvanometer-type recorder. Several companies are producing direct readout recorders that use an ultraviolet light source and sensitive paper. The Honeywell line of "Visicorder" is available with six or more channels. Miniature galvanometers having a frequency response to 4800 Hz are used to deflect the light beams onto the moving paper. Among others, a galvanometer with a response of 1000 Hz is available; this is good for most EMG applications, being considerably faster than ink-writing pen recorders.

The galvanometers may be set to produce a trace on any portion of the sensitized paper. Amplifiers for use with the various Visicorders are available from Honeywell (for example, the Accudata 108 a.c. amplifiers and 109 d.c. amplifier). These two must be used in series for proper drive of the fluid-damped galvanometers in the Visicorder.

Many investigators will find these components compatible with their analog computers, averagers, etc. The PDP series of computers have proved particularly useful in our experience. Of course, ink-writing oscillographs will have special application in certain studies as a final output.

However, the use of standard EEG equipment without adjuncts makes SMUT—which is quite easy otherwise—an unnecessarily unreliable procedure.

Obvious problems such as poor grounding, surrounding 60 Hz and radio frequency interference, poor connections, and short-circuiting of electrodes can be obviated or cured by appropriate precautions familiar to all workers in bioelectronics.

B. Electrodes

For the excellent recordings required in motor-unit training, the difficulties arising from surface electrodes appear to be prohibitive (but see below). Moreover, inserted electrodes are no longer as forbidding as they once were. Our fine-wire electrodes (Basmajian & Stecko, 1962) are as easy to use and as easy to tolerate as are surface electrodes (Fig. 5-3). They are (1) extremely fine and, therefore, painless, (2) easily injected and withdrawn, (3) as broad in pickup from a specific muscle as the best surface electrodes, and (4) give excellent sharp motor-unit spikes with fidelity. With 1 mm of their tip exposed, such electrodes record the voltage from a muscle much better than surface electrodes (Sutton, 1962). Bipolar fine-wire electrodes isolate their pickup either to the whole muscle being studied or to the confines of the compartment within a muscle if it has a multipennate structure. Barriers of fibers connective tissue within a muscle or around it act as insulation. Thus, one can record all the activity as far as such a barrier without the interfering pickup from beyond the barrier (such as there always is with surface electrodes).

Bipolar fine-wire electrodes are made from a polyurethane- or nylon-insulated Karma alloy wire only 25 or 75 μ in diameter or any other similar

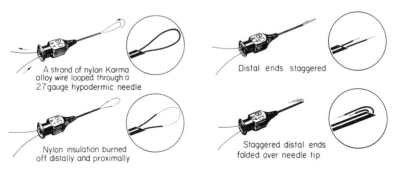

FIG. 5-3. The construction of bipolar fine-wire electrodes. The needle carries the wires into the muscle and is withdrawn. The turnbacks prevent the wire from slipping out. Because a slight tug will pull out the wires, they are taped to the skin, allowing for a liberal loop between the point where they emerge and the adhesive.

material now available (e.g., Teflon-coated stainless steel or platinum). They are very simple to make and their preparation is described by us in detail in the literature (Basmajian & Stecko, 1962; Basmajian, 1967). Care must be taken to avoid short-circuiting between the two electrode wires within the tissues; this can be assured by precise production methods.

C. Surface Electrodes as Necessary Substitutes

As Kahn, Bloodworth and Woods (1971) have pointed out, there are circumstances in which any injection of an electrode must be avoided because of the inherent bias it may introduce in certain special types of psychophysiological experiments with SMUs (i.e., where the covert nature of the phenomenon must be masked by the use of many dummy electrodes to mislead the subject). Basing their work on that of Bruno, Davidowitz, and Hefferline (1970), they have shown that when the surface-electrode response is recorded on magnetic tape at high speed and then played back at low speed through a standard EMG–EEG oscillograph, pen records of good resolution were obtainable. Of necessity this technique relies upon the recording of obtrusive units which, of course, may be the requirement of a specific project as in our study of SMUT in young children (Simard, Basmajian, & Janda, 1968). The exclusive use of surface electrodes should be avoided except where the experimental design demands them (e.g., where many dummy electrodes are placed on a subject and he is kept uninformed about the aim of an experiment either to study or to condition covert activity).

Green *et al.* (1969) have found alternate ways in which to minimize the problems inherent in the use of skin electrodes with standard polygraph equipment. They developed a rectifier circuit for detection of SMU firing with a feedback meter. Input through standoff capacitors allows the use of preamplifiers with high-voltage d.c. outputs. Green and his co-workers have used whatever combination of amplifiers was available. Details are presented in their paper (Green *et al.*, 1969).

III. Basic Motor Unit Training

A. Preliminaries

Fundamentally the technique consists of providing human subjects with the best possible feedback of the activity of his motoneurons through the monitoring of EMG signals. The general experimental approach is not complex and different laboratories have already devised their own routines. The following is a general design:

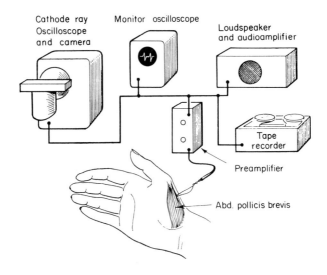

FIG. 5-4. A basic setup for SMUT experiment.

After the aseptic injection of fine-wire electrodes and routine testing, a subject needs only to be given general instructions. He is asked to make contractions of the muscle under study while listening to, and seeing, the motor unit potentials on the monitors (Fig. 5-4). A period of up to 15 min is sufficient to familiarize him with the response of the apparatus to a range of movements and postures.

Subjects are invariably amazed at the responsiveness of the loudspeaker and cathode-ray tube to their slightest efforts, and they accept these as a new form of "proprioception" without difficulty. It is not necessary for subjects to have any knowledge of electromyography. After getting a general explanation they need only to concentrate their attention on the obvious response of the electromyograph. With encouragement and guidance, even the most naive subject is soon able to maintain various levels of activity in a muscle on the sensory basis provided by the monitors. Indeed, most of the procedures he carries out involve such gentle contractions that his only awareness of them is through the apparatus. Following a period of orientation, the subject can be put through a series of tests for many hours.

Several basic tests of criteria may be employed. Since people show considerable difference in their responses, adoption of a set routine is difficult. In general, they are required to perform a series of tasks. Usually the first is to isolate and maintain the regular firing of a single motor unit from among the 100 or so that a normal person can recruit and display with this technique. When he has learned to suppress all the neighboring motor units completely, he is asked to put the unit under control through a series of tricks including speeding up its rate of firing, slowing it down, turning it "off" and "on" in various set patterns and in response to commands.

After acquiring good control of the first motor unit, a subject may be asked to isolate a second with which he then learns the same tricks; then a third, and so on. In a serial procedure, his next task may be to recruit, unerringly and in isolation, the several units over which he has gained the best control.

Many subjects then can be tested at greater length on any special skills revealed in the earlier part of their testing (for example, either an especially fine control of, or an ability to play tricks with, a single motor unit). Finally, the best performers may be tested on their ability to maintain the activity of specific motor-unit potentials in the absence of either one or both of the visual and auditory feedbacks. That is, the monitors are gradually turned off and, after a "weaning" period, the subject must try to maintain or recall a well-learned unit without the artificial "proprioception" provided earlier (Basmajian *et al.*, 1965).

Any skeletal muscle may be selected. The ones we have used most often are the abductor pollicis brevis, abductor digiti minimi, tibialis anterior, biceps and brachialis brachii and the extensors of the arm and forearm. But our group has also successfully trained motor units in back muscles, shoulder and neck muscles, tongue muscles and others; there appears to be no limit.

B. Sample Routines for Single Motor Unit Training

1. FIRST ROUTINE

The routine that follows has been used in a number of studies (Simard & Basmajian, 1967) and is offered here as an example. The essential features are: (a) clear-cut criteria of success or failure and (b) a stepwise program of training. Ninety percent of human subjects can be trained by such a procedure. Following the preliminary training, rhythm control tests are applied in progressively more difficult forms: thus, at each test, a subject is required to produce controlled sets of three twitches of his isolated motor unit in response to different types of demand situations. These are:

Test a. The subject is asked to give a set of three spike potentials of his controlled motor unit while silently counting; one, two, three. Thereupon he is asked to repeat this attempt a minimum of ten times.

Test b. The subject is asked to think of the sound of a horse galloping ("tip-a-tap" or "clippity-clop") and, using this rhythm to produce sets of three twitches not less than ten times.

Test c. The investigator plays musical notes (do-re-mi) on the keyboard of a transistor audio oscillator. The duration of each note is approximately

½ sec and each interval approaches ¼ sec. The subject tries to give a set of three spike potentials of his controlled unit immediately after hearing all three notes. This attempt is repeated not less than ten times.

Test d. The investigator plays the same note three times in ten sets. (The duration of each note is ½ sec, the internal between each note, 1 sec, and the interval between each set of three notes, approximately 2 sec). The subject tries to give one spike potential of his controlled unit immediately after each note heard.

Test e. The investigator produces ten sets of three consecutive sharp taps by striking a pencil against a table. After each set of three taps the subject is allowed a period of approximately 2 sec to produce three successive motor unit spike potentials. This procedure is repeated ten times.

The FM tape provides a source of data which allows us both to replay and to manipulate the data with our LAB-8 analog-to-digital conversion for calculation.

2. SECOND ROUTINE

Another sample technique for conditioning of SMUs is provided by Lloyd and Leibrecht (1971). Information about his performance is presented to a subject by a panel of three lights mounted in front of his chair. Onset of a white light indicates the beginning of each trial and remains on throughout the trial. A second white light informs the subject of a correct response while an amber light indicates an incorrect response. Trial length is controlled by peripheral timing equipment. Response times are recorded by clock counters interfaced with the timing equipment.

PROCEDURE. Prior to the beginning of the experiment, the group of subjects observes a demonstration of SMU isolation and control by an experimenter. The general nature and purpose of the experiment are described and the subjects are allowed to question the demonstrator. This procedure is necessary since none of the subjects have observed SMU activity or experienced the implantation of fine-wire electrodes.

Each subject is seated comfortably in the reclining chair. The fine-wire bipolar electrodes are injected aseptically into the belly of the right tibialis anterior muscle. The barbed wires remain in the muscle tissue while the free portions are taped to the skin. The oscilloscope display faces away from the subject so that no cues other than the lights are available.

The subject is instructed to activate a SMU at a steady frequency for as long as possible during each trial. The relationship between his response and the lights is then explained. The first white light indicates the beginning of each trial and remains on for the duration of the trial. If no EMG response is made by the subject within the first 5 sec, the amber light is turned on for 800 msec. Onset of the incorrect response light terminates the trial. If an incorrect response is made by the subject (two or more motor units in the EMG), the experimenter terminates the trial by activating the incorrect response light. If a SMU appears within the first 5 sec of a trial, the experimenter turns the correct light on which extends the trial to a total duration of 10 sec. The correct response light remains on for as long as a SMU continues to fire. If the SMU ceases firing, or if additional motor units appear, the correct response light is turned off and is reactivated only if the SMU reappears. The criterion for a correct response is the presence of any SMU. Therefore, a subject can switch from one SMU to another during the course of the experiment.

Each trial is separated by a rest interval of approximately 15 sec during which time the subject is instructed to relax. The subject is run to a performance criterion of five successive trials in which continuous activation of a SMU occurs for a minimum of 5 sec. Each session consists of 100 trials and requires approximately 35 min. A maximum of 500 trials is provided in five sessions which are administered in one day with the same set of fine wire electrodes. There is a rest period of approximately 45 min between sessions. The subject is instructed to remain in the testing area during the rest periods and to keep relatively inactive in order to minimize changes in the position of the electrodes.

IV. Further Psychophysiologic Techniques Employing SMUT

In this section, only techniques for exploring psychophysiological phenomena are described. Readers desiring further information of general motor and proprioceptive research with SMUs should consult Basmajian (1963, 1967, 1972), Basmajian et al. (1965), Carlsöö and Edfeldt (1963), Basmajian and Simard (1967), Simard and Basmajian (1967), Simard et al. (1968), Scully and Basmajian (1969), Powers (1969), Jacobs and Felton (1969), Hardyk, Petrinovich, and Ellsworth (1966), Wagman, Pierce, and Burger (1965), Zappalá (1970), Simard and Ladd (1969), Petajan and Philip (1969), and McGuigan (1970).

A. Reaction-Time Experiments

Following a routine of training SMUs, a fairly routine technique of reaction time (RT) experimentation may be performed.

1. THYSELL TECHNIQUE

In the procedure described by Thysell (1969), the RT stimuli consisted of a red warning light and a white reaction light. The onset of the red warning light was controlled by a manually operated switch. Its duration and offset were controlled by a Hunter timer. A second timer activated the white light and a Hunter Klockounter simultaneously with the offset of the warning light. The white light and Klockounter circuits were integrated with that of the amplifier-loudspeaker system so that a SMU response turned off the white light and stopped the Klockounter. For the Overt RT test a small, rigidly fixed contact button replaced the electrodes. A slight, 2–3 mm, abduction of the small finger brought the subject into contact with the button, triggering the offset of the Klockounter and light through the same circuit employed for the MU response.

PROCEDURE. The electrodes were inserted and the subject was seated before a small table in a shielded cage. The oscilloscope and loudspeaker were located directly in front of him at a distance of about 2 ft. Following a brief description of the nature of MU action potentials, the investigator instructed the subject to abduct his small finger and then very slowly relax the abductor. The movement produced repeated firing of many units, causing a burst of activity on the oscilloscope screen and from the loudspeaker. As the subject relaxed the muscle this activity decreased. The investigator monitored the oscilloscope and loudspeaker feedback with the subject, explained to him what it meant, and gave him continual instructions to relax further or maintain the current degree of tension. Guided by the visual and auditory feedback, all subjects quickly learned to discriminate among action potentials of different SMUs.

With repetition of the abduction-relaxation sequence, under continual verbal guidance, the subject would eventually succeed in producing the repetitive firing of an SMU, in the absence of potentials from any other unit. He was then instructed to vary the rate of firing of the unit by very minute changes in tension in the muscle until he could slow down, stop, and resume firing of the unit. The subject then practiced until he was able to fire a SMU once, twice, then three times on command for three successive trials. At that point he was considered to have mastered the task, his total learning time was recorded, and practice was terminated.

After a short rest the subject began the RT task. Each trial was signaled by the onset of a red warning light with a duration of 500, 1000, or 1500 msec. At the onset of the white reaction light he responded by firing the trained SMU once. During the intertrial interval, which was varied randomly from 10 to 30 sec, he practiced firing the unit. A 2 min rest period was given after every 24 trials. A total of 120 trials was administered in each session. Throughout the testing sessions the subject received feedback via the oscilloscope and loudspeaker. In addition the investigator informed him immediately on any trial on which an unacceptable response was given. Unacceptable responses included multiple firing of the unit, the firing of more than one unit, and firing during the warning light interval. One week after the first test session, a second test session was administered. The procedure was the same in all respects as that employed during the first test session.

After another week the overt RT test was given. The overt response required was a 2–3 mm abduction of the small finger. This response was chosen because it involved the same muscle group as that from which SMU RT had been recorded.

In his earlier reaction time work, Thysell scrupulously trained subjects to a criterion of SMU control. (The criterion was the ability to fire a unit in a "once, twice, thrice" sequence for three consecutive trials without error.) Under this criterion many subjects required 30–60 min, or more, of training before beginning reaction time trials. In his most recently completed experiment, Thysell trained subjects for a maximum of only 20 min and then initiated reaction time testing regardless of whether the subject had reached criterion. Reaction time performance in this experiment was superior to that in earlier experiments. Subjects performed with fewer errors and had faster mean reaction times (Thysell, personal communication, 1971).

Thysell believes that the frustration engendered by a long training period leads to poor reaction time performance. Minimizing the frustration by shortening the training period leads to better performance even though the subject has achieved a poorer level of control. Another possibility is that the 1, 2, 3 pattern required during training is irrelevant to reaction time performance. Training subjects to a high level of proficiency on this criterion may actually be training in a pattern that later interferes with good reaction time performance.

Thysell has carried out a series of reaction time experiments comparing overt (key press) and electrode (untrained, multiunit potentials) reaction times taken simultaneously. In a typical experiment the subject sits with both hands palm down in front of him, an electrode inserted in the abductor digiti V of one hand and a microswitch placed just outside the little finger of the other hand. At the "go" signal the subject abducts both little fingers, the reaction time of each being recorded independently. The mean electrode reaction time is considerably faster than the mean overt reaction time, but the variances of the two responses are about equal and the correlation between them is quite high.

W. D. McLeod and R. V. Thysell (personal communication, 1972), are currently employing a square-wave pulse, triggered by a SMU firing for audio feedback. This feedback seems as effective as that from the MU itself. It has the advantage that the characteristics (for example, voltage and duration) of the pulse can be varied and specified precisely.

2. SUTTON AND KIMM TECHNIQUE

A variation of the Thysell technique was described by Sutton and Kimm (1969, 1970). The subject faced a half-silvered mirror placed so that motor unit activity from a CRO display could be readily viewed, and through which a visual stimulus for the RT procedure could be presented, superimposed on the oscilloscope trace of motor activity. Thus, the subject could continuously observe his own unit activity as well as monitor the RT signals (light flash).

Amplified electrical signals from the muscle were displayed both visually and aurally. The subject was instructed to use the visual and auditory signals as information to help him achieve voluntary control over the firing of a motor unit such that a single spike could be produced on command. When the subject reported he was able to produce a single spike voluntarily, a short practice run involving simple RT to a flash of light was begun. No warning signal was presented, and the interstimulus interval was 8 sec. Termination of this practice session was followed immediately by the data collection period unless the subject had difficulty in controlling the unit during the practice run. In such cases, further practice was permitted and a rerun of the practice series was given.

PROCEDURE. A sequence of 50 RT trials was completed during each session. Each subject participated in a minimum of six sessions, during which data were collected from equal numbers of biceps and triceps motor units. One session per day was conducted. Electromyographic RT data were collected in approximately one-third of the sessions. These data were obtained under identical conditions employed for motor unit data acquisition, using records derived from the intramuscular electrode from which unit data also were obtained.

The RT signal (light flash) originated from a Grass PS-2 photic stimulator located outside the experimental room. Auditory cues were thereby reduced to a minimum. The light was masked so that a 25 mm diameter beam was transmitted through the fixation screen. Between trials, this masked spot was dimly illuminated and provided a fixation point for the subject.

A tape-recorded series of 50 intermittent pulses controlled the RT signals. All subjects viewed the same series of flashes during each experimental session.

A synchronous output from the photic stimulator initiated the timing cycle of an electronic timer (Hewlett-Packard 5223L) and the spike from a motor unit terminated the cycle.

Reaction time to each photic stimulus was recorded in milliseconds and displayed trial-by-trial on the screen. The subject was required to call out the RT value achieved on each trial. This provided immediate information regarding RT latency. The subject was requested to make his RT as brief as possible.

All data, including trigger pulses, motor unit responses, and subject's voice, were recorded on magnetic tape for later analysis.

The experimenter monitored each trial, either on-line or from the recorded sessions. Responses were counted as successful only if a single motor unit spike occurred within 2 sec after the flash. Trial failures consisted of multiple spikes or of no spikes within the above time limit.

The RT data were processed offline with a LINC-8 computer using a program for computing medians and interquartile values. The RT histogram printouts were also obtained from the program. Inasmuch as the distributions of RT were usually asymmetrical, statistical analyses were conducted on medians.

B. Studies of Covert Operant Conditioning

Hefferline and Perera (1963), Bruno et al. (1970), and Hefferline et al. (1971) have applied localized EMG to the discrimination of a covert operant. Their specific objective was to train a subject to report a covert twitch in a thumb muscle detectable only through EMG monitoring. The subject sat in a reclining chair within a dimly lit, shielded enclosure. An intercom provided communication with the experimenter, and a one-way mirror permitted observation of the subject. After preparation of the skin, surface electrodes were applied to the thenar eminence formed by the muscles of the thumb, and to the hypothenar eminence formed by the muscles of the little finger. A ground electrode was attached to the earlobe, and additional sets of dummy electrodes were applied to confuse the subject about the actual recording site. Action potentials appearing between the thumb and little finger muscles were differentially amplified by a factor of one million, and then rectified and displayed on a vacuum-tube voltmeter. A permanent record of the rectified signals was made using either magnetic tape or a recording light-beam galvanometer. The system, which had a time constant of 0.1 sec, was calibrated at the beginning and end of each session.

The response in each study consisted of a ballistic deflection of the voltmeter needle or photorecorder beam which rose sharply above and then returned quickly to the resting tension level. Response magnitudes were categorized according to peak needle or beam excursions, and frequency distributions for most of their subjects were compiled after their sessions by examining the tape or photorecorder records. At the beginning of each experimental run, a response magnitude which occurred not more than once in 1 or 2 min was selected for later reinforcement. This response was always much smaller than that which produced observable movement in the least responsive subject (about 300 μV).

When the effect of positive reinforcement was under study the subject was informed that "we are measuring your ability to relax ... after thirty or forty minutes numbers will begin to appear on the lighted box in front of you. These numbers represent your total score. For each increase in score you will receive a nickel." When negative reinforcement was used, some subjects were told that the study concerned the effects on body tension of noise superimposed on music; others were told that a response, so small as to be invisible, would temporarily turn off the noise, or, when the noise was not present, postpone its onset; and still others were informed that the effective response was a tiny twitch of the left thumb.

Each session began with a 10 min base-line period during which the spontaneous rate of responding was determined and a criterion response magnitude selected. In the subsequent conditioning period (60 min or more), the experimenter pressed a key which produced the reinforcement whenever a criterion response was observed. For subjects receiving positive reinforcement, the experimenter's key press advanced the displayed score by one digit, representing another nickel earned. For subjects working under a negative reinforcement contingency, the key press turned off, for 15 sec, the aversively loud, 60 Hz hum which, during conditioning, was superimposed on the music presented via earphones throughout the entire session, When the noise was already off, a key press postponed its resumption for 15 sec. For all subjects the experimental run terminated with 10 or more minutes of extinction during which criterion responses no longer produced reinforcement.

C. *Studies of Personality Factors Affecting SMU Control*

1. Effects of Previous Training and Skill

Various criteria of successful SMU performance may be matched against previous skills (Scully & Basmajian, 1969). The simplest criterion may be the raw time required to bring an SMU up to a specific criterion.

2. Effects of Handedness

Powers (1969) demonstrated a technique for investigation of cross transfer and handedness. A long series of subjects were studied under identical conditions on two occasions using a different hand each time. Significant differences were found, suggesting that this type of investigation deserves widespread adoption in psychophysiology laboratories.

3. Effects of Internal States

At the time of writing, a number of laboratories are exploring the use of EMGs of localized or SMU contractions for studying covert responses to emotional states. Fundamentally, there is no training of units but rather the use of motor unit responses for externalizing internal events. Equipment and general methodology are similar except that there is no direct feedback for the subject. In one paradigm being tested by S. D. Kahn (personal communication, 1971) the EMG response is processed electronically to manipulate visual displays offered to the subject as reinforcing or negative stimuli.

References

Adrian, E. D., & Bronk, D. W. The discharge of impulses in motor nerve fibres. Part II. The frequency of discharge in reflex and voluntary contractions. *Journal of Physiology*, 1929, **67,** 119–151.

Basmajian, J. V. Control and training of individual motor units. *Science*, 1963, **141,** 440–441.

Basmajian, J. V. *Muscles alive: Their functions revealed by electromyography.* (2nd ed.) Baltimore, Maryland: Williams & Wilkins, 1967.

Basmajian, J. V. Electromyography comes of age. *Science*, 1972, **176,** 603–609.

Basmajian, J. V., Baeza, M., & Fabrigar, C. Conscious control and training of individual spinal motor neurons in normal human subjects. *Journal of New Drugs*, 1965, **5,** 78–85.

Basmajian, J. V., & Simard, T. G. Effects of distracting movements on the control of trained motor units. *American Journal of Physical Medicine*, 1967, **46,** 1427–1449.

Basmajian, J. V., & Stecko, G. A new bi-polar indwelling electrode for electromyography. *Journal of Applied Physiology*, 1962, **17,** 849.

Bruno, L. J. J., Davidowitz, J., & Hefferline, R. F. *Behavioral Research Methods and Instrumentation,* 1970, **2,** 211–219.

Carslöö, S., & Edfeldt, A. W. Attempts at muscle control with visual and auditory impulses and auxiliary stimuli. *Scandinavian Journal of Psychology*, 1963, **4,** 231–235.

Gasser, H. S., & Newcomer, H. S. Physiological action current in the phrenic nerve. An application of the thermionic vacuum tube to nerve physiology. *American Journal of Physiology*, 1921, **57,** 1–26.

Gilson, A. S., & Mills, W. D. Single responses of motor units in consequence of volitional effort. *Proceedings of the Society of Experimental Biology and Medicine*, 1940, **45,** 650–652.

Gilson, A. S., & Mills, W. D. Activities of single motor units in man during slight voluntary efforts. *American Journal of Physiology*, 1941, **133,** 658–669.

Green, E. E., Walters, E. D., Green, A. M., & Murphy, G. Feedback technique for deep relaxation. *Psychophysiology*, 1969, **6,** 371–377.

Hardyk, C. D., Petrinovich, L. F., & Ellsworth, D. W. Feedback of speech muscle activity during silent reading: Rapid extinction. *Science*, 1966, **154,** 1467–1468.

Harrison, V. F., & Mortensen, O. A. Identification and voluntary control of single motor unit activity in the tibialis anterior muscle. *Anatomical Record*, 1962, **144,** 109–116.

Hefferline, R. F., Bruno, L. J. J., & Davidowitz, J. E. Feedback control of covert behavior. In K. J. Connolly (Ed.), *Mechanisms of motor skill development*. New York: Academic Press, 1971.

Hefferline, R. F., & Perera, T. B. Proprioceptive discrimination of a covert operant without its observation by the subject. *Science*, 1963, **139**, 834–835.

Henneman, E., Somjen, G., & Carpenter, D. O. Excitability and inhibitibility of motoneurons of different sizes. *Journal of Neurophysiology*, 1965, **28**, 599–620.

Jacobs, A., & Felton, G. S. Visual feedback of myoelectric output to facilitate muscle relaxation in normal persons and patients with neck injuries. *Archives of Physical Medicine and Rehabilitation*, 1969, **50**, 34–39.

Kahn, S. D., Bloodworth, D. S., & Woods, R. H. Comparative advantages of bipolar abraded skin electrodes over bipolar intramuscular electrodes for single motor unit recording in psychophysiological research. Unpublished manuscript, 1971.

Lindsley, D. B. Electrical activity of human motor units during voluntary contraction. *American Journal of Physiology*, 1935, **114**, 90–99.

Lloyd, A. J., & Leibrecht, B. C. Conditioning of a single motor unit. *Journal of Experimental Psychology*, 1971, in press.

McGuigan, F. J. Covert oral behavior as a function of quality of handwriting. *American Journal of Psychology*, 1970, **83**, 377–388.

Petajan, J. H., & Philip, B. A. Frequency control of motor unit action potentials. *Electroencephalography and Clinical Neurophysiology*, 1969, **27**, 66–72.

Powers, W. R. Conscious control of single motor units in the preferred and non-preferred hand: An electromyographic study. Ph.D. Thesis. Queen's University, Kingston, Ontario, Canada, 1969.

Scully, H. E., & Basmajian, J. V. Effect of nerve stimulation on trained motor unit control. *Archives of Physical Medicine and Rehabilitation*, 1969, **50**, 32–33.

Simard, T. G., & Basmajian, J. V. Methods in training the conscious control of motor units. *Archives of Physical Medicine and Rehabilitation*, 1967, **48**, 12–19.

Simard, T. G., Basmajian, J. V., & Janda, V. Effect of ischemia on trained motor units. *American Journal of Physical Medicine*, 1968, **47**, 64–71.

Simard, T. G., & Ladd, W. L. Conscious control of motor units with thalidomide children: an electromyographic study. *Developmental Medicine and Child Neurology*, 1969, **11**, 743–748.

Smith, O. C. Action potentials from single motor units in voluntary contraction. *American Journal of Physiology*, 1934, **108**, 629–638.

Sutton, D. L. Surface and needle electrodes in electromyography. *Dental Progress*, 1962, **2**, 127–131.

Sutton, D., & Kimm, J. Reaction time of motor units in biceps and triceps. *Experimental Neurology*, 1969, **23**, 503–515.

Sutton, D., & Kimm, J. Alcohol effects on human motor unit reaction time. *Physiology and Behavior*, 1970, **5**, 889–892.

Thysell, R. V. Reaction time of single motor units. *Psychophysiology*, 1969, **6**, 174–185.

Wagman, I. H., Pierce, D. S., & Burger, R. E. Proprioceptive influence in volitional control of individual motor units. *Nature*, 1965, **207**, 957–958.

Zappalá, A. Influence of training and sex on the isolation and control of single motor units. *American Journal of Physical Medicine*, 1970, **49**, 348–361.

Chapter 6

Electromyography: Human and General

Joseph Germana

Department of Psychology
Virginia Polytechnic Institute and State University
Blacksburg, Virginia

I. Introduction 155
II. The Coordination of Behavior and Response Uncertainty 156
III. The Effects of Response Uncertainty on Behavior 158
IV. Autonomic–Somatic Integration 159
V. Conclusion 161
    References 162

I. Introduction

The contents of this chapter represent a departure from the material typically offered in treatments of a circumscribed scientific methodology. Conventional presentations are usually restricted to matters of research apparatus, the techniques of measurement, and the mathematical-statistical analysis of experimental data. Such efforts are worthwhile to the extent that they do not merely duplicate technical information presently available in readily accessible form.

However, the fact concerning general electromyographic (EMG) recording is that these purely methodological considerations have been given extensive and adequate treatment in previous works. Certainly, no investigator intending to use surface electrodes to record EMG activity can overlook the basic and comprehensive discussion of methodology provided by Davis (1959) in the *Manual of Surface Electromyography*. The more recent works of Basmajian (1967) and Lippold (1967) have additional content of interest and value. In this brief chapter, I have instead attempted to outline the psychophysiological significance of general EMG activation.

II. The Coordination of Behavior and Response Uncertainty

One major principle concerning the relationship of brain and behavior would seem so apparent as to require only infrequent mention. And yet, it is precisely this functional point of view which deserves consistent emphasis because it has provided a coherent framework for many apparently diverse sets of neurobehavioral phenomena. Sperry (1952) has directly stated this view in the following manner: ". . . the principal function of the nervous system is the coordinated innervation of the musculature. Its fundamental anatomical plan and working principles are understandable only on these terms [p. 298]." Many of the classic neuroscientists including Sechenov, Jackson, Sherrington, and Pavlov offered the same opinion. For example, a consideration of the evolution of the nervous system led Herrick (1961) to conclude that the most primitive systems ". . . form a network that serves to keep all parts of the body in communication and so facilitates orderly coordination of the bodily movements. In higher animals with more complicated structure and behavior special collections of nerve cells are set apart to provide more efficient coordination and integration [p. 255]."

At the same time, an objective study of the coordination and regulation of motor activities reveals the inherent complexity and significance of behavior. Bernstein (1967) identified the essential problem of behavioral regulation in the following way: "The first clear biomechanical distinction between the motor apparatus in man and the higher animals and any artificial self-controlling devices . . . lies in the enormous number . . . of *degrees of freedom* which it can attain [p. 125]." Kinematic analyses of some elementary behaviors reveal the anatomical, mechanical, and physiological sources of these degrees of freedom, the types of error with which they can be associated, and therefore the complexity of the spatiotemporal problems involved in the regulation of even such simple behaviors. Coordination has been defined by Bernstein (1967) as ". . . *the process of mastery of redundant degrees of freedom* [p. 127]."

The "functional" or "system" view of the interrelation between physiological and behavioral actions has served as a valuable methodological paradigm. Although it can be supplemented by more abstract or "formal" paradigms, the functional view is consistently applied in this treatment.

If the regulatory functions of the central nervous system are primarily concerned with the coordination of behavior and this process involves the progressive mastery of the degrees of freedom associated with behavior, then the essential problem confronting the nervous system can be described as indeterminacy or response uncertainty. Consequently, response uncertainty would seem to precede and accompany the performance of a behavioral response that has *not* been transformed into an automatic or reflexive sequence of coordinated events. In addition, any situation that increases the complexity of the behavioral condition through manipulation of stimulus or response events can enhance or prolong response uncertainty.

Thus, the uncertainty derived from the inherent complexity of behavior can be augmented by several conditions which produce additional sources of indeterminacy. In general, these additional degrees of freedom may be generated by increasing stimulus uncertainty, by modifying the nature or complexity of a behavior, or by changing the conditional significance of a response. In all cases, however, the additional response uncertainty that results ultimately concerns the basic problem of selecting appropriate or adaptive behavior which can be produced in an efficient manner (Germana, 1972).

On the other hand, response uncertainty can be reduced through repeated occurrences of the stimulus and behavioral events, and through the adjustment of behavior to the contingencies of reinforcement. The coordination processes of the nervous system are critically dependent on the afferent feedback produced by behavior and the external consequences of behavior. Again, Bernstein (1967) has described the situation quite well: "Like every other form of nervous activity which is structured to meet particular situations, motor corrdination develops slowly as a result of experiment and exercise . . . in reality, we observe as a rule that improvement in coordination is achieved by utilizing all possible roundabout methods in order to reduce the number of degrees of freedom at the periphery to a minimum [pp. 107–108]."

III. The Effects of Response Uncertainty on Behavior

As behaving organisms with efficient nervous systems, our own conscious perspective and feelings of effort become withdrawn in those well-coordinated behaviors that consistently produce their intended results, that is, those responses satisfactorily adjusted to the existing external contingencies. It is therefore difficult for us to subjectively recognize the inherent complexity of those automatic or reflexive behaviors that probably constitute the greater proportion of our repertoire. In addition, the phasic, goal-directed aspects of behavior tend to subjectively overshadow the basic postural activities or attitudes that serve as preparations for these specific operant responses. However, these anticipatory and supporting aspects of behavior become particularly significant when the effects of response uncertainty are investigated.

In fact, response uncertainty appears to be most immediately and clearly demonstrated in preparations for overt behavior. At this time, it would

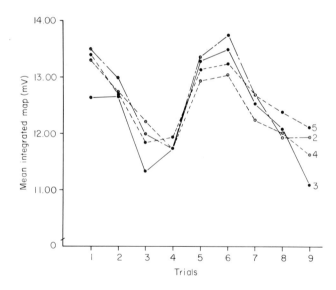

FIG. 6-1. Mean integrated muscle action potentials recorded from the brachioradialis region of the forearm during habituation (trials 1–3) and subsequent conditioning of a motor response (trials 4–9) in ten human subjects. Each trial consisted of five segments, but the data from segment 1 were directly confounded by the occurrence of the overt behavioral response. The results clearly demonstrate the habituation and activational peaking effects (Germana, 1969a).

seem most profitable to regard these preparations as anticipatory and concomitant, frequently covert responses which facilitate the performance of overt operant responses. The preparatory responses associated with well-established and coordinated behaviors may be specifically localized to the familiar pattern involving facilitation of primary and synergistic effector mechanisms—inhibition of antagonistic mechanisms.

Under conditions of greater response uncertainty, however, more extensive or general preparations for overt behavior are produced, and this broad form of response generalization involves not only somatic, but autonomic and central mechanisms as well. The covert events that comprise these extensive preparations, and which can reflect response uncertainty, include increased, widely generalized EMG activity, increased skin conductance level, heart and respiration rates, peripheral vasoconstriction, pupillary dilatation, EEG desynchronization, and other responses originally treated by Berlyne (1960), Duffy (1962), and Malmo (1959, 1962) as representing an increased state of "activation" or "arousal."

Thus, proceeding in a different manner, that is, from a strictly neurobehavioral view, a familiar conclusion can be reached: general EMG activity, as recorded by surface electrodes from widely diffuse and functionally disparate sites, reflects part of a general activation pattern. In addition, the present analysis establishes the basic significance of these responses as extensive preparations for overt behavior, and identifies the essential nature of the conditions that produce these events as response uncertainty.

IV. Autonomic–Somatic Integration

It may be observed that the several situations previously listed as conditions involved response uncertainty have been shown to produce activational responses. However, some of these effects have been more widely replicated than others. In particular, two conditions have been replicated across organisms, situational specificities, and psychophysiological measures. The two effects occur with novel stimuli and with conditional stimuli during the initial stages of conditioning.

The effects are clearly demonstrated with EMG (or muscle action potentials, MAP) in Fig. 6-1. The subjects in this experiment (Germana, 1969a) were administered nine trials, each trial consisting of five segments. The "motor-speech" method of conditioning was employed by pairing the termination of the fifth segment (an asterisk slide) with the verbal command "press." Each subject pressed a button in his left hand when the command

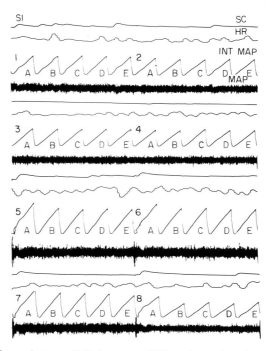

FIG. 6-2. The skin conductance (SC), heart rate (HR), and muscle action potential (MAP—raw and integrated) activity of an individual subject in the experiment. Increased MAP is clearly demonstrated after the onset of conditioning (trial 5).

was delivered on the last six trials. The command was not given during the first three trials.

The results indicate an habituational decrement in EMG from the brachioradialis region of the left forearm on the first four trials (before the first command had been delivered). During subsequent conditioning trials, EMG activity first increased and then again habituated. This last effect, as obtained with psychophysiological measures, has been termed "activational peaking" (Germana, 1968). Activational peaking is clearly demonstrated in the EMG activity of Fig. 6-2, a record from an individual subject in the experiment.

The phenomena of habituation and activational peaking, on the psychophysiological level, appear homologous to the two basic interneuronal processes of habituation and sensitization recently discussed by Groves and Thompson (1970). Although these data are far removed, they do suggest a strong neurophysiological and anatomical coherence received by the present approach to EMG activity and other activational events. I have previously presented (Germana, 1969b) some data which indicate a central integration of autonomic and somatic events. Two major facts emerge from these data.

First, although a classical distinction between the autonomic and somatic nervous systems can be maintained peripherally, the general plan within the central nervous system involves the integration of autonomic and somatic processes. The integration is frequently so salient that it has led some investigators to suggest a reorganization of our conceptual framework. Hess (1954) and Gellhorn (1967) have consistently urged and supported the concept that the functions of several brain structures should be described as "ergotropic" and "trophotropic," indicating response complexes of primarily sympathetic–somatic events, on the one hand, and parasympathetic–somatic effects, on the other.

Second, the structures that most clearly demonstrate this integration of autonomic and somatic processes are those that have been regarded as constituting a core arousal system (for example, regions of the brainstem reticular formation and its forward thalamic and hypothalamic projections), and other structures traditionally treated as components of the extrapyramidal system (for example, premotor cortex, basal ganglia). The fact is that these structures have long been implicated in the form of tonic, background control of behavior that constitutes preparatory, postural, attitudinal responses.

Some recent treatments of hippocampal theta activity suggest similar functions. For example, Black (1970) has concluded that hippocampal theta probably represents a central behavior state that reveals both autonomic (for example, heart rate) and somatic (general EMG and gross behavioral) consequences when the discriminative stimulus used to condition theta in curarized dogs is presented to the animals in a subsequent noncurarized stage. Vanderwolf (1971) has studied hippocampal theta activity in free-moving rats and has suggested that theta reflects preparations to perform "voluntary," as contrasted with automatic, responses. It would appear that hippocampal theta activity may represent a central indicant of activation, an extensive preparation for overt behavior produced by conditions of response uncertainty. It is interesting to note that hippocampal theta and other activities in the core arousal system demonstrate both the habituation and activational peaking effects (for example, see Grastyan, Lissak, Mandarsz, & Dunhoffer, 1959).

V. Conclusion

In this chapter, I have provided only a very brief sketch of one functional context for general EMG activation—one in which it is categorically associated with other nonspecific activities of the organism. Greater empirical and hypothetical detail in the application of a "system" paradigm to these psychophysiological response complexes has been provided in a recent treatment of learning and conditioning (Germana, 1973).

In this last source, it is suggested that these, general or activational responses occur during periods of behavioral uncertainty, and that they may be regarded as a "variational" system of covert responses in the sense that they affect both a greater equalization within the distribution of behaviorally selective probabilities, and increase variability in the dynamic or coordinational aspects of behavior. As such, they can be contrasted with "specificational" patterns of covert events that act as highly differentiated "signals" for behavioral responses.

References

Basmajian, J. V. *Muscles alive: Their functions revealed by electromyography.* (2nd ed.) Baltimore, Maryland: Williams & Wilkins, 1967.
Berlyne, D. E. *Conflict, arousal, and curiosity.* New York: McGraw-Hill, 1960.
Bernstein, N. *The coordination and regulation of movements.* Oxford: Pergamon, 1967.
Black, A. H. Mediating mechanisms of conditioning. *Conditional Reflex,* 1970, **5,** 140–152.
Davis, J. F. *Manual of surface electromyography.* Springfield, Virginia: Clearinghouse for Federal Scientific and Technical Information, 1959. (Originally prepared in 1952.)
Duffy, E. *Activation and behavior.* New York: Wiley, 1962.
Gellhorn, E. *Autonomic–somatic integrations.* Minneapolis: Univ. of Minn. Press, 1967.
Germana, J. Psychophysiological correlates of conditioned response formation. *Psychological Bulletin,* 1968, **70,** 105–114.
Germana, J. Patterns of autonomic and somatic activity during classical conditioning of a motor response. *Journal of Comparative and Physiological Psychology,* 1969, **69,** 173–178. (a)
Germana, J. Central efferent processes and autonomic-behavioral integration. *Psychophysiology,* 1969, **6,** 78–90. (b)
Germana, J. Response uncertainty and autonomic-behavioral integration. *Proceedings of the New York Academy of Sciences,* 1972, **193,** 185–188.
Germana, J. Psychophysiology of learning and conditioning. In F. J. McGuigan & D. B. Lumsden (Eds.), *Contemporary views of conditioning and learning.* Washington, D. C.: H. V. Winston Press, 1973.
Grastyan, E., Lissak, K., Mandarsz, I., & Dunhoffer, H. Hippocampal electrical activity during the development of conditioned reflexes. *Electroencephalography and Clinical Neurophysiology,* 1959, **11,** 409–430.
Groves, P. M., & Thompson, R. F. Habituation: A dual-process theory. *Psychological Review,* 1970, **77,** 419–450.
Herrick, C. J. *The evolution of human nature.* New York: Harper, 1961.
Hess, W. R. *Diencephalon: Autonomic and extrapyramidal functions.* New York: Grune & Stratton, 1954.

Lippold, O. C. J. Electromyography. In P. H. Venables & I. Martin (Eds.), *A manual of psychophysiological methods.* New York: Wiley, 1967.

Malmo, R. B. Activation: A neurophysiological dimension. *Psychological Review,* 1959, **66,** 367–386.

Malmo, R. B. Activation. In A. J. Bachrach (Ed.), *Experimental foundations of clinical psychology.* New York: Basic Books, 1962.

Sperry, R. W. Neurology and the mind-brain problem. *American Scientist,* 1952, **40,** 291–312.

Vanderwolf, C. H. Limbic-diencephalic mechanisms of voluntary movement. *Psychological Review,* 1971, **78,** 83–113.

Chapter 7

Electrocardiogram: Techniques and Analysis

Neil Schneiderman

*Department of Psychology
and Laboratory for Quantitative Biology
University of Miami
Coral Gables, Florida*

George W. Dauth[1]

*Department of Behavioral Physiology
New York State Psychiatric Institute
and Columbia University
New York, New York*

David H. VanDercar[2]

*Laboratory of Physiological Psychology
Rockefeller University
New York, New York*

I. Introduction	166
II. The Heart	167
A. Structure	167
B. The Cardiac Cycle	167
C. The Electrocardiogram	168
D. Normal and Abnormal Patterns in the Electrocardiogram	170
III. Regulation of Cardiac Responses	172
A. Relationship of Heart Rate to Other Cardiovascular Responses	172
B. Neural and Hormonal Control of the Heart	172

[1] Present address: Department of Neurology, College of Physicians and Surgeons, Columbia University, New York, New York.

[2] Present address: Department of Psychology, University of South Florida, Tampa, Florida.

        C. Cardiac–Somatic Coupling 175
        D. Heart Rate and Respiration 175
        E. Curarization... 177
   IV. Basic Instrumentation 179
        A. Electrodes.. 179
        B. Reduction of Cable Artifacts... 181
        C. Biotelemetry.. 182
        D. Amplifiers and Polygraphs... 184
        E. Cardiotachometers 186
        F. Triggering... 186
        G. Data Reduction 188
    V. Data Analysis.. 191
   VI. Operant Conditioning and Biofeedback.................................... 193
        References 197

I. Introduction

The heart is the pump of the cardiovascular system. Its function is to provide the power for circulating the blood. In order to perform this task efficiently in an organism whose blood needs change continually, the heart receives feedback from various chemical and pressure receptors. These sensors form part of a regulatory mechanism which adjusts heart rate (HR), contractile force, cardiac output, and the distribution of blood to various vascular (blood vessel) networks. The easiest of these variables to record is HR. Because HR is sensitive to the changes in energy requirement associated with different behavioral states, it is used widely in behavioral research as an index of attention, arousal, emotion, and conditioning. It is also used as an indicant of the organism's well-being during various physiological interventions.

Although the electrocardiogram (ECG)[3] is a useful tool in behavioral research, its limitations as well as advantages should be kept in mind. The ECG, for instance, provides information about only one aspect of the organism's functioning. Consequently, identical changes in HR can be correlated with markedly different physiological and behavioral states. An increase in HR may be associated with a fall in blood pressure when we simply assume a standing position or it may accompany a rise in blood pressure during sympathetic arousal. Conversely, a decrease in HR may be part of a general parasympathetic adjustment or it may reflect a compensatory reaction to a sympathetically induced rise in blood pressure. In order to obtain a better understanding of the relationships between the ECG and

---

[3] The more frequently used abbreviation, EKG, comes from the German word Elektrokardiogramm.

7. ELECTROCARDIOGRAM: TECHNIQUES AND ANALYSIS

behavior, investigators have begun increasingly to monitor other physiological responses along with HR.

In the present chapter we shall briefly describe the physiological basis of the ECG, the manner in which it is recorded, and the relationship of ECG responses to the total physiological functioning of the organism. We shall also discuss the measurement of ECG responses and describe the instrumentation used to record and analyze them. The procedures and instrumentation used in such behavioral research as the instrumental modification of HR will be presented and attendant problems such as curarization abnormalities and controls for respiration examined.

II. The Heart

A. Structure

The mammalian heart is a four-chambered muscular organ. Most of the heart wall consists of specialized muscle tissue called myocardium. The two upper cavities of the heart are called atria and the two lower ones are called ventricles. The ventricles have larger and thicker walls than the atria. This is associated with the heavier pumping burden of the ventricles. In similar fashion the walls of the left ventricle are thicker and larger than those of the right ventricle. This is associated with the right ventricle having only to supply the lungs, whereas the left ventricle must supply the rest of the vasculature. The left and right sides of the heart are separated completely by a septum.

The conduction system of the heart consists of four structures. These are the sinoatrial (SA) node, the atrioventricular (AV) node, the atrioventricular bundle (bundle of His and its branches) and the Purkinje fibers. A schematic representation of the conduction system is shown in Fig. 7-1. It can be seen that the SA node is located at the junction of the superior vena cava and the right atrium. The AV node is located in the posterior portion of the right atrium proximal to the interatrial septum. It is continuous with the bundle of His, whose branches continue to the ventricular muscle via the Purkinje system.

B. The Cardiac Cycle

Some heart tissue is able to depolarize spontaneously. Such tissue is referred to as pacemaker tissue. At any instant, the portion of the heart having the highest rate of depolarization is referred to as the pacemaker.

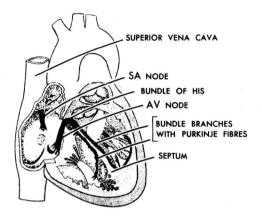

FIG. 7-1. Schematic representation of the conduction system of the heart.

Depolarization spreads from the pacemaker to other cardiac regions before they can respond spontaneously. For this reason the rate of discharge of the cardiac pacemaker determines the rate at which the heart beats. In the mammalian heart the normal pacemaker is the SA node. It is located in the upper right-hand corner of the right atrium.

Although the SA node is the normal pacemaker, several other areas of the heart also are capable of depolarizing spontaneously. These include the junction of the SA node and atrial muscle, the AV node, and the bundle of His. All of these sites are characterized by low resting membrane potentials and small action potentials. The locations besides the SA node which have latent pacemaker capability are known as *ectopic foci*. Cardiac beats beginning at one of these latent pacemaker sites are called ectopic beats.

Changes in HR can be mediated by altered SA node activity, by shifts in the site of pacemaker activity, or by a combination of these. A decrease in the rate of depolarization at the SA node, for instance, leads to a decrease in HR. Slowing of HR could also occur in response to the slowing or complete suppression of SA node activity such as may occur during direct vagal stimulation.

Once electrical activity is initiated in the SA node, a wave of excitation spreads through the atrial walls and through the interatrial septum in an orderly fashion. After a relatively long latency the AV node depolarizes. The excitation then spreads down the bundle of His and its branches to the Purkinje system. From there the electrical wave travels through the ventricular muscle.

C. *The Electrocardiogram*

As excitation spreads across the heart, the cells that are depolarized become electrically negative with respect to regions still in the resting state.

The sequences of depolarization and repolarization which occur as excitation spreads across the heart during successive cardiac cycles are recorded as the ECG. A normal cardiac cycle, shown in Fig. 7-2, begins with the P wave, which reflects the depolarization of the atria. Excitation reaches the AV node before the end of the P wave. The excitation travels through the bundles of His and the Purkinje system during the P–R segment. Deflections corresponding to ventricular excitation are recorded as the QRS complex. Repolarization of the atria, which follows the P wave and continues during the QRS complex produces an ECG wave which is usually too small to be detected with conventional leads. In contrast, the T wave which represents ventricular repolarization is quite conspicuous.

The spread of electrical excitation across the heart triggers a sequence of contractions in the myocardium. This has an important influence on the diastole (filling stage) and systole (ejection stage) of the heart cavities. Thus, atrial systole beings shortly after the onset of the P wave, ventricular systole begins near the end of the R wave, and ventricular systole ends just after the T wave. It can be seen in Fig. 7-2 that the QRS complex terminates just before the onset of the systolic phase in the blood pressure recording from the carotid artery.

Although the form of the ECG is very similar among mammals, HR differs considerably among species. In general, larger mammals have slower HR than smaller ones. In the resting state a horse's HR may be approximately 35–40 beats per minute (bpm), whereas in the mouse resting HR varies between 600 and 650 bpm. The mean bpm for man, rabbit, and rat are roughly 70, 225, and 375, respectively.

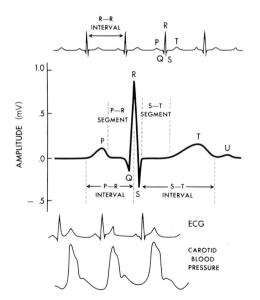

FIG. 7-2. Complexes, intervals, and segments of the electrocardiogram (ECG). The two lowermost tracings show the temporal relationship between the ECG and blood pressure recorded from the carotid artery.

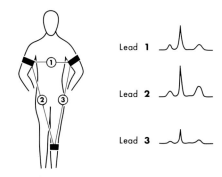

FIG. 7-3. The conventional bipolar limb leads of the electrocardiogram. To record from each lead shown, separate electrodes from each of the two limbs specified are each connected to the electrocardiograph.

An ECG may be recorded by measuring the potential difference between any two sites on the surface of the body. The electrodes may be placed on the body surface because body fluids are good conductors. In general, the extremities can be considered as lead wires connected to the body since the configurations and magnitudes of the ECG waves are not significantly altered if the electrodes are moved along the extremities. In contrast, the magnitude and configuration of individual ECG waves vary with the location of the electrodes placed on the trunk. Although the configuration of individual ECG waves changes with bodily location of the electrodes, the individual components are time-locked with each other.

Any ECG recording between two sites constitutes an electrocardiographic lead. The conventional bipolar limb leads are shown in Fig. 7-3. It can be seen that lead I is between the two arms, lead II is between the right arm and left leg, and lead III is between the left arm and left leg. Because lead II provides the largest and clearest representation of the QRS complex, it is the most used lead in behavioral research. In medical practice, the ECG has been standardized so that the recording paper is moved at 25 mm/sec and a 1 μV input to the amplifier produces a 1 cm deflection of the pen. To the extent that behavioral research has emphasized changes in HR to the exclusion of changes occurring in the general form of the ECG, the amplifier value and paper speed conventions have not been closely followed. Instead, because of the requirements of automatic data processing equipment, an attempt is usually made to provide the best signal-to-noise ratio for the R wave.

D. *Normal and Abnormal Patterns in the Electrocardiogram*

During the past few years behavioral research using the ECG has expanded into a number of fields. Investigators, for instance, have begun to use behavior modification techniques to alter abnormal as well as normal aspects of the ECG. In addition, experimenters placing cardiovascular stress upon their subjects as well as investigators using curariform drugs or artificial pacemakers have a need to know about the abnormalities which can be detected in the ECG.

The term *normal sinus rhythm* indicates that the pacemaker is within the SA node, that the sequence of electrical events in the ECG is normal and constant, and that HR is within normal limits. In human children and young adults as well as in several other species such as the dog, HR changes phasically during the respiratory cycle. This condition, referred to as *sinus arrythmia* is not a sign of pathology.

When HR falls markedly below the normal rate the condition is called *bradycardia*. In contrast, when a pronounced increase in HR occurs, the condition is known as *tachycardia*. Differences in basal HR among species and lack of information about the typical ranges of HR in different behavioral states has made it difficult to precisely define bradycardia and tachycardia.

Abnormal contractions of the heart are referred to as *extrasystoles*. When the abnormal impulses arise in the SA node, the beat is referred to as an *interpolated premature beat*. Other extrasystoles are paced outside of the SA node. The most common ectopic beats are *atrial premature beats* which are paced within the atrium and *ventricular ectopic beats* which are paced in the ventricular system. Atrial premature beats are characterized by an abnormal P wave followed by a normal QRS complex and T wave. In contrast, when an ectopic beat arises in a branch of the bundle of His or in the ventricular myocardium, the QRS complex is abnormal in shape and duration. Occasionally a ventricular extrasystole arising from a single focus discharges rhythmically with constant relation to the QRS complex. The extra beat, called a *coupled* beat, may occur regularly or after every second, third, or fourth beat.

An episode of very rapid beating with sudden onset and offset is referred to as *paroxysmal tachycardia*. At extremely high rates atrial flutter may occur in which the atria depolarize at such a fast rate that the AV conduction system cannot respond. In this situation every P wave is not followed by a QRS complex. At rates exceeding 350–400 bpm in man *atrial fibrillation* may develop in which the ventricle still beats but fine oscillations take place in the rest of the record. In contrast to atrial fibrillation, *ventricular fibrillation* is usually a terminal event in which recognizable beats do not occur. Although ventricular fibrillation may be produced by electrocution, its treatment also consists of electric shock.

III. Regulation of Cardiac Responses

A. Relationship of Heart Rate to Other Cardiovascular Responses

Although the ECG provides important information about cardiac function, the activity of the heart can only be understood properly within the context of the total cardiovascular system. The function of the cardiovascular system is to provide the proper quantity of oxygenated blood to the various vascular networks in response to the demands of various tissues and organs. This typically requires changes in the relative distribution of blood from one vascular bed to another as well as changes in cardiac output.

Cardiac output is defined as the quantity of blood ejected each minute by one of the ventricles. This quantity is a function of HR multiplied by cardiac stroke volume. Stroke volume represents the volume of blood in the ventricle at the end of diastole minus the ventricular volume at the end of systole. Cardiac output can be modified by changes in diastolic or systolic volume as well as by changes in HR.

Blood forced into the aorta during systole exerts considerable pressure. As the blood passes from the larger arteries to the smaller arteries and arterioles, blood pressure falls rapidly. The fall in pressure occurs because the small vessels offer greater resistance to flow. Blood pressure in the cardiovascular system is thus a function of cardiac output (HR × stroke volume) and peripheral resistance. Since the diameter of the arterial lumen is subject to neural control, resistance may change also as a function of behavioral manipulations. The relationship between blood flow, pressure, and peripheral resistance may be expressed as

$$\text{resistance} = \frac{\text{flow}}{\text{pressure}}.$$

Our present exposition indicates that knowledge derived solely from the ECG provides only a small part of the total information about the way in which the cardiovascular system ensures that the proper quantity of blood is pumped to specific tissues and organs. For this reason investigators interested in the physiological concomitants of behavioral states have increasingly examined such measures as blood pressure and regional flow as well as the ECG. At the same time it should be recognized that for some specific problems the HR measure alone may suitably serve as an index of change in behavioral states.

B. Neural and Hormonal Control of the Heart

Although cardiac tissue is intrinsically rhythmic, its activity is modulated bilaterally by neural inputs from the sympathetic and parasympathetic divisions of the autonomic nervous system. The cardiac branch of each vagus nerve provides a cholinergic, parasympathetic input to the heart which is responsible for decreasing HR. In contrast, the cardiac accelerator nerves provide adrenergic, sympathetic inputs which can accelerate HR and increase the force of cardiac muscle contraction.

In most circumstances HR is influenced by both the cardiac accelerator and the vagus nerves. The specific contribution of each neural component is usually dependent upon the situation and the species. Severing the vagus nerves bilaterally or injecting an appropriate parasympathetic blocking agent such as atropine in the resting, nonstressed animal causes a pronounced increase in HR in most mammalian species. In contrast, severing the cardiac sympathetic nerves or administering a beta-adrenergic blocking agent such as propranolol ordinarily decreases HR in the resting animal. A recent study by Pappas, DiCara, and Miller (1971), however, found that for curarized rats no changes occurred in basal HR as a function of sympathectomy by 6-hydroxydopamine injection.

Increases in HR during exercise in cats, dogs, and humans appear to be primarily mediated by a decrease in vagal activity accompanied by a smaller increase in sympathetic activity (for example, Gasser & Meek, 1914; Robinson, Epstein, Beiser, & Braunwald, 1966; Samaan, 1935). Conditioned tachycardia in dogs also has been presumed to be mediated primarily by a decrease in vagal restraint (Dykman & Gantt, 1959), although Cohen and Pitts (1968) have provided evidence that the sympathetic nerves make a larger contribution than the vagus nerves in conditioned cardioacceleration of the pigeon.

In the intact moving subject general sympathetic activation causes an increase in HR and blood pressure. If a sympathetically aroused animal shows behavioral freezing, however, the increase in blood pressure will be accompanied by a decrease in HR. In this case, the increase in blood pressure stimulates baroreceptors in the carotid sinus and aortic arch, resulting in impulses being sent to the vasomotor center in the medulla. This in turn triggers an increase in vagal activity which originates in the medulla, causing a slowing of HR. Because of the vagal compensatory response, the ECG alone cannot be used to distinguish between a sympathetically aroused organism making a freezing response and an organism showing general parasympathetic activation.

The important integrating role of the medulla in cardiovascular regulation was shown in Ludwig's laboratory in the 1870's. It was found that successive transections of the brainstem begun in the diencephalon and moving caudally only induced hypotension when the lesion was below the lower pons. In 1916 Ranson and Billingsley used electrical stimulation to locate pressor and depressor zones in the medulla. Subsequent work has established that these zones exert their influence through the sympathetic nervous system, and that they influence HR as well as blood pressure responses. The pressor and depressor areas, known as the vasomotor center, are intimately related to the dorsal motor nucleus of the medulla (nucleus ambiguous in the cat), which exercises control over the vagal cardiac innervation.

Although the medulla plays an important role in the regulation of cardiovascular responses, considerable integration appears also to occur at supramedullary levels. Much of this integration appears to occur in the neocortex, limbic system, and hypothalamus and seems related to (a) energy requirements of the organism, (b) autonomic functions appropriate to behavioral states such as rage, and (c) regulation of such internal processes as heat dissipation. While most cardiovascular responses elicited from higher centers relay in the medulla, Lindgren, Rosen, Strandberg, and Uvnas (1956) have provided evidence for a sympathetic, cholinergic, vasodilator outflow to skeletal muscle which bypasses the cardiovascular centers of the medulla.

Heart rate and other cardiovascular changes have been elicited by electrical stimulation of the neocortex, limbic system and hypothalamus (e.g., Ban, 1966; Covian & Timo-Iaria, 1966; Hess, 1949; Hoff & Green, 1936; Kaada, 1951). Descending cardiovascular pathways have been traced from the hypothalamus through the midbrain and pons (for example, Chai & Wang, 1962; Enoch & Kerr, 1967a, b). Electrical stimulation of the brain also has been used to motivate HR classical conditioning by Elster, VanDercar and Schneiderman (1970), Malmo (1965), VanDercar, Elster, and Schneiderman (1970), and Sideroff, Schneiderman and Powell (1971). Central lesions have been used to abolish HR conditioning (for example, Durkovic & Cohen, 1969a, b).

In addition to the direct neural control of the heart, the endocrine system also appears to play a role in the regulation of cardiovascular activity. Sayers and Solomon (1960), for instance, showed that the work performance of the rat heart–lung preparation was increased when either aldosterone, corticosterone, or hydrocortisone is added to the perfusing blood. In contrast, work performance was decreased when the preparation was perfused with the blood from adrenalectomized rats.

Secretions of norepinephrine and epinephrine by the adrenal medulla appear to have identical effects upon the heart as does the secretion of norepinephrine at the sympathetic neuroeffector junctions within the heart. In contrast, the nature of the influence of the adrenal cortical steroids upon the myocardium is a more controversial matter.

Thyroid hormones as well as adrenal hormones affect the heart. The HR of patients with *hypothyroidism*, for example, is slowed and cardiac output is diminished. Conversely, patients with *hyperthyroidism* exhibit tachycardia, increased cardiac output and arrhythmias such as atrial fibrillation. Studies conducted upon isolated cardiac muscle preparations have indicated that thyroxine has direct effects upon the heart.

C. Cardiac–Somatic Coupling

The relationship between cardiovascular activity and muscular exertion has been investigated extensively. During tasks involving muscular exertion, increases occur in both HR (Gasser & Meek, 1914; Krogh & Linhard, 1913) and strength of cardiac contraction (Sarnoff, 1955). In a more recent experiment, Webb and Obrist (1967) made food reinforcement contingent upon an increase followed by a decrease in somatic activity. They found that HR accelerated when somatic activity increased and decelerated below base line during decreases in somatic activity.

Although the HR changes elicited in conditioning situations are intimately related to gross motor activity and the energy requirements of the organism, they are not always importantly related to more restricted skeletal responses such as the eyeblink or finger-withdrawal. In addition, experimental evidence suggests that while HR is highly correlated with gross motor activity, it is not directly mediated by it. Black (1965), for example, has demonstrated that both accelerative and decelerative HR CRs in dogs can each remain qualitatively the same whether the animals are run under a paralyzing drug or in the normal state. When these data are considered together with the known positive correlation between HR and general skeletal activity, it appears that a single central state may concomitantly influence the central control of both HR and gross movement. The nature and extent of this coupling requires further elucidation.

D. Heart Rate and Respiration

Changes in respiration frequently have been shown to have a systematic relationship to changes in HR. In sinus arrythmia, for example, heart rate increases during inspiration and decreases during expiration. In part, the arrythmia seems to be caused by impulses from the inspiratory center being sent to the cardiac center of the medulla. In addition, inflation of the lung excites vagal stretch receptors which inhibit the dorsal motor nucleus and excite the cardioaccelerator center.

The influence of respiratory changes upon HR has been of considerable concern in experiments assessing HR conditioning. In animal studies, of course, the effects of respiration are ruled out by curarization. This has not been feasible thus far in the HR conditioning of humans. Instead, most studies conducted on humans have used the technique of paced respiration. Under conditions of paced respiration, the subjects are typically taught to match their breathing rate to the beat of a metronome.

In an early experiment, Westcott and Huttenlocher (1961) trained subjects to breath at a high but constant rate during the presentation of the conditioned stimulus. Although the forms of the classically conditioned HR conditioned responses were similar under paced and unpaced respiration, several changes were noted in the latency and amplitude of the HR responses. In another classical conditioning experiment, Wood and Obrist (1968) used the paced respiration technique and found that it abolished the accelerative component of the HR conditioned response, while leaving the decelerative component intact.

A relationship between respiration and HR also has been observed during attempts to instrumentally control HR. Brener and Hothersall (1966) and Hnatiow and Lang (1965) reported that the HR changes they obtained were correlated with respiratory changes. In a subsequent experiment, Brener and Hothersall (1967) examined the instrumental conditioning of HR under conditions of paced and unpaced respiration. They found that the paced respiration technique minimized the respiratory differences between groups instructed to increase or decrease their HR. Moreover, reliable HR conditioning occurred with or without paced respiration.

The paced respiration technique ensures that breathing rate is held constant across conditions. In addition, most investigators using the technique have been conscientious in eliminating gross differences in respiratory amplitude. Nevertheless, subtle differences in respiratory amplitude, slope, and the parameter values of the inspiration–expiration ratio do occur under paced respiration. Consequently, attempts have been made to condition uncurarized human subjects whose breathing is controlled by a respirator.

In one instrumental conditioning experiment VanDercar, Feldstein, and Solomon (in preparation) used a Bennett MA I human respirator. This instrument can be used in either an assist or control mode. When the respirator is functioning in the assist mode, inspiration is initiated by the subject, but a fixed preset volume of air is delivered during each inspiratory cycle. In this mode the subject has the ability to control breathing rate to some degree, but has no control over the volume of air delivered. When the respirator is functioning in the control mode, both the rate and volume of air delivered are held constant at predetermined values.

By carefully adjusting the rate and volume controls on the respirator, a subject can be maintained for a long period of time fairly comfortably. In order to detect possible changes in the subject's respiratory pressure due to resistance against the respirator, respiratory pressure is monitored. This is accomplished by monitoring the pressure in the system by means of a strain gauge pressure transducer.

In the experiment by VanDercar, Feldstein, and Solomon human subjects were instrumentally conditioned to increase heart rate in the presence of one stimulus and to decrease it in the presence of another. Most subjects first were conditioned without the respirator, then trained further with the respirator in the control mode, and finally trained again without the respirator. Original training occurred for some subjects while they were being respirated in the control mode. The results indicated that when subjects were placed in the control mode the ability to control heart rate usually did not exist even after multiple training sessions. In the few instances in which operant control of heart rate was manifested in the control mode, the heart rate conditioning was accompanied by changes in respiratory pressure and/or other somatic activity.

E. Curarization

In order to eliminate the effects of respiration and skeletomotor activity upon HR responses, several investigators have curarized their subjects. Classical conditioning of HR has thus been demonstrated in curarized dogs (Black, 1965), rabbits (for example, Yehle, Dauth, & Schneiderman, 1967) and rats (for example, DiCara, Braun, & Pappas, 1970). Instrumental conditioning of HR in curarized animals has similarly been demonstrated in rats (for example, Trowill, 1967; Miller & DiCara, 1967) and dogs (Black, 1967). In all of these experiments the animals were curarized and then run while being artificially ventilated.

Subjects to be artificially respirated are either tracheotomized, intubated with an endotracheal cannula, or fitted with a face mask. The latter two procedures avoid the trauma involved in tracheotomy and facilitate running subjects over multiple sessions.

Most contemporary studies examining HR responses in curarized subjects have induced paralysis by intravenous injection of d-tubocurarine chloride or gallamine triethiodide (Flaxedil). Although injection of these drugs has proved useful, infusion provides a more even drug effect during prolonged sessions. With increased use of the constant infusion technique, the short-acting curarizing agent succinylcholine is likely to be used more widely. Succinylcholine has the advantage over Flaxedil or d-tubocurarine chloride of wearing off more rapidly once the infusion is terminated.

Curarizing drugs have helped to elucidate the nature of HR responses, but the obtained data must be interpreted cautiously. In order to ensure that curarization is sufficiently deep, EMG activity should be monitored continuously. The stability of the HR base line should be examined and respiratory p_{CO_2} should be measured to ensure that the animal is properly ventilated. Several good p_{CO_2} monitors are available commercially. In addition to monitoring respiratory p_{CO_2}, analyses of blood gases can be carried out using a micro-pH and gas analyzing system.

Inadequate ventilation of an animal may lead to respiratory acidosis. In this case underventilation leads to an increase in arterial p_{CO_2} which in turn causes the acidosis. Symptoms of respiratory acidosis include bradycardia, increased duration of diastole, and decreased strength of heartbeat.

Not all of the difficulties associated with curarization are caused by inadequate ventilation. High doses of curariform drugs have been reported variously to debilitate sensory nerve endings and receptors, central synapses, autonomic ganglia, and the vagal outflow (for example, Black, 1967; Galindo, 1971; Gellhorn, 1958; Guyton & Reeder, 1950; Koelle, 1962, 1965). Sometimes in curarized animals an increase in the level of plasma potassium induces a condition known as hyperkalemia. The early stages of hyperkalemia are marked by the appearance of tall, peaked T waves. At higher potassium levels the atria become paralyzed and the QRS complex is prolonged. Finally, as the extracellular level of potassium continues to increase, the resting membrane potentials of the muscle fibers of the heart decrease until they become inexcitable and death occurs.

An unusual aspect of curarization is that it appears easier to operantly condition autonomic responses in the curarized than in the normal state. Miller (1969) has offered the plausible explanation that this may be caused by reduction of proprioceptive "noise" in the curarized preparation. Alternatively, it is possible that under curarization the cardiac innervation may be altered in such a way as to facilitate the learning of an HR response. Most studies presumably demonstrating operant conditioning of HR, for example, have reported elevated HR baselines. Black (1967) also has provided evidence that the parasympathetic but not the sympathetic innervation of the heart is blocked by high doses of d-tubocurarine. Thus, the improved operant conditioning of HR in curarized relative to noncurarized animals could conceivably be facilitated by an autonomic imbalance.

Although several studies have successfully demonstrated HR conditioning under curarizing drugs, the successful use of these agents may be less straightforward and simple than has been commonly supposed. Several investigators, for instance, have had difficulty obtaining adequate conditioned or even unconditioned HR responses in animals which were presumably respirated adequately. This suggests that further research may be needed to determine how HR responses to a variety of stimuli in different species are influenced by (a) different curarizing drugs, (b) depth of curarization, (c) mode of curariform administration, and (d) the kind of experimental situation in which the curare is administered.

IV. Basic Instrumentation

A. Electrodes

There are two basic types of ECG electrodes, subdermal and surface. Subdermal electrodes are most often used with infrahuman, and surface electrodes with human subjects. Surface electrodes are easy and painless to apply and there is no danger of their causing infection. For this reason they have been used also on primates run over many sessions.

In selecting electrode sites for humans, the least hairy places on the wrists and ankles are used. Before the electrodes are applied, contact is sometimes improved by scrubbing the skin surface with alcohol until it is slightly red. This ensures that the horny layer of dead cells from the epidermis is removed.

Surface electrodes may be either polarizing or nonpolarizing. Polarizing electrodes are made from a nonchlorided metal such as gold, silver, platinum, or stainless steel. Nonpolarizing electrodes use a metal/metal chloride pellet as the transducing element. Figure 7-4 provides a schematic diagram

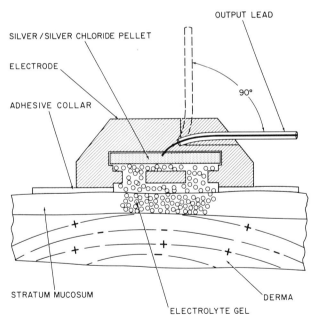

FIG. 7-4. Schematic diagram of a Beckman nonpolarizing surface electrode with disposable adhesive collar. (Courtesy of Spinco Division of Beckman Instruments, Inc.)

of a Beckman nonpolarizing surface electrode. This electrode is held in place by a disposable adhesive collar. The electrode is constructed from a plastic disk in which is embedded a silver/silver chloride pellet. In order to provide a conducting medium, the electrode is filled with a commercially available electrolyte gel. When the electrode is applied to the skin surface, the electrolyte gel provides an ionic conduction path between the electrode and the underlying volume conductor.

When an electrode is placed in a electrolyte medium, a half-cell voltage develops across the interface between the electrode and the electrolyte as a consequence of the oxidation and reduction reactions taking place. In the case of the nonpolarizing electrode, the half cell voltage is stable and relatively independent of the oxidation-reduction reactions taking place at the electrode–electrolyte surface. In the case of the polarizing electrode, however, the half-cell voltage is not stable and the fluctuations are recorded as a base-line shift.

Electrode polarization does not usually present a problem in ECG recording, because a.c. amplifiers are normally used. The time constant of these amplifiers is about 0.2 sec so that slower potential shifts such as those caused by polarization are eliminated. Electrode polarization only presents a problem when long time constant a.c. amplifiers or d.c. amplifiers are used.

Most studies of HR in infrahumans use subdermal electrodes. Basically, these electrodes consist of steel or platinum alloy pins or wires. With this type of electrode care must be taken to ensure that the electrodes are not inserted into muscle since otherwise the ECG would be contaminated by the EMG. Actually, with all ECG electrodes, it is desirable to avoid placements over or very near large muscles.

Electrode pins are simply inserted under the skin in acute, immobilized or anesthetized animals. In animals which do not attempt to dislodge the electrodes, such as the rabbit, stainless steel safety pins may be chronically inserted through a fold of skin. The safety pins are approximately 1 inch long and have a small connector soldered to them. This permits the leads to the amplifier to be easily disconnected from the animal between sessions. Usually, one electrode is placed above the right shoulder and the other electrode is placed above the left rear haunch. When a balanced amplifier is used to record HR, it is usual to ground the animal by means of a third electrode.

Although safety-pin electrodes are suitable for rabbits, animals such as rats and cats would rip the electrodes from their bodies. Consequently, in these latter animals the electrodes are completely implanted under the skin and the electrode leads are made to exit through an Amphenol connector which is fastened to the skull with dental cement.

Briefly, the electrode consists of the exposed portion of an insulated wire. The wire consists of .005 inch diameter stainless steel and the insulation consists of a nontoxic coat such as Teflon. Under general anethesia a small incision is made over the lowest rib on each side. The uninsulated loop is sutured or tied to the rib. At its free end the wire is passed under the skin and a path for it is made through the fascia with a blunt probe. An incision is made in the skin covering the back of the skull and the free end of the wire is brought out. Enough slack is left in the wire so that a strain is not placed on the electrode or the rib when the animal moves.

The free end of the wires exiting at the skull are fastened to a crimp type connector (Amphenol reliatac) and attached to a socket. The socket assembly is then attached to the skull using acrylic dental cement. Small stainless steel screws are embedded in the skull near the socket to provide an anchor for the cement.

The technique of tying or suturing the electrodes to the ribs prevents the electrodes from working free. In addition, the method produces a respiration artifact in the ECG record which permits the experimenter also to monitor breathing rate.

B. Reduction of Cable Artifacts

In recording the ECG or other bioelectrical potential from a free-moving animal, a commonly encountered source of electrical noise consists of cable artifacts. These artifacts are produced whenever the animal's movements cause the wires leading to the recording equipment to cut through electromagnetic force fields. In order to eliminate these artifacts, a unity gain voltage follower can be used. This permits the recorded signal to be changed from high to low impedance without changing the strength or form of the signal.

The assembled unit and circuit diagram are shown in Fig. 7-5. Briefly, the assembled unit consists of two parts. One of these is a miniature dual field effect transistor which is cemented to a small male Amphenol plug. This plug connects to a female Amphenol socket cemented to the animal's skull. The second part of the assembled unit, which is placed in a small box away from the animal, consists of the remaining circuitry.[4]

The unity gain voltage follower eliminates cable artifacts when ECG, EEG, or single unit recordings are taken from free-moving animals in the laboratory. Several other things also can be done to facilitate recording in this situation. Thin, flexible wire should be used to prevent leads from twisting. A slip ring or commutator permits unencumbered axial rotation by the subject. Finally, a counterweight system is desirable to ensure that sufficient tension is placed on the leads to keep them from the animal's reach.

---

[4] Mike Rosetto, an electrical engineer at Rockefeller University collaborated in the design and development of this device.

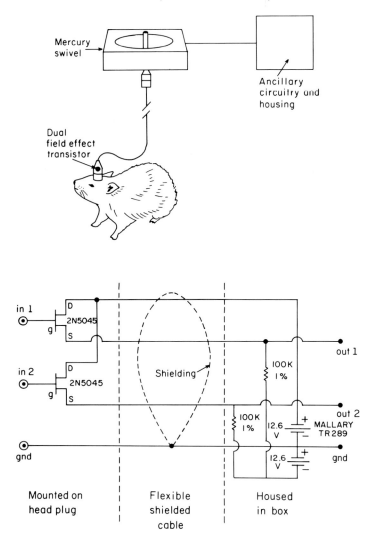

FIG. 7-5. Unity gain voltage follower including assembled unit (top) and circuit diagram (bottom).

C. Biotelemetry

In naturalistic or social situations requiring extended mobility of the subject, the leads from the animal to the amplifier may be replaced by a biotelemetry system. The system consists of three main components. These are (a) a preamplifier, which amplifies the signal from the electrodes to a usable level, (b) a small radio transmitter, and (c) a radio receiver. The preamplifier and transmitter are mounted on or implanted in the subject. Transmissions are picked up by the receiver and then are amplified and recorded in the usual manner.

The most commonly used method of transmitting biotelemetry information is by frequency modulation (FM). In FM modulation a carrier frequency is modulated as a function of the amplitude and frequency of the signal which is to be transmitted. The amplitude of the carrier frequency does not change, but rather modulation by the signal results in a change in the carrier frequency. An increase in signal amplitude is coded as an increase in carrier frequency, and an increase in signal frequency is coded as the number of times that the shift in the carrier frequency occurs. Since the receiver only decodes frequency variations in the carrier while ignoring carrier amplitude, the FM system is relatively immune to interference from static, electrical discharges, and power line interference.

Most biotelemetry systems use the standard FM radio broadcast band. Since most systems available have a range of only 20–50 ft, they do not require a license for operation. However, there are larger, more powerful transmitters available which enable the investigator to receive telemetry information from a considerable distance.

The standard FM broadcast band comprises the spectrum between 88 and 108 MHz. While commercial FM stations use this frequency band, there are enough unused portions to permit investigators to work without interference from commercial stations. Use of the standard FM band is economical, because it permits utilization of an ordinary FM radio as the receiver.

Many factors must be taken into account when selecting a particular biotelemetry system. These include size and weight of the preamplifier and transmitter, whether they are implantable, whether the transmitter can be tuned in the field, preamplifier sensitivity, transmission range, and total cost.

Size and weight often are important considerations. Transmitters vary in size from that of a common transistor to that of a matchbox. Most often, the size and weight of the preamplifier–transmitter system is determined by the batteries needed for operation. In the case of smaller animals the size and weight of the unit may restrict recording to a single channel.

Figure 7-6 shows a commercially available implantable transmitter, which complete with power supply weighs about 20 gm. The unit cannot be tuned in the field, however, and uses a receiver which is permanently tuned to a given transmitter frequency. For most research purposes, these limitations pose no problem.

Battery failure is sometimes a problem in telemetry systems implanted in the animal. In general, 100–200 hr of continuous service can be expected from implantable systems before battery replacement becomes necessary. In implanted animals, battery replacement requires removal of the trans-

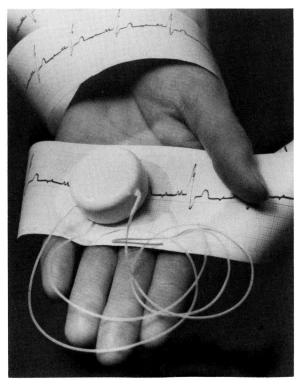

FIG. 7-6. Implantable transmitter complete with power supply weighs about 20 gm. (Courtesy of American Electronic Laboratories, Inc.)

mitter and reimplantation of the subject. The transmitter shown in Fig. 7-6 has a small magnetic switch incorporated into it. If the transmitter is implanted, the experimenter can turn it on or off by holding a magnet next to the animal. This means that the transmitter will be using the battery only during the time that data are to be recorded.

D. *Amplifiers and Polygraphs*

A differential, a.c. coupled amplifier is usually used for ECG recording. The time constant of the amplifier is usually about .2 sec. This means that the amplifier will amplify a signal having a frequency as low as .5 Hz without distortion or attenuation. At the high end of the frequency spectrum, the ECG amplifier should provide linear responses to about 20 kHz. Since the ECG signal is in the millivolt range, a low noise amplifier is not required and an amplifier which has an inherent noise level of less than 50 μV peak to peak is adequate. Amplification of the ECG requires medium gain amplification. Since the ECG signal is usually several millivolts, an amplifier with a total gain of about 3000 will provide adequate amplification.

The usual method of recording the ECG is on a polygraph. A polygraph is a multichannel device which can amplify and record several different physiological events at once. The usual method of recording these signals is to use a moving chart paper on which the polygraph pens reproduce the signal.

In addition to the chart drive and a power supply, the polygraph consists of three modules per recording channel. These are the preamplifier, the pen driver amplifier, and the oscillograph pen. The preamplifier for recording the ECG is the differential, a.c. coupled, medium gain amplifier previously described. It performs the initial voltage amplification of the ECG potential. The driver amplifier is a special amplifier that provides the power necessary to drive the oscillograph pens recording the deflections on the chart paper.

The sensitivity of polygraph and oscilloscope amplifiers are frequently defined in terms of voltage per centimeter of deflection. Thus, if the polygraph amplifier is set to a sensitivity of 1 mV/cm, a 1 mV signal at the amplifier input will cause a 1 cm deflection of the pen-writer. Similarly, for an oscilloscope set at the same sensitivity, a 1 mV signal at the input will cause a 1 cm deflection of the cathode ray beam.

Many modern polygraphs are modular in design and use interchangeable plug-in preamplifier units. While most manufacturers offer special ECG preamplifiers, more sensitive and versatile EEG amplifiers may also be used for ECG recording.

In order to record the ECG on magnetic tape or to feed the signal into a computer, the signal must be taken from one of the output stages in the amplifier system. Some polygraphs have provisions for connecting devices such as computers, signal averagers, and tape recorders. This output is a low impedance output which conforms to IRIG standards used in the design of many tape recorders.

The driver amplifiers of several polygraph models have auxiliary high and low impedance reports which accept signals up to 3.6 V rms (10 V peak to peak) and convert them to IRIG standard voltages at the auxillary output. Thus, high or low level inputs from such devices as telemetry receivers can be converted to standard IRIG output levels for recording.

If the polygraph being used does not have a special output for tape recorders, care should be taken in determining the amplification stage at which the signal should be taken. Thus, the impedance should be low in comparison to the input impedance of the recorder or computer. A mismatch will distort the signal or induce a malfunction of the amplifier. Care also should be taken to ensure that the amplitude of the signal from the amplifier is in the input range of the recorder.

E. Cardiotachometers

The cardiotachometer functions as a time to amplitude converter. It measures the time between successive R waves and provides a voltage output proportional in amplitude to the interval between two successive R waves. Thus if the interbeat interval (IBI) increases (decreases), the output signal will also increase (decrease) linearly. The output of the cardiotachometer can be displayed on a meter, recorded on analog tape, or printed on a polygraph.

Some manufacturers offer cardiotachometers for use in their polygraphs, while others manufacture units which can be used independently. Circuit diagrams also have been described in the literature. Of these, a circuit by Swinnen (1968) has the advantage of incorporating operational amplifiers, thus eliminating much of the work associated with building from individual components.

In choosing a cardiotachometer, careful attention should be paid to the range of HR which will be investigated. A cardiotachometer designed exclusively for humans may have an upper range (for example, 200 bpm) which is too low for work on rats and rabbits. Several of the commercial models as well as the Swinnen device have overlapping ranges. Their absolute ranges are from about 6 to 600 bpm.

F. Triggering

A simple, inexpensive method of R wave detection involves the use of a polygraph pen-writer in conjunction with a pulse-former and electromechanical counter. This system is shown in Fig. 7-7. Briefly, an adjustable steel contact wire is isolated electrically from the polygraph frame, whereas the polygraph pen and the relay equipment are grounded together. When the pen makes its upward excursion during the R wave, contact is made. This grounds the pulse-former, causing it to generate an output which advances the counter. The pulse-former is interposed between the contact wire and counter to (a) eliminate multiple triggering caused by contact bounce, and (b) ensure that the duration of contact is sufficient to activate the counter. In order to work efficiently the relay equipment should be arc-suppressed.

A more widely used method of detecting R waves is with an amplitude discriminator called a Schmitt trigger. This is a device used to detect whether or not the voltage of a signal exceeds a particular level determined by the investigator. When the voltage of the signal exceeds the predetermined level, the Schmitt trigger generates an output pulse. The trigger level is adjusted by the experimenter so that the R, but not the T or P wave will trigger the amplitude discriminator.

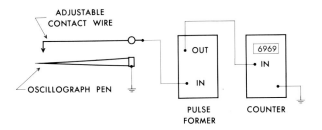

Fig. 7-7. A simple, inexpensive method of *R*-wave detection and counting.

The Schmitt trigger has been used in conjunction with counters and also has been incorporated into the design of cardiotachometers to ensure that only the R wave is detected. Both the Schmitt trigger and the contact methods of R wave detection, however, are vulnerable to artifacts introduced into the record by subject movement, electrical noise (60 Hz line noise), and some slow wave baseline changes. In order to eliminate these artifacts, additional devices have been used.

In contrast to the amplitude discriminator which is triggered by any voltage change above a predetermined level, a *peak detector* or *amplitude window* only triggers when the output voltage is between two preset levels. At the outset of the experiment the investigator determines the normal voltage level of the R wave and sets limits. Consequently, when movement artifacts or transients induce pulses larger than those of the R wave, the peak detector does not fire.

Discriminations between signal and artifacts can be made on the basis of temporal as well as amplitude characteristics. Thus, for example, a time window in conjunction with a Schmitt trigger or amplitude window can be used to discriminate the R wave from transients, movement and respiration induced artifacts. When the time window is used, signals shorter and longer than those specified by the experimenter preclude the triggering of an output pulse.

Besides the amplitude and temporal duration of the R wave, this component of the ECG can be discriminated automatically on the basis of the temporal aspects of the IBI. In this method a timing circuit called a *one-shot multivibrator* is triggered by the R wave. Once triggered, the one-shot changes state for a preset time before reverting to its stable state.

Sometimes, in ECG recording, the T wave has approximately the same amplitude as the R wave. In order to prevent triggering from occurring twice during each cycle of the ECG, the one-shot can be set for a delay period greater than the R–T interval. When it is unlikely that stimulation will induce very large changes in IBI, the one-shot can be adjusted so that the duration of the changed state is just less than the minimum IBI. This has the benefit of eliminating transient and movement artifacts as well as false triggering from the T or P wave.

In addition to its use in reliably triggering from the R wave, the one-shot multivibrator can be used to selectively trigger just the P or T wave. In this instance, the one-shot is triggered by one of the smaller components of the ECG, and the device remains in the nonfiring changed state during the occurrence of the R wave.

In order to adjust the Schmitt trigger, peak detector, time window, or one-shot, it is necessary to display the signal and the output. This is usually performed on an oscilloscope, while a visual check is made to be sure that the discriminator is adjusted properly. Once adjusted, every R wave should trigger an output, whereas all other signals should be rejected by the discriminator.

In general, the devices described in this section together with the filtering circuits built into most modern polygraphs provide adequate criteria for triggering components of the ECG. Grieve (1960) has provided still another mechanism for detecting R waves, however, which specifically eliminates contamination from muscle noise during violent exertion. In Grieve's method, signals from three parts of the body are fed through a coincidence gate. Only those time-locked signals (that is, the ECG cycle) common to all three inputs constitute the output.

Besides triggering R waves from the ECG, HR can be triggered by the systolic blood pressure wave. This method prevents false triggering from other components of the ECG, and frequently permits a measure of HR to be obtained even when the subject is receiving peripheral electric shock. Because the entire blood pressure base-line shifts up or down as pressure increases or decreases, slow wave base-line shifts must be filtered out. Fortunately, most polygraph amplifiers and even some detectors have high pass filter circuits which limit low frequency responses.

G. Data Reduction

Once an amplified signal is fed to a peak detector or other discriminating and pulse-shaping device, a single electronic or electromechanical counter in conjunction with a timer is capable of providing an on-line measure of HR. In many instances, however, the investigator is interested in the difference in HR between a prestimulus base-line period and a poststimulus period. In this case two counters may be used in conjunction with a stepper and recycling timer.

7. ELECTROCARDIOGRAM: TECHNIQUES AND ANALYSIS

For a finer grain analysis of HR following stimulus onset, a bank of 10 counters may be employed. This, for example, would permit a comparison of HR during a 10 sec base-line period with the HR occurring during nine successive 10 sec periods following stimulus onset. If the counts in each counter are allowed to accumulate over several trials, a poststimulus time histogram of heartbeats is generated. A poststimulus time histogram represents the probability density function of heart beats occurring during successive temporal intervals.

In addition to electromechanical counters, electronic counters having printing capability are also marketed. For more complex analyses, or analyses requiring greater temporal resolution, other instrumentation is available. These include wired program averaging computers or minicomputers which are software programmable.

There are numerous advantages in using a computer for analyses of HR. Both the wired program and software program computers provide several types of output. These may include punched paper tape, magnetic tape, printed output or analog plots of various histograms.

In addition to poststimulus time histograms, the computers have the capability and speed to calculate beat-by-beat intervals for great numbers of heartbeats. A successive beat-by-beat histogram in which the ordering of the original IBI is preserved is called a *sequential histogram*.

Because the sequential histogram preserves the order of IBI, the investigator may apply statistical tests of time (autocorrelation) and serial dependence (serial correlation). The sequential histogram also permits the investigator to cross correlate HR with some other event. A detailed discussion of the methods of auto-, serial-, and cross-correlation may be found in papers by Moore, Perkel, and Segundo (1966) and Perkel, Gerstein, and Moore (1967).

Another type of histogram which is often used is the interval histogram representing the probability density function across time of different temporal intervals. In contrast to poststimulus and sequential histograms, the interval histogram is not usually related to a discrete stimulus. Figure 7-8 represents stylized, fictional interval histograms following the administration of atropine, saline, and propranolol. Thus, these data could indicate that atropine speeded up and propanolol slowed down HR in comparison with the saline treatment.

In addition to providing histogram plots, the wire program and minicomputers are capable of signal averaging. By averaging 100 waveforms before and after a given experimental treatment, for example, an investigator can accurately assess differences in the latency or amplitude components of the ECG waveform.

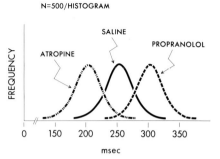

Fig. 7-8. Stylized, fictitious interval histograms indicating possible results following administration of atropine, saline, and propranolol.

In averaging the ECG waveform, the analog signal must first be transformed into digital information. This is accomplished by an analog-to-digital (A/D) converter. The A/D converter is built into the wired program computer, but is an accessory item for most minicomputers.

In addition to standard detecting and timing interfaces offered by the manufacturers of minicomputers, several investigators have developed their own devices. Kerr, Tobin, Milkman, Djoleto, Khachaturian, Williams, Schacter, and Lachin (1970), for instance, have described an interface system for online acquisition of HR data using the PDP-12. This system is based upon an R-peak detector and R-R interval counter developed by Tobin, Djoleto, Milkman, Kerr, Williams, Khachaturian, and Schachter (1970). The device, when interfaced to the PDP-12, will detect R waves and compute $R - R$ intervals with an error of only $+1$ msec. In addition, the detection system will function adequately with amplitude variations of 50% in the R wave due to 60 Hz noise and other base-line shifts. Information is also provided to the computer when an R wave is missed.

The wired program and software program minicomputers have advantages and disadvantages with respect to each other. Wired program computers, for instance, are easier to use and do not require program writing or debugging. They are incapable, however, of performing such elementary statistical manipulations as calculating means and variances.

In contrast to the wired program computer, the software program minicomputer with appropriate interfacing to recording equipment is capable of performing statistical analyses. In addition, it has the capability of performing experimental programming functions based on online data analysis. Thus, for example, we shall see in the last section of this chapter, how a LINC-8 computer has been used to set criteria for operantly conditioning either the RR, PR, or PP intervals of the rat's ECG (Fields, 1970).

The main disadvantage of the software program minicomputer is that it must be programmed. Much of this is in machine language. In general, the user of the general purpose minicomputer can expect to spend from several months to a year developing and testing his machine language programs for data acquisition and analysis. It seems to be the first law of programming that nothing ever works right the first time! However, once a library of programs is developed, the minicomputer is very effiicient.

It is possible to purchase a minicomputer with a 4K memory and a real time clock interface for about $6000. This compares favorably with the cost of the average wired program computer. Bush (1969) lists over 40 companies that manufacture minicomputers with a price range from $3000 to $25,000.

V. Data Analysis

The devices we have described provide considerable latitude in the automatic reduction of HR data. Analyses of this data are of course dictated by the considerations of particular experiments. In studying the effects of drugs upon the HR base line, for example, it may be sufficient to measure mean HR periodically. Alternatively, it might be desirable to examine HR interval histograms which provide measures of dispersion and skewness as well as mean HR. In studying reaction time and conditioning it might be sufficient to merely examine the gross change in HR before and after stimulus onset or it may be necessary to conduct a fine-grain sequential analysis of the data.

Chase, Graham, and Graham (1968), for instance, examined the HR response topography occurring during a 4 sec READY-GO interval in a human reaction time task. They succeeded in identifying three sequential components of the HR response. Immediately after the onset of the READY signal, Chase *et al.* (1968) observed a decrease in HR, which they interpreted as an orienting response. The second component consisted of a rate decrease or increase depending upon the energy requirements of the subsequent GO task. Finally, a decrease in HR immediately preceeding the GO signal was interpreted as being part of an attentional process. The important point for our discussion is that the sequence of HR decrease, increase, and decrease could have been obsured by averaging HR throughout the 4 sec READY-GO interval. In reaction time and conditioning experiments examining changes in HR, it is important for the investigator to attend to possible temporal sequences in the response following stimulus onset.

Another problem of data analysis consists of comparing responses among systems whose measures are ordinarily defined differently. Thus, in conditioning experiments using a discrete system such as eyeblink, most data are derived from frequency and latency measures. In contrast, most data from a continuous response system such as HR are based upon magnitude measures.

Kimble (1961) examined the problem of comparing performance among response systems on the basis of different response measures. He concluded that "the correlations among various measures have usually been too low to support a common-process view [p. 111]." Although this suggests that conclusions from experiments comparing response system performance on the basis of different measurement criteria should be interpreted cautiously, the problem is not insurmountable. In many cases, experimental manipulations can be arranged so that reliable conditioning is observed in one response system, but is not detectable in another system on the basis of any response measure.

Another technique which can be used to facilitate comparisons between discrete and continuous response systems is to define both responses according to frequency criteria. In an experiment exploring interstimulus interval functions, VanDercar and Schneiderman (1967) defined both nictitating membrane and HR according to frequency criteria. In this case membrane extensions less than 1 mm were considered as "noise," whereas extensions greater than 1 mm were defined as responses. Frequency measures of HR responding were similarly computed by defining a conditioned response on each trial as a change from the preconditioned stimulus baseline exceeding either 0, 1, 2, 3, 4, or 5%. Provided that minima of 3–5% changes from base line were used to define the conditioned response, the HR frequency and amplitude data provided virtually identical interstimulus interval functions. Although frequency comparisons between HR and discrete response systems may prove useful occasionally, their value is limited. This is chiefly because (a) the signal-to-noise ratio differs among systems, (b) changes in the conditioning situation may alter the direction or topography of the HR response, and (c) the HR measure may be influenced by the law of initial value (for example, Lacey, 1956; Wilder, 1967).

The law of initial value says that the magnitude of the response in a physiological system depends upon the initial value of the baseline. Thus, for example, if an animal's base-line HR is 180 bpm and its biological floor is 175 bpm, the animal would ordinarily be incapable of decreasing HR by more than 5 bpm. In contrast, if the same animal had an HR base line of 250, relatively larger HR decreases might occur in the same stimulus situation.

In recording HR, difference scores and percentage changes from base-line have both been used. The difference score carries the implicit assumption that a decrease from 290 to 280 bpm is basically the same as a decrease from 190 to 180 bpm. In contrast, the percent change from baseline score treats a larger absolute decrease in HR from a high rate the same as a somewhat smaller decrease from a lower base line.

In most experimental situations, at least for classical conditioning of the rabbit, it has not mattered very much which measure we have used. This is probably because most of our experimental manipulations have begun from fairly restricted, intermediate HR baseline values. In reanalyzing the data from more than a dozen conditioning studies conducted on rabbits, we have found no substantial disparities between results based upon the percentage change or difference score measures.

For the restrained rabbit, changes in initial baseline HR do not seem to be markedly related to changes in response magnitude (Manning, Schneiderman, & Lordahl, 1969). In contrast, changes in initial baseline appear to be of much greater importance in other species. We have found, for example, that baseline HR in the restrained rhesus monkey is extremely labile and varies tremendously as a function of eating, sleeping, and other activities. We have found also that when the HR base line in the rhesus exceeds 200–225 bpm, conditioned and unconditioned HR increases to intracranial stimulation do not occur.

In view of the problem of selecting an appropriate measure of HR change, the law of initial value deserves further attention. The function, or functions, describing relationships between baseline HR and HR response magnitude to stimulation in different species need to be determined. There is little a priori reason to suppose that the function is linear throughout the ranges of HR that occur in different species.

VI. Operant Conditioning and Biofeedback

During recent years considerable attention has focused upon the conditioning of autonomic responses. In several of the studies from Neal Millers' laboratory (for example, Miller & DiCara, 1967), reinforcement was delivered on a contingent basis using two predetermining counters in conjunction with a timing circuit. Briefly, if the number of beats recorded on one counter during a fixed period reached a criterion value set on the predetermining counter, reinforcement occurred.

An alternative system for reinforcing HR and providing feedback to the subject is based upon the use of a cardiotachometer and an external Schmitt trigger. An adaptation of the method used by Engel and Chism (1967) will be described. The left-hand side of Fig. 7-9 depicts this method for reinforcing HR on a beat-by-beat basis. Polygraph tracings obtained with the instrumentation are shown on the right-hand side.

Initially, the ECG leads are fed into an a.c. preamplifier and into a cardiotachometer. Channels 1 and 2 of Fig. 7-9 provide an accurate picture of the subject's HR performance. Output of the cardiotachometer is fed into one d.c. amplifier which drives the pen of channel 2 and slave-drives a second amplifier which is used to control reinforcement. The Schmitt trigger shown

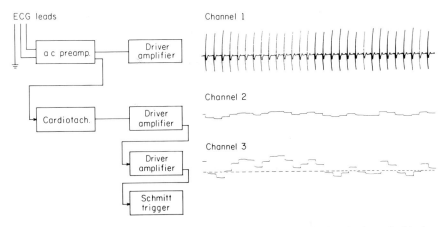

FIG. 7-9. Method for reinforcing changes in heart rate on a beat-by-beat basis. Left side depicts arrangement of instrumentation. Right side shows polygraph tracings obtained with this instrumentation. The predetermined triggering value of the Schmitt trigger is indicated by the dotted line superimposed on the tracing of channel 3.

in Fig. 7-9 receives its voltage from the slave-driven amplifier. Although the Schmitt trigger need not be built into the polygraph, this feature has recently been added to some machines.

The pen tracing of channel 3 reflects the changes in voltage obtained from the slave-driven amplifier. By moving the base-line adjust knob of the slave-driven amplifier, the cardiotachometer tracing can be raised above or below the predetermined triggering value of the Schmitt trigger (indicated by the dotted line of channel 3). In this way the criterion can be made more or less difficult.

The system described in Fig. 7-9 is suitable for providing feedback to a subject as well as for activating a brain stimulator, electronic shocker, or other instrument providing reinforcement. In the case of feedback the subject is informed of his progress in increasing or decreasing HR. Initially, the trigger level can be set so that approximately one half of all IBI are on one side of the triggering level. The IBI above this level are used to trigger one signal light, whereas the IBI below this level trigger another. As a greater number of longer or shorter IBIs fall on one side of the trigger level than the other, this is reflected in the proportion of time that each of the signal lights is illuminated.

In the typical experiment in which HR is operantly conditioned, a shaping procedure is used. Thus, the experimenter often must change the HR criterion leading to reinforcement. This is usually done by (a) periodically recomputing HR between trials and setting the criterion level at a fixed percent above or below this level, or (b) gradually increasing or decreasing the criterion in accordance with the success of the subject in changing HR.

Although each of these methods has proved useful, neither takes into account changes occurring in base-line HR just prior to the beginning of a trial. For example, suppose the subject is being reinforced for decreasing HR. If on a particular trial his HR is much higher than the originally computed baseline value, he will not receive reinforcement even if a relatively large HR decrease is made to a discriminative stimulus. The solution to this problem consists of developing a procedure for continually modifying the HR criterion on the basis of continual changes occurring in the HR base line.

In a rather ingenious experiment, Fields (1970) used a LINC-8 computer to operantly condition either the RR, PR, or PP interval of the rat's ECG. Monitoring of these intervals was performed for 512 consecutive intervals. As each new interval entered storage the oldest interval (now 513) was dropped.

During conditioning a running histogram of intervals was calculated. A percentage of either the upper or lower tail of the distribution of either R–R, P–R, or P–P intervals was selected for reinforcement. Intervals falling within the tail (for example, upper 10%) were reinforced.

Although the LINC-8 method provides an excellent method for continually adjusting the HR reinforcement criterion on the basis of changes in the HR base line, it requires the availability of an expensive computer. Consequently, one of us (DHV)[2] has developed a simple, inexpensive, analog device which permits trial-to-trial changes in HR to be incorporated into the reinforcement criterion.

The device is essentially a floating reference level detector which receives its input directly from a cardiotachometer. It computes mean HR for a specified period just prior to a trial and automatically triggers if at any time during the trial HR changes by a fixed amount above (below) a predetermined criterion.

Computation of the pretrial HR is accomplished by integrating the voltage output of the cardiotachometer for a fixed period of time (.1, .2, .5, 1, 2, 5, or 10 sec) as specified by the experimenter. This is done by setting a rotary switch to the desired interval and by switching the gate input voltage from ground to -12 V for the period during which integration is to take place. Then, during the trial, the voltage input from the cardiotachometer is continually compared with the mean voltage value of the pretrial period. If at any time during the duration of the trial HR exceeds (falls below) the mean HR of the pretrial period by a predetermined specified amount, reinforcement is delivered. For a trial period of 3 min or less the device is accurate to within 1%. A schematic diagram of the floating reference detector is shown in Fig. 7-10.

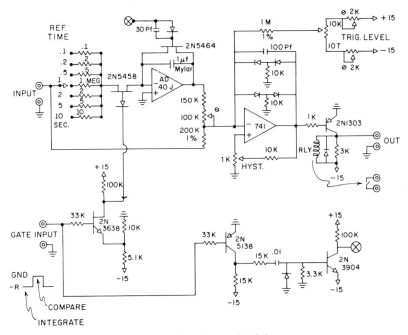

FIG. 7-10. Floating reference level detector.

In our previous discussion of feedback we described how differential stimuli can be presented whenever HR is above or below a preset criterion value. Although this method has proved to be quite valuable, an experimenter may sometimes wish to inform the subject concerning the *extent* to which his performance is above or below criterion level. In this case a tone can be presented to the subject, the frequency of which is a function of HR. This is easily accomplished by feeding the voltage output of a cardiotachometer directly into a voltage to frequency converter. In this way, the frequency of a tone can be made to increase and decrease concomitantly with HR.

An alternative method for providing continuous feedback involves the use of a light which changes in intensity as a function of HR. In this instance the potentiometer of an inexpensive dimmer switch can be replaced with a photoresistive cadmium sulfide cell which is optically coupled to a large photoemitting diode.

The output of a cardiotachometer is delivered to the photoemitting diode through an appropriate resistor so that voltage changes associated with changes in HR modulate the intensity of the light-emitting diode. These changes in the illumination of the light-emitting diode in turn alter the resistance of the photoresistive cell, increasing and decreasing the intensity

7. ELECTROCARDIOGRAM: TECHNIQUES AND ANALYSIS

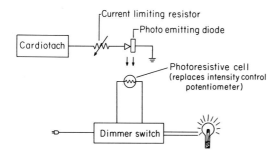

FIG. 7-11. Schematic diagram of a device which provides continuous feedback about changes in heart rate through variations in signal light intensity.

of a standard light bulb. A schematic diagram of this instrumentation is shown in Fig. 7-11.

Most of the instrumentation described in this chapter is available from commercial sources. Much of the remainder is easily constructed. Increasingly, investigators examining ECG variables are turning to automatic programming and data processing devices. This has been accompanied by a growing emphasis upon examining the methodological problems encountered in the attempt to understand the physiology and the behavior of the response. The result has been a gain in rigor and sophistication that promises to make the ECG an even more useful research tool than it has been in the past.

Acknowledgments

The research from our laboratories described in this chapter was supported by National Science Foundation Grant GB-24713, by U.S. Public Health Service Grants (National Institute of Mental Health) MH 13189 and MH 19183 to Rockefeller University, MH 10315 to New York State Psychiatric Institute of Columbia University, and (NIH Child Health and Human Development) HD 00187 to the University of Miami. We thank Bruce Pappas and Roger Ray for their helpful comments.

References

Ban, T. The septo-preoptico-hypothalamic system and its autonomic function. In T. Tokizane & J. P. Schade (Eds.), *Progress in brain research: Correlative neurosciences,* Part A. *Fundamental mechanisms.* Amsterdam: Elsevier, 1966.

Black, A. H. Cardiac conditioning in curarized dogs: The relationship between heart rate and skeletal behavior. In W. F. Prokasy (Ed.), *Classical conditioning.* New York: Appleton, 1965. Pp. 20–47.

Black, A. H. Operant conditioning of heart rate under curare. Technical Report No. 12, 1967, Department of Psychology, McMaster University, Hamilton, Ontario.

Brener, J. M., & Hothersall, D. Heart rate control under conditions of augmented sensory feedback. *Psychophysiology,* 1966, **3,** 23–28.

Brener, J. M., & Hothersall, D. Paced respiration and heart rate control. *Psychophysiology*, 1967, **4**, 1–6.

Bush, R. C. Which minicomputer do you want? *Industrial Research*, 1969, **11**, 49–51.

Chai, C. Y., & Wang, S. G. Localization of central cardiovascular control mechanism in lower brain stem of the cat. *American Journal of Physiology*, 1962, **202**, 25–30.

Chase, W. G., Graham, F. H., & Graham, D. T. Components of heart rate response in anticipation of reaction time and exercise tasks. *Journal of Experimental Psychology*, 1968, **76**, 642–648.

Cohen, D. H., & Pitts, L. H. Vagal and sympathetic components of conditioned cardioacceleration in the pigeon. *Brain Research*, 1968, **9**, 15–31.

Covian, M. R., & Timo-Iaria, C. Decreased blood pressure due to brain septal stimulation: Parameters of stimulation, bradycardia, baroreceptor reflex (cat). *Physiology and Behavior*, 1966, **1**, 37–43.

DiCara, L. V., Braun, J. J., & Pappas, B. A. Classical conditioning and instrumental learning of cardiac and gastrointestinal responses following removal of neocortex in the rat. *Journal of Comparative and Physiological Psychology*, 1970, **73**, 208–216.

Durkovic, R. G., & Cohen, D. H. Effects of rostral midbrain lesions on conditioning of heart and respiratory rate responses in the pigeon. *Journal of Comparative and Physiological Psychology*, 1969, **68**, 184–192. (a)

Durkovic, R. G., & Cohen, D. H. Effects of caudal midbrain lesions in conditioning of heart and respiratory rate responses in the pigeon. *Journal of Comparative and Physiological Psychology*, 1969, **69**, 329–338. (b)

Dykman, R. A., & Gantt, W. H. The parasympathetic component of unlearned and acquired cardiac responses. *Journal of Comparative and Physiological Psychology*, 1959, **52**, 163–167.

Elster, A. J., VanDercar, D. H., & Schneiderman, N. Classical conditioning of heart rate discrimination using subcortical electrical stimulation as conditioned and unconditioned stimuli. *Physiology and Behavior*, 1970, **5**, 503–508.

Engel, B. T., & Chism, R. A. Operant conditioning of heart rate speeding. *Psychophysiology*, 1967, **3**, 418–426.

Enoch, D. M., & Kerr, F. W. L. Hypothalamic vasopressor and vesicopressor pathways. I. Functional studies. *Archives of Neurology*, 1967, **16**, 290–306. (a)

Enoch, D. M., & Kerr, F. W. L. Hypothalamic vasopressor and vesicopressor pathways. II. Anatomic study of their course and connections. *Archives of Neurology*, 1967, **16**, 307–320. (b)

Fields, C. Instrumental conditioning of the rat cardiac control systems. *Proceedings of the National Academy of Sciences*, 1970, **65**, 293–299.

Gasser, H. S., & Meek, W. J. A study of the mechanism by which muscular exercise produces acceleration of the heart. *American Journal of Physiology*, 1914, **34**, 48–72.

Galindo, A. Prejunctional effect of curare. Its relative importance. *Journal of Neurophysiology*, 1971, **34**, 289–301.

Gelhorn, E. The influence of curare on hypothalamic excitability and the electroencephalogram. *Electroencephalography and Clinical Neurophysiology*, 1958, **10**, 697–703.

Grieve, D. W. Heart rate detection in exercise. *Proceedings of the Second International Conference on Medical Electronics*. London: Iliffe, 1960.

Guyton, A. C., & Reeder, R. C. Quantitative studies on the automatic actions of curare. *Journal of Pharmacology and Experimental Therapeutics*, 1950, **98**, 188–193.

Hess, W. R. *Das Zwischenhirn. Syndrome lokalisations funktioner*. Basle: Schwabe, 1949.

Hnatiow, M., & Lang, P. J. Learned stabilization of cardiac rate. *Psychophysiology*, 1965, **1**, 330–336.

Hoff, E. C., & Green, H. D. Cardiovascular reactions induced by electrical stimulation of the cerebral cortex. *American Journal of Physiology*, 1936, **117**, 411–422.

Kaada, B. R. Somato-motor, autonomic and electrocorticographic responses to electrical stimulation of "rhinencephalic" and other forebrain structures in primates, cat and dog. *Acta Physiologica Scandinavia*, 1951, **24**, 1–285.

Kerr, J., Tobin, M., Milkman, M., Djoleto, B. D., Khachaturian, Z., Williams, T., Schachter, J., & Lachin, J. A PDP-12 System for on-line acquisition of heart rate data. *PDP-12 user application report*. Digital Equipment Corp., 1970.

Kimble, G. A. *Hilgard and Marquis' Conditioning and Learning*. New York: Appleton, 1961.

Koelle, G. B. A new general concept of the neuro-humoral functions of acetylcholine and acetylcholinesterase. *Journal of Pharmaceutics and Pharmacology*, 1962, **14**, 65–90.

Koelle, G. B. Neuromuscular Blocking Agents. In L. S. Goodman & A. Gilman (Eds.), *The pharmacological basis of therapeutics*. New York: Macmillan, 1965.

Krogh, A., & Linhard, J. The regulation of respiration and circulation during the initial stages of muscular work. *Journal of Physiology*, 1913, **47**, 112–136.

Lacey, J. I. The evaluation of autonomic responses: Toward a general solution. *Annals of the New York Academy of Sciences*, 1956, **67**, 123–164.

Lindgren, P., Rosen, A., Strandberg, P., & Uvnas, B. The sympathetic vasodilator and vasoconstrictor outflow—a cortico—spinal autonomic pathway. *Journal of Comparative Neurology*, 1956, **105**, 95–104.

Malmo, R. Classical and instrumental conditioning with septal stimulation as reinforcement. *Journal of Comparative and Physiological Psychology*, 1965, **60**, 1–8.

Manning, A. A., Schneiderman, N., & Lordahl, D. S. Delay versus trace heart-rate classical discrimination conditioning in rabbits as a function of inter-stimulus interval. *Journal of Experimental Psychology*, 1969, **80**, 225–230.

Miller, N. E. Learning of visceral and glandular responses. *Science*, 1969, **168**, 444–445.

Miller, N. E., & DiCara, L. Instrumental learning of heart rate changes in curarized rats: Shaping and specificity to discriminative stimulus. *Journal of Comparative and Physiological Psychology*, 1967, **63**, 12–19.

Moore, G. P., Perkel, D. H., & Segundo, J. P. Statistical analysis and functional interpretation of neuronal spike data. *Annual Review of Physiology*, 1966, **28**, 493–522.

Pappas, B., DiCara, L. V., & Miller, N. E. Effects of acute sympathectomy by 6-hydroxydopamine on the classical conditioning of blood pressure and heart rate. Paper presented at the meetings of the American Psychological Association, Washington, 1971.

Perkel, D. H., Gerstein, G. L., & Moore, G. P. Neuronal spike trains and stochastic point processes. *Memorandum RM-4816-PR*, March, 1967, Rand Corporation. Santa Monica, California.

Ranson, S. W., & Billingsley, P. Vasomotor reactions from stimulation of the floor of the fourth ventricle. *American Journal of Physiology*, 1916, **41**, 85–99.

Robinson, B. F., Epstein, S. E., Beiser, G. D., & Braunwald, E. Control of heart rate by the autonomic nervous system. *Circulation Research*, 1966, **19**, 400–411.

Samaan, A. Muscular work in dogs submitted to different conditions of cardiac and splanchnic innervations. *Journal of Physiology (London)*, 1935, **83**, 313–331.

Sarnoff, S. J. Myocardial contractility as described by ventricular function curves. *Physiological Reviews*, 1955, **35**, 107–122.

Sayers, G., & Solomon, N. Work performance of a rat heart-lung preparation: Standardization and influence of corticosteroids. *Endocrinology*, 1960, **66**, 719–725.

Sideroff, S., Schneiderman, N., & Powell, D. A. Motivational properties of septal stimulation as the US in classical conditioning of heart rate in rabbits. *Journal of Comparative and Physiological Psychology*, 1971, **74**, 1–10.

Swinnen, M. T. The design of biomedical instrumentation made easy through the use of operational amplifiers. *Psychophysiology,* 1968, **5,** 178–187.

Tobin, M., Djoleto, B. D., Milkman, M., Kerr, J., Williams, T., Khachaturian, Z., & Schachter, J. The R peak detector and R-R interval counter: a new interface to the PDP-12 computer for on-line analysis and processing of heart rate data. *Proceedings of the Decus Spring Symposium,* 1970, 265–271, Digital Equipment Corporation.

Trowill, J. A. Instrumental conditioning of the heart rate in the curarized rat. *Journal of Comparative and Physiological Psychology,* 1967, **63,** 7–11.

VanDercar, D., Elster, A., & Schneiderman, N. Classical heart rate conditioning motivated by hypothalamic or septal US stimulation. *Journal of Comparative and Physiological Psychology,* 1970, **72,** 145–152.

VanDercar, D. H., & Schneiderman, N. Interstimulus interval functions in different response systems during classical discrimination conditioning of rabbits. *Psychonomic Science,* 1967, **9,** 9–10.

Webb, R. A., & Obrist, P. A. Heart-rate change during complex operant performance in the dog. *Proceedings of the 75th Annual Convention of the American Psychological Association,* 1967, 137–138.

Westcott, M. R., & Huttenlocher, J. Cardiac conditioning: The effects and implications of controlled and uncontrolled respiration. *Journal of Experimental Psychology,* 1961, **61,** 353–359.

Wilder, J. *Stimulus and response: The law of initial value.* Baltimore, Maryland: Williams & Wilkins, 1967.

Wood, D. M., & Obrist, P. A. Effects of controlled and uncontrolled respiration on the conditioned heart rate response in humans. *Journal of Experimental Psychology,* 1968, **77,** 468–473.

Yehle, A. L., Dauth, G., & Schneiderman, N. Correlates of heart-rate classical conditioning in curarized rabbits. *Journal of Comparative and Physiological Psychology,* 1967, **64,** 98–104.

Chapter 8

The Cardiac Response during Infancy

Michael Lewis

Educational Testing Service
Princeton, New Jersey

I. The Use of Physiological Recordings in Infancy 201
II. Cardiac Response as a Measure of Psychophysiological Responsivity 203
    A. General Laboratory Considerations 204
    B. The Measurement of Heart Rate 207
    C. Data Analysis 208
    D. Parameters of Heart Rate Response 211
    E. Initial Level Effects 213
III. Studies of Heart Rate Response in Infants 215
IV. Developmental Issues in Heart Rate Response 219
    A. State 220
    B. Nature of the Stimulus 221
    C. Stimulus Intensity and Other Parameters 222
V. Heart Rate Response and Cognitive Functions 222
VI. Discussion and Summary 225
    References 226

I. The Use of Physiological Recording in Infancy

Two major aspects of inquiring motivate investigators of infant psychophysiological behavior. These approaches lead to somewhat different studies of infant behavior and it is important for any discussion to specify them. The first approach centers around the study of the psychophysical response of the young child to a variety of sensory experiences and is characterized by such questions as: "What is the effect of an auditory signal on the galvanic skin response (GSR)?" or, "What is the nature of the function of the cardiac rate response when the energy level of the environment changes?" It emphasizes the infant's psychophysiological behavior as a consequence of certain kinds of stimulation.

The second approach is characterized by the attempt to find psychophysical responses that measure specific infant processes. In some sense the psychophysical data are just another response used to infer something about the infant's capacity, most often some cognitive capacity. These studies are often concerned with answering questions such as: "Is one stimulus more interesting than another to the infant?" If this can be determined, then the investigator may postulate something about the organism's motivational or cognitive systems, or both. While at first glance it would appear that both these approaches have much in common, they do often lead to different kinds of research strategies and, as a consequence, different kinds of conclusions. For example, if one were interested in heart rate (HR) change as a consequence of an auditory signal, the auditory signal would most likely be well defined in hertz and intensity but might be relatively meaningless to the organism. If, on the other hand, one were interested in using a psychophysiological response as an index of some cognitive aspect, then one might use meaningful stimuli such as human voices. That different stimuli can lead to different conclusions about function seems clear in light of recent research evidence. For example, it is possible to demonstrate a developmental change in the ability to show differential HR response to two meaningless auditory stimuli; whereas a 12-week-old infant shows little or no differential HR response, older infants are able to do so. One might conclude erroneously that the 12-week-old is incapable of differentiating sound. Experiments using meaningful sound with the same age infants, however, produce results that indicate their ability to differentiate (Eimas, Siqueland, Jusczyk, & Vigorito, 1971; Moore, 1971). Any discussion of research data on infants' psychophysical response must come to grips with the problem that different results are often the consequence of these different types of stimuli, themselves functions of the experimental question asked. It seems unrealistic to investigate the purely psychophysiological responses independent of the type of stimulus. Perhaps it is not possible to describe *the* psychophysiological responses for a particular sensory modality. The organism is constantly interacting with its environment; not even the newborn exists in a "pure" environment.

We have suggested that different strategies may be employed by different experimenters, all using the same psychophysiological measure. Still a more basic question is the value of using psychophysiological measures in infancy, especially for those who are interested in the personal, social, and cognitive development of the infant. Isn't this an area in which psychophysiological data would be of limited use? Moreover, employing psychophysiological measures may be moving infant behavior too quickly to a molecular level of analysis when molar levels have yet to be explored. The latter point is a general criticism often leveled at psychophysiological research at any age: the belief of many investigators that if satisfactory answers on a more molar behavioral level are not available, then perhaps study at a molecular level will provide the solution. In general, psychophysiological data do not provide solutions to problems at a behavioral level. It is interesting to note that even in discussion of such concepts as state (surely we would consider psychophysiological data relevant to this construct) such physiological data are often misleading. In fact, in a recent presidential address before the Society for Psychophysiological Research, Laverne Johnson cautioned his listeners that determination of state from physiological data is useless (Johnson, 1970). The question remains of why anyone interested in infancy should be concerned with such physiological techniques if the motive for the study of infants is not psychophysiological reactions *per se*. The rationale lies in the nature of the organism under study. The infant can be characterized by the absence of motoric control. Thus, while he is like the nonhuman organism in the sense that he cannot understand our verbal instructions, he is unique in that even if he could understand, he could not motorically carry them out. Thus, the investigator trying to get at such concepts as the intellectual or affectual development of the infant is forced to explore other approaches. It is here that psychophysiological techniques find their proper place. If indeed there are physiological correlates of behavior such as cognition, attention, and affect, then psychophysiological techniques become appropriate in the study of organisms relatively unable to behave motorically.

II. Cardiac Response as a Measure of Psychophysiological Responsivity

Of the various psychophysiological responses available for study in the infant, none is more widely used than the autonomic nervous system (ANS) heart rate response (HR). As Ellingson's (1967) review makes clear, central nervous system (CNS) research with infants and young children is extremely difficult. The study of ANS is consistently easier and, coupled with the theorizing of Lacey, has resulted in a preponderance of work using HR. Lacey (1959) has argued that HR is related to the organism's intended transaction with his environment. Stimulus intake is related to HR deceleration while stimulus rejection is related to HR acceleration. This theorizing, along with some empirical work by Lacey, Kagan, Lacey, and Moss (1963), introduced those interested in the study of cognition to the possibility of the use of HR response to study infant behavior. Graham and Clifton (1966), in a review of the orienting reflex literature made popular by Sokolov (1963), argued that the HR deceleration should be a component of the orienting reflex (OR). Others, for example, Obrist, Webb, and Sutterer (1969) and Obrist, Webb, Sutterer, and Howard (1970), have taken exception to Lacey's position. Lacey's theorizing led the way for those investigators interested in studying mental processes in infants to investigate psychophysiological and specifically HR response.

This chapter will first discuss general laboratory problems in obtaining infant HR responses, followed by measurement and methodological considerations. The final section will deal with some of the results to date indicating the relationship between HR response and infant behavior.

A. General Laboratory Considerations

Research in infancy can be divided into that dealing with the neonatal period and "other"—the other comprising the rest of the first 2 years of life. Most research on infancy is conducted in the neonatal period; this because the subjects are easily available in the hospitals. Research with neonates has the great advantage of convenience in obtaining large numbers of subjects easily and in having a rather compliant and easy-to-work-with subject. One disadvantage of the neonate is the relatively short period of time he is awake and alert. Many of the techniques which one can use with the newborn become increasingly difficult as the infant becomes older. Electrode placement, for example, is relatively easy in the newborn child, but another matter in the case of a struggling, active, and frightened infant of 9 months. Thus, any comments about infancy research must take into consideration the relatively wide age period in terms of development covered by the term infancy.

The electrodes used in infancy research have to be small and able to withstand a great deal of movement. Probably the most serious laboratory concern for the investigator of HR response is the movement artifact problem. It is clear that unlike those of adult subjects, infants' electrode leads cannot be attached to his extremities. There is a considerable amount of movement and thrashing which renders a large part of the records unreadable. The alternative is to attach the electrodes to the infant's body.

8. THE CARDIAC RESPONSE DURING INFANCY 205

For this purpose miniature skin electrodes are used, along with standard electrode paste. These electrodes are fastened to the child's body by double sticky-sided doughnuts that secure the electrodes firmly. The two cardiac leads are placed as follows: the first directly under the left nipple, and the second on the back. It is important to note that to avoid the movement artifact, the second electrode should not be placed on the shoulder blade. The ground electrode is placed upon the child's abdomen. We have found that this placement results in the smallest movement artifact. Moreover, the double sticky-sided doughnut adheres the electrode firmly to the child's body. The removal of this attachment is momentarily painful for the infant, but this slight discomfort is often compensated for by the relatively clean recording available. It is necessary on rare occasions to move the chest electrode further toward the center of the infant's chest, especially when a rather weak polygraph record is obtained. In order to determine placement before applying the electrode, the source of the strongest auditory cardiac beat should be found and the electrode placed at that point.

In general laboratory operating procedures, the parents are informed and their consent obtained for every experiment, specifically for the electrode attachment. Most often the parents' concern centers around the question of whether the electrodes will shock or hurt the child in any way. After assuring the parents that there is no chance of causing the child any harm, we are always able to obtain their cooperation. However, even so, many parents are rather apprehensive about wiring their child. Great care and consideration must be taken in order to reduce their anxiety. One anxiety reduction method involves the design of the laboratory and the hiding of wires. Care is taken so that the electrode wires are hidden under the child's clothes. Moreover, the electrode leads to the terminal box are also hidden. In this way wires are not readily visible to the parents.

The child's anxiety about electrode placement is, of course, another issue. In the first 6–9 months of life, children show very little discomfort. After 6 months and around 9 months stranger anxiety usually appears. At this point the approach of a stranger and the touching of the child by the stranger often evokes severe anxiety, exhibited in the child's attempts to tear the electrodes from his body. The solution is to hide them under the child's garments, leading the wires out the child's back in a taillike fashion. Also of help are rather long leads connecting into an unseen terminal box. The general stranger anxiety cannot be eliminated. The solution which we have used is to make sure that the experimenter attaching the electrodes is not the one who will subsequently have an interaction with the child. Moreover, we often allow a considerable amount of adaptation to the electrodes so that by the time the child is ready to participate in the experiment, he has for the most part forgotten about the electrodes. The electrode

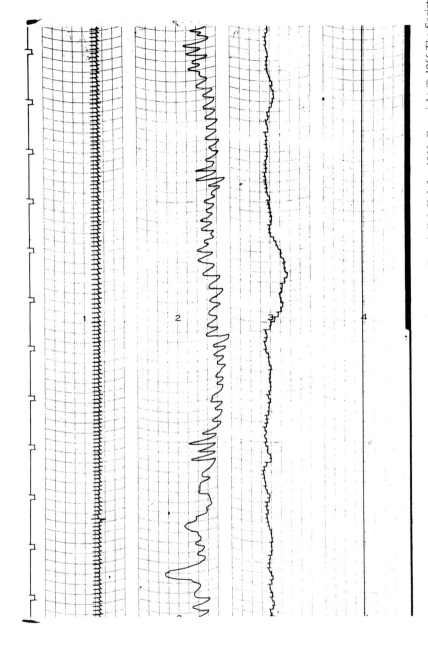

FIG. 8-1. A portion of an individual infant's polygraph record. (From Lewis, Kagan, Campbell & Kalafat, 1966. Copyright © 1966 The Society for Research in Child Development.)

placement constitutes a stress situation for the 6-month-old-plus infant. For its parents, it constitutes a stress situation no matter what the age of the child. What effect this stress has upon subsequent psychophysiological data is unclear. It is clear, however, that an adaptation period is necessary before the experimenter attempts to obtain data. It should be noted that there are marked individual infant differences in response to the stress. Not to be ignored as well are the individual parent differences in response to their child's distress. This often varies from the mother physically holding and rocking and comforting the child, to a mother essentially ignoring and chastizing the child for being upset over such an incidental matter. In our laboratory we have been recording incidents of infant upset and we find that, in general, boys seem more upset by the experience than girls. Moreover, we have found that the mother's response to the child's distress in terms of nature and amount of her comforting is directly related to other measures of the mother's attachment relationship with the child. Mothers, observed in their homes, who show a great deal of touching and proximal behaviors toward their children are the same mothers who comfort their children by holding under this stress. This is mentioned both to alert the reader to the possible "risks" involved in electrode attachment and to make clear that stress, especially for the young, can be found in the most innocuous of circumstances. Finally, it is to be noted that experimental procedures for the production of distress and individual differences in responsivity can also be found in most situations.

B. The Measurement of Heart Rate

The recording of infant cardiac data is much the same as that for adults. The electrode leads terminate in an electrocardiograph channel of the polygraph which generates a typical kind of EKG output (see Fig. 8-1). The spike pattern is characteristic of the EKG recording, with the spike referred to as the "R wave." Investigators interested in the efficiency of the heart as a muscle study the shape of the particular curve. However, those interested in HR data are concerned with the number of R waves per unit of time or the time between R waves. A device often used is a cardiotachometer, the function of which is to convert the R–R interval into a rate measure. In Fig. 8-1 the EKG can be seen on channel 1, respiration on channel 2, and the rate measure on channel 3. The function of the cardiotachometer is to convert the R–R interval into a rate measure. It does this for each and every R–R interval. It should be pointed out that the R–R interval (time in milliseconds) and rate (in beats per minute, bpm) are not identical. Investigators have hand-scored either the R–R interval (time in milliseconds) or rate (beats per minute). This, as one can imagine, is both a laborious and rather inaccurate method of data recovery. Hand computations have been utilized, however, and can be used in the absence of other kinds of data reduction devices. It should be understood, however, that there are error factors which will affect absolute HR scores. If one is concerned with difference scores rather than absolute scores, then the error of hand scoring becomes less serious, although by no means eliminated.

Most often the EKG recording is transmitted to the polygraph as well as some other device, either directly to a computer facility or to an FM tape which will then be used with a computer. It is from the EKG signal that the more accurate measures of HR are obtained. At the moment in our laboratory we are using a PDP-8 computer with an A to D converter. The A to D converter determines the spike of the R wave and the computer both clocks the R–R interval and converts the time into a rate measure (bpm). Many devices using this type of procedure have been developed— for example, Welford's SETAR (Welford, 1962) which uses a punchpaper tape recording system, which can then be fed into a computer.

What quickly becomes clear is the abundance of data which is generated using the HR response. This is true regardless of whether it is an infant or adult subject. However, a unique feature of the infant subject is the speed of its heart action, varying from a neonatal rate of about 150 bpm on the average to a 1 or 2-year-old rate of 120–100 bpm. Infants generate two to three times as many bits of information per unit of time as adults. An adult with a HR of 60 bpm collects 3600 R–R intervals over an hour while an infant with an average of 200 bpm would generate 12,000 intervals. Some kind of data reduction should be available for HR research; otherwise, one is often forced to use sampling procedures. Whereas sampling procedures are quite legitimate, full response data is invaluable for alternative analysis.

C. Data Analysis

When data have been obtained and recorded in some form, there are a variety of data analysis procedures which can and have been used.

One of the first procedures employed by Lacey was a mean response measure (see for example, Lacey *et al.,* 1963), in which the mean HR x seconds prior to a stimulus situation is compared to pre- and poststimulation levels. The fact that the stimulation situation or experimental condition might last for several minutes is neglected and the average HR during this entire period is obtained. Clearly, this procedure is the simplest available to the experimenter. However, there are several risks. Consider, for example, a response which shows an initial increase followed by a decrease in HR when compared to the prestimulus level. One might imagine such a cardiac response to be a consequence of a child's sudden movement readying himself to attend to a visual array. If one sums over the entire stimulus period, the average score one obtains eliminates the multiphasic response. However, this procedure does have the advantage of eliminating changes which have little psychophysiological meaning. On the other hand, it also eliminates the kinds of HR change which occur over time. It becomes especially questionable when the stimulus situations involve more than a second or two of time and when the experimenter has reason to suspect a multiphasic response.

The second type of analysis of HR data is the observation of each point of the response. "Point" here can be defined in several ways: either R–R interval in terms of seconds or rate in terms of beats per minute. It is to be noted that these measures, although related, are not synonymous. [See Lacey (1962) and Khachaturian, Kerr, Kruger, and Schachter (1972) for a full discussion of this point.] Having decided on one or the other measurement form, the experimenter must select either the absolute or change scores (from some prestimulus level). There are thus several different variations the HR data can take.

Still another set of possibilities has to do with the time dimension. These are the observation of "psychophysiological" time or "real" time. In the first condition, the important time dimension is said to be the number of R–R intervals from the onset of the stimulus event, independent of real time. For example, two subjects, one having 120 bpm and the other 180 bpm, will be matched on the first, second, etc. R–R interval. Of course, the subject with 180 bpm will have more R–R intervals than the subject with only 120 bpm. Underlying this procedure is the assumption that real time is relatively less important than the number of times the heart has beaten from the onset of the stimulus event. An alternative procedure is to have a real time base, for example, a second-by-second scale. In this case, the subject who has 120 bpm will have an average of two scores in order to obtain his second-by-second score, whereas the subject with 180 bpm will have as his average three scores. The advantage here is that there is a real time base and it is possible to compare subjects with differing HR along a constant time dimension. The disadvantage is that the number of R–R intervals contributing to the average at a particular time varies from subject to subject as a function of his rate. More important is the real possibility that HR responsivity occurs very quickly and that single intervals may be important for analysis. If this in fact is the case—and the work of Lipton and his associates suggest it to be (Lipton, Steinschneider, & Richmond, 1961a, b)—then the averaging of several scores over time, even as small as a second, may be misleading. In fact, this is one explanation why

Schachter and his associates (see, for example, Schachter, Williams, Khachaturian, Tobin, Kruger, & Kerr, 1971), repeatedly get a biphasic response of initial deceleration followed by acceleration to auditory stimuli in the neonate, while Graham and her associates (see Graham & Jackson, 1970) get simple acceleration. Schachter uses a smaller time interval than Graham. If the deceleration is only two or three beats, these data can be lost by averaging over a second. Whatever the procedure for the response or time scales, the information obtained from an analysis of all the data enables one to look at the function characteristics of the HR response to a stimulus situation. Obtaining this function allows the investigator a variety of options for subsequent analysis. These data are quite open to descriptive analysis. Unfortunately, relatively little work has been done analyzing these functions. (See Schachter, Williams, Khachaturian, Tobin, & Druger, 1968, for an exception.) By and large, the functions are most used to describe the general characteristics of the HR response. A much more popular procedure is to observe certain characteristics of the function rather than utilizing the function itself. The advantage of analyzing curve data should not be lost to the reader. There are many problems in psychophysiological research that lend themselves to this type of analysis. For example, it has been widely held that a biphasic response of initial HR deceleration followed by acceleration is the result of a homeostatic process; the acceleration component represents an overshooting of the cardiac response in an attempt to regain base level. In an analysis of curve data, evidence was suggested to indicate that biphasic responses do not always represent an overshooting phenomenon, but rather an important and independent characteristic of the HR response itself (Schachter *et al.*, 1968). This kind of analysis is best performed by observing the function.

While observation of the fit of general mathematical functions to cardiac HR curve data are valuable, subtle mathematical differences in the HR curve may not be detectable using curve-fitting techniques. Moreover, individual subject data may have little resemblance to the mean HR curve. For example, the data of a subject who shows a biphasic response of deceleration followed by acceleration may be lost in averaging with a subject who shows only a decelerative reaction. Such considerations might suggest that more subtle descriptors of the cardiac response be observed.

If, in addition, the problems posed have to do with individual differences in HR response, it becomes essential to obtain descriptors of the response for each subject. While theoretically curve description for each subject is possible, individual variability within the response curve makes curve fitting extremely difficult. Descriptors of the individual subjects' response are helpful and have been used in a variety of studies.

D. Parameters of Heart Rate Response

A wide variety of descriptors of HR response are possible. In the present discussion we will list just a few and some of the results obtained with these. It must be kept in mind that other measures can be considered. The first, *peak magnitude*, is measured as the difference between the mean of the three R–R intervals immediately prior to stimulation and the mean of the three shortest R–R intervals. Lipton, Steinschneider, and Richmond (1961a), have also used a peak magnitude measure. However, in their case, the single shortest R–R interval was used. Many arguments can be offered on the advantages and disadvantages of using single R–R intervals to describe a particular HR parameter. In our work, for example (Lewis, Dodd, & Harwitz, 1969), we have chosen to use the mean of three highest consecutive beats in order to avoid the spurious change of a single interval.

A second measure, *trough magnitude*, is the difference between the mean of the three R–R intervals immediately prior to the stimulation and the mean of the three longest consecutive R–R intervals. The third measure is *HR range*, or the difference between the peak and the trough magnitude. This can be used as a measure of variability or lability (Lewis, Wilson, Ban, & Baumel, 1970). A fourth measure which we have called *initial response* is the difference between the mean of the three R–R intervals immediately prior to stimulation and the mean of the R–R intervals, usually five, immediately following the onset of stimulation. In addition, two latency measures are available. These are *latency-to-peak magnitude*, which is the number of R–R intervals from stimulus onset to the shortest single interval in the mean peak magnitude. *Latency-to-trough magnitude* is the number of R–R intervals from stimulus onset to the longest single R–R in the mean trough magnitude.

Various other measures dealing with the termination of HR response to stimulation—referred to as return levels—have been suggested by Lipton *et al.* (1961a), Clifton, Graham, and Hatton (1968), and Lewis (1971). No systematic study has been conducted relating these various measures of cardiac response. Lipton and his associates have suggested that some of the responses for an individual are quite stable (Lipton, Steinschneider, & Richmond, 1961b); however, there is little evidence indicating any consistent relationship between peak and trough magnitude or latency measures. In fact, Lewis *et al.* (1969) found little relationship between HR response parameters to tactile stimulation in sleeping and waking infants. In terms of the habituation data, several findings were observed. First, there were state differences which varied as a function of the nature of the response parameter studied. For example, in that study three HR parameters, trough magnitude and the latencies-to-peak and trough magnitudes, failed to evidence habituation, while peak magnitude and the range score showed habituation clearly. Initial response showed a trial effect which could not be interpreted as habituation. It is clear, at least from these results, that not all parameters of the HR response yield similar results and that the response parameter that the experimenter chooses will affect the results. The results of this study force us to conclude that in any study of the HR response a variety of response parameters is necessary in order to generalize about the HR response *per se*. Even so, if the response parameters do not covary, the only statements that can be made will be limited to specifying the response parameter.

As a corollary to this problem of response parameter, it is also important to keep in mind the fact that results from a single response measure may be quite misleading and may generate the wrong conclusions. Assume a situation in which the first few trials of stimulation result in a monophasic cardiac response of acceleration with a subsequent return to baseline. This could occur because the initial trials might startle the infant, resulting in an HR response of acceleration. Imagine further that as one presents more trials, the defensive reflex or startle alters to one of orienting where the acceleration component becomes smaller and the deceleration component becomes greater. If in a measurement procedure only the acceleration component were observed, one would conclude that the response habituated and would report that there was HR habituation to repeated stimulation, when in fact there was only habituation of the acceleration component. This is not an imaginary situation. For example, Raskin, Hattle, Harris, and De Young (1967) have shown that initial response of cardiac acceleration to a 40-dB tone tends to alter to cardiac deceleration over repeated presentations, an instructive instance of the above difficulty. In order to control for this measurement problem, multiple parameters of HR response should be analyzed. Unfortunately, very few studies have done so.

There are several problems in estimating HR response parameters that should be considered. The first is related to sampling procedures in which only trough magnitude is considered. Because some investigators are primarily interested in degree of deceleration, they consider any amount of acceleration to be zero deceleration (see McCall & Kagan, 1967). Thus, the scale contains true decelerative scores (R–R intervals are slower during the stimulus period than during the prestimulus period), but no accelerative scores. The error of this procedure is obvious. Consider six trials of acceleration and one of deceleration. Under this system all accelerations are scored as zero and the mean deceleration score would be based on the single deceleration value. This type of statistical bias is possible when sampling procedures are used without first observing the entire response.

A second problem has to do with using uneven time periods in a sampling procedure. Graham and Jackson (1970) have demonstrated that if one chooses to look at the three shortest R–R intervals of a 5 sec prestimulus period and compare it to the three shortest R–R intervals of a 10 sec stimulus period, acceleration or deceleration could be inferred when in fact there is no real rate change between these periods. An alternative problem in terms of sampling the extreme intervals in a prestimulation and stimulation period is the difficulty that one might conclude that there is directional HR change when in fact there is nothing more than HR variability change.

All these problems suggest that observation of the total HR response function is necessary in order to prevent error in choosing a particular response parameter.

E. Initial Level Effects

The "law of initial values" (LIV) was first introduced by Wilder (1950), and more recently expounded by Lacey (1956). This problem has been reviewed by many investigators, for example, most recently by Graham and Jackson (1970), and by Benjamin (1967), to mention a few. In general the law is related to the relationship between the organism's prestimulation HR and its HR response to a stimulus. More precisely, it states that "high autonomic excitation preceding a stimulus is correlated with low autonomic reactivity upon stimulus presentation and vice versa" (Woodcock, 1971). Thus, if the prestimulation HR is high, deceleration upon stimulus presentation is more likely than if the HR is low. This may be more than a simple scaling problem, but one that reflects the organism's homeostatic mechanism (see Lacey, 1956). In a recent review, Graham and Jackson (1970) hold that the LIV is applicable both for infant and adult HR response. They present data which suggest that the regression coefficients average between .63 and .92, indicating a relationship between the prestimulus level and stimulation response. However, others have argued that there is, especially among infants, little effect of the law of initial value. Eisenberg (1967) found no LIV effect to auditory signals, and Bartoshuk (1964) also failed to find any effect. Moreover, some people have found that the effect of prestimulation level is not linear but curvilinear [see Lazarus, Speisman, & Mordkoff (1963) as an example]. The data are discrepant. Since the prestimulation effect does seem at times to affect the response, it is necessary to consider this problem in data analysis, especially if the independent variable is also associated with the prestimulation level.

For any particular study the effect of initial level can easily be determined by obtaining and plotting the correlation between the prestimulation value on the one hand and the response parameter on the other. If there does appear to be an effect as a consequence of the initial level, then it is necessary to introduce controls. One control entails experimental manipulation of initial level where the experimenter waits for the HR to reach certain levels before introducing the stimulus. This presents particular difficulties in that one often observes slowly increasing base levels over repeated presentation of stimulus events; or, changes in base level can occur as a consequence of changing the state of the child from one of quiescent wakefulness to more restless wakefulness.

Waiting for HR to attain certain levels generally proves unsatisfactory, however, and it therefore becomes necessary statistically to control for the effects of prestimulation levels. To summarize the procedure simply: in eliminating the effect of initial level it is necessary to remove from the response that portion which is predicted or related to the initial level. The techniques for doing this have been stated several times and for a more detailed and authoritative discussion of the exact procedure the reader is referred to Benjamin (1963, 1967) and to Lacey (1956). It is important to note that in not all cases can regression analyses be used. For example, in a study reported by Lewis, Bartels, and Goldberg (1967), two stimulation conditions for each subject were used, these being the waking and sleeping states. It was found that these two conditions had significantly different prestimulation levels. In order to compare response parameters for these conditions, it was necessary to remove the prestimulation level. However, the prestimulation level for the two conditions did not form a single distribution. If this is the case, a statistical attempt to eliminate initial level effects cannot be computed and the observation of state differences, independent of prestimulation level, is not possible. The problem awaits the solution of this kind of statistical difficulty.

The solution used by Lewis, Bartels, and Goldberg (1967) was to obtain for each subject his mean HR and standard deviation for condition 1. Then, using each subject as his own control, each trial of the subject in condition 2 was compared to a trial in condition 1. Any trial during condition 2 for which the prestimulation HR was within a standard deviation of the prestimulation HR in condition 1 was viewed as having equal prestimulation levels. Only these trials were analyzed in an attempt to observe the effect of state on HR response independent of prestimulation levels. Whereas this technique has some validity, it must be remembered that most trials were eliminated and not all the data could be used.

The LIV is a rather complex phenomenon interfering with the observation of the HR response to a stimulus condition. It should be noted that some investigators are interested in the effects of prestimulation levels in terms of the definition of state. State has been defined as an arousal continuum, one measure of which might be the prestimulus HR level. In these problems the prestimulus level represents an experimental variable to be investigated. However, in most cases the experimenter is interested in the effect of the stimulus on the HR response. In those situations the prestimulation level must be controlled. Finally, it must be pointed out that the difference score, that is, the difference between prestimulation and the stimulation levels, is not a satisfactory way of eliminating the effect of initial level.

III. Studies of Heart Rate Response in Infants

In the early 1960s using Lacey's argument that changes in HR are related to the organism's intended interaction, we undertook a series of studies to investigate the relationship of attentional processes to cognition and included in our measure of attention HR response. It was reasoned that if HR was related to the organism's intended transaction with the environment—HR deceleration associated with intake—then HR deceleration should be related to other measures of attention. Moreover, we reasoned that greater amounts of attention should be associated with greater degrees of HR deceleration.

In the first study by Lewis et al. (1966), two samples of infants about 6 months of age were presented with a series of visual stimuli. The data indicated that the predominant response was one of HR deceleration. In order to relate this more closely to fixation data, that is, the amount of time the subject oriented his eyes toward the array, trough and peak magnitude scores were obtained. The trough magnitude data in relation to fixation time are presented in Fig. 8-2A and indicate that the amount of HR deceleration was related to the amount of fixation such that the more the child oriented toward the visual array, the greater the decelerative response. Figure 8-2B presents these data for peak magnitude which also indicate a relationship with fixation time. That both trough and peak magnitude responses showed deceleration indicates that the entire response was one of deceleration rather than variability increases.

Subsequent to this, other studies in a variety of situations over age ranges from 3 months through 3 years indicated a positive relationship between HR deceleration and other measures of attention, for example, fixation time, activity cessation, and decreases in vocalization. In a study by Lewis and Goldberg (1969), it was found that for the 3½-year-old, the habituation patterns for the measures fixation time and trough magnitude were quite similar (see Fig. 8-3). In fact, the correlation between fixation and deceleration was quite high, +.70.

More recently, in a study of attention distribution as a function of complexity and congruity in 24-month-old children, Lewis, Wilson, and Baumel (1971) demonstrated a significant correlation between fixation time, HR

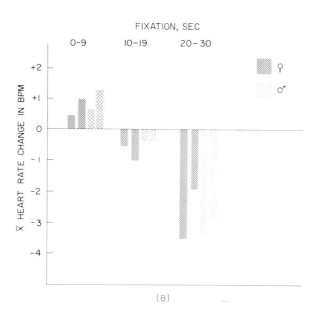

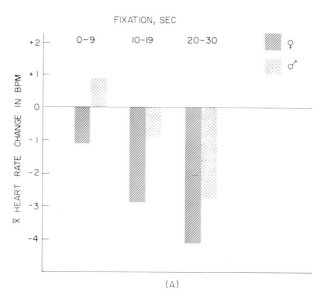

Fig. 8-2. (A) Heart rate change (trough magnitude) as a function of the duration of fixation on a visual array. (From Lewis et al., 1966.) (B) Heart rate change (peak magnitude) as a function of the duration of fixation on a visual array. (From Lewis et al., 1966.)

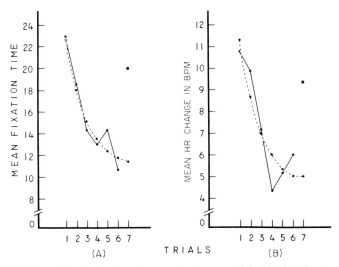

FIG. 8-3. (A) Mean total fixation time in seconds for each trial. (B) Mean heart rate change in beats per minute for each trial. Since the cardiac response (trough magnitude) was a deceleration in heart rate, the Y axis indicates that the number of beats decrease with larger numbers, indicating greater decelerations. In both graphs, the solid line indicates observed points with trial 7 (violation) shown as an isolated dot. The dotted line represents the points predicted by the regression equation including trial 7. (From Lewis & Goldberg, 1969.)

response, smiling, pointing, vocalization, and activity. The data indicate that attention in infants and young children is characterized by long fixation time, greater HR deceleration as measured by trough magnitude, decreases in arm movements and vocalization, increases in smiling and pointing behaviors. Thus, throughout the first $3\frac{1}{2}$ years of life the response of HR deceleration is related to other measures of attention.

Most of the research with infants in our laboratory has been concerned with attentional response to visual stimuli. In several papers (Lewis, 1971; Lewis & Spaulding, 1967), we have found that infants as early as 3 months of age show monophasic cardiac deceleration to the onset and presentation of complex auditory stimuli of 30 sec durations. This holds across a variety of auditory stimuli ranging from human voices, intermittent tones, and rock music. In one study, Lewis and Spaulding (1967) compared the HR response to visual and auditory stimuli in the same 6-month-old children. Figure 8-4 presents the heart rate curves averaged over all trials of the visual and auditory stimuli. The vertical line indicates the onset of the stimulus event. The results indicate that a monophasic response of HR deceleration most accurately describes the behavior of these 6-month-old infants. If one accounts for the immediacy of the stimulation—that is, auditory stimuli

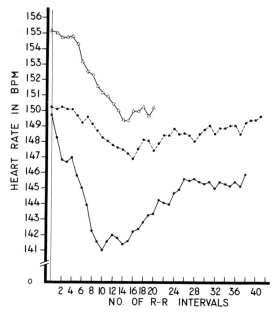

FIG. 8-4. A beat-by-beat analysis of the cardiac response to the onset of visual and auditory presentation as well as the cardiac response to the onset of orientation to the visual stimuli: (●——●) auditory stimuli; (●---●) visual stimuli; (△——△) visual orientation. (From M. Lewis and S. J. Spaulding, Differential cardiac response to visual and auditory stimulation in the young child, *Psychophysiology,* 1967, **3,** 229–237. © 1967 The Williams & Wilkins Co., Baltimore, Maryland.)

are more immediate than visual—there is little difference between the HR response to either modality.

Lewis (1971) has recently explored the HR response to both onset and termination of a complex auditory stimulus in infants 3, 6, and 12 months of age. The results indicate a predominantly monophasic HR decelerative response to stimulus onset and offset, indicating that energy change in either direction produces HR deceleration. Moreover, the data demonstrate that infants as early as 12 weeks show monophasic HR deceleration response to auditory stimuli. Berg and Graham (1970) and Hatton (1969) have also observed deceleration to the offset of an auditory signal in young infants. Kagan and Lewis (1965) have also shown that HR decelerative response (trough magnitude) is related to the cessation of arm movement in the case of the presentation of auditory stimuli.

In a developmental study of attention using a multivariant approach, Wilson and Lewis (1971) reported data on the same infants seen at 6, 13, 25, and 44 months. Looking at a variety of attentional measures—looking time, heart rate, activity, vocalizing, and smiling—a component analysis indicated two major components. The first component was consistently characterized by much looking, decreases in HR—particularly at the older ages—and little activity. This component includes those responses involved in receptor and body orientation and has been termed an orienting factor. In contrast, the second component is generally characterized by many discrete looks, much vocalizing or smiling—and was interpreted as representing an affect factor in attention. Thus, two components of attention have been identified. The question arises, however, of why two components are necessary to describe adequately a child's response to a visual stimulus and moreover why HR deceleration should appear in both components. Closer analysis of the process of attending provides an answer. Imagine a sequence of responses when a child looks at his mother's face. The initial reaction is one of orientation—intense looking, depression of HR, inhibition of ongoing behaviors such as activity, smiling, and vocalizing—in fact, all the responses which appear in the first component or orienting factor. After recognition takes place, the affect element of the stimulus—the face and especially the mother's face—comes into play and evokes affectual responses in the child. Thus, the child's response to his mother's face is in fact two clusters of responses—the initial orienting response and the subsequent affectual responses. This analysis suggests several points, the first of which is that the HR response does seem to be a component of general attending responses in both visual and auditory stimuli. However, it also suggests that the HR response may be in the service of other kinds of response systems such as the affect component. It is important to note that the HR response is in the service of many systems, only one of which may be the subject's intended transaction with the environment. Thus, it is quite possible to see a child deeply attending and at the same time kicking and thrashing about in excitement. Under these circumstances one might find the vector on the HR response between deceleration associated with the orienting and acceleration associated with the movement of the child's behavior. What the consequences of these divergent vectors are is still unclear. Perhaps these two divergent vectors balance each other, resulting in no observable HR response (see Jennings, Averill, Opton, & Lazarus, 1970). Perhaps they are sequential, resulting in some multiphasic response such as initial deceleration as a consequence of orientation followed by acceleration as a consequence of movement.

IV. Developmental Issues in Heart Rate Response

While the data are unequivocal that HR deceleration is a primary response to attending to auditory and visual stimuli in infants 12 weeks and over, there is a body of literature which suggests that HR acceleration is the primary response of newborn and very young infants (for example, Bartoshuk, 1962a,b, 1964; Bridger, 1962; Keen, Chase, & Graham, 1965; Gray & Crowell, 1968). However, there are exceptions to this work. Schulman (1968a,b) in several studies has demonstrated that premature and newborn infants can show HR deceleration to auditory stimuli. It is interesting to note that most of the newborn research is concerned with auditory and tactile stimulation, while most of the research with older infants has included visual stimuli as well. The discrepancy between newborn and older infant research, the former showing HR acceleration and the latter HR deceleration, has suggested to many that there may be a developmental change in the organism's ability to decelerate to discrete stimulation. Obviously, this would be quite an important developmental change, given that the HR deceleration has been used as a measure of attending. This age difference suggests the possibility that newborn and very young infants are less able to attend than older infants. The nonphysiological data, however, do not seem to support the belief that newborns cannot attend. For example, the Brown studies on neonatal conditioning strongly suggest that the newborn can attend and in fact can learn (see Lipsitt, 1963). If the HR response is one of only acceleration and deceleration is necessary for attending, then we have a discrepancy between the theoretical position and the fact that the newborn infant can learn. Whereas it is possible that part of the age difference is related to a developmental shift—this position is taken by Lipton, Steinschneider, and Richmond (1966)—it is necessary to consider at least three other variables which might be used to explain the age differences.

A. State

Lewis, Bartels, and Goldberg (1967) and Lewis, Dodd, and Harwitz (1969) have attempted to investigate the effects of state on the HR response. Observation of the experimental procedures of those working with neonates suggested that the state of the newborn child was not carefully controlled; and alert (see Lipton, Steinschneider, & Richmond, 1966). Since it has been postulated that stimulus rejection should result in HR acceleration, sleeping newborns would be expected to reject the stimuli and remain asleep. Under these conditions one might expect to see HR acceleration. However, if they were awake and alert one might expect them to be interested in taking in the stimulation and thus we would observe HR deceleration. One explanation, then, for the difference between the newborn data and infants at older ages may be related to the state differences between these two age groups. Clearly, the newborn is asleep most of the time (see Parmalee,

Wenner, & Schulz, 1965), whereas the older infant is more likely to be awake during the experimental situation. In the series of experiments Lewis and his associates found, in fact, HR acceleration in the infant when asleep and HR deceleration or, on occasion, no HR response when the infant was awake. Similar results were obtained by Berg, Berg, and Graham (1968) and Hatton (1969). Thus, until state is carefully controlled we cannot rule this out as a possible cause for the difference between ages.

B. Nature of the Stimulus

The nature of the stimulus presented is, of course, another important factor. In general, investigators have concerned themselves with observing the effect of "pure" auditory signals. That the infant quite early in its life—by four weeks—can differentiate "ga" and "ba" sounds would suggest that the ability to attend in the auditory mode is available quite early, especially with meaningful signals. Whereas it is possible that pure tones would produce HR acceleration in the newborn, it is not at all clear what the effect of human voices might be. Thus, the nature of the stimulus must be taken into account. Social stimuli having more meaning might be attended to, that is, elicit HR deceleration, whereas nonsense stimuli such as tones and beeps, may be excluded (elicit HR acceleration) possibly quite intentionally. This is not to be ignored. Much, if not most, of psychophysiological research, even on the infant, is predicated on presenting meaningless stimuli. The effects of meaningful stimuli are relatively unexplored. Eisenberg (1965), however, has found that newborns are especially attentive to sounds within the speech range and to speechlike patterns. Moreover, Moffett (1968) found very large decelerative response—20 bpm—to human voices.

This raises the whole issue of stimulus relevance. It would seem reasonable to assume some stimuli are more biologically meaningful than others. Human voices should be more meaningful than square waves of some "pure" tone, or movement to the upright more meaningful than an air blast to the abdomen. In fact, data suggest that the upright position facilitates attentive behavior in infants (Korner & Grobstein, 1966). If certain kinds of stimuli have more relevance than others, we would expect them to elicit attentive behavior. Moreover, and an important corollary, uninteresting or irrelevant stimuli would be expected to elicit not only nonattentive behavior but even an attempt to exclude these events. In the former case HR deceleration would be expected, while in the latter HR acceleration.

In order to conclude that newborns and the very young cannot decelerate their HR, it is essential to observe their reaction to stimuli which could be considered meaningful. In a recent paper by Malcuit and Clifton (1971), stimuli which could be considered more meaningful than those usually employed with the newborn were used. One consisted of moving the infant from a prone position 2 inches up toward a more upright one. Another stimulus, much like the one employed by Lewis *et al.* (1967) with older children, was tactile and consisted of stroking the infants' foreheads with a small paintbrush. The third stimulus was an auditory tone. The data revealed that sustained (more than 1 or 2 beats) HR deceleration was possible in neonates when stimuli more salient were obtained. The authors also observed a rather interesting state interaction, which only reinforces the effect of state mentioned earlier.

The data are impressive and should put to rest the belief that HR deceleration is not possible in the newborn. The data indicate that deceleration is possible when certain types of stimuli are presented. That the newborn may have a smaller range of stimuli that interest it or are relevant to it than older infants is not to be denied. The developmental function, then, is not the ability or lack thereof to show HR deceleration, but rather in the number and type of stimuli which interest the organism—the interacting with the organism's ability to stay interested, namely its state.

C. Stimulus Intensity and Other Parameters

A study by Jackson, as referred to in Graham and Jackson (1970), varied rise time and found somewhat inconclusive differences in the HR response as a consequence. For two stimuli with rapid rise time there was no clear HR effect, whereas slow rise time resulted in acceleration, but this HR acceleration was delayed. It is clear that neonates startle easily (see Berg & Beebe-Center, 1941) as a result of an immature neuronal organization. Since startling does result in HR acceleration, it may be that stimuli which appear abruptly or which are initially intense produce a startle response. The newborn, unable to dampen the startle, shows an acceleration rather than the deceleration response. This argument suggests that it is not the ability to decelerate which is missing in the newborn, but rather the inability to suppress startle to stimulus onset.

V. Heart Rate Response and Cognitive Functions

If HR deceleration is associated with environmental intake, then on those cognitive tasks for which environmental intake is crucial, the results should indicate two things: (1) HR deceleration will accompany the performance on the cognitive task, and (2) differential HR deceleration should be associated with cognitive efficiency. Most of the studies used to test this hypothesis employ the reaction time paradigm (see, for example, Boyle,

8. THE CARDIAC RESPONSE DURING INFANCY

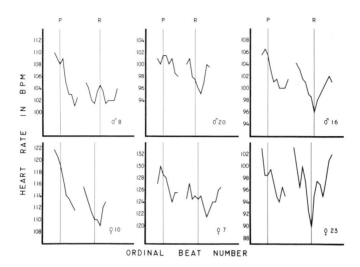

FIG. 8-5. Six individual median cardiac response curves during the matching figures task. (From M. Lewis and C. D. Wilson, The cardiac response to a perceptual cognitive task in the young child, *Psychophysiology,* 1970, 6(4), 411–420. © 1970 The Williams & Wilkins Co., Baltimore, Maryland.)

Dykman & Ackerman, 1965; Dureman & Saaren-Seppälä, 1964; Lacey & Lacey, 1958; Obrist, Hallman, & Wood, 1964). While reaction time has often been considered a cognitive task, there were a few studies to indicate, however, that the accuracy of perception and response on a perceptual task should be accompanied by HR deceleration.

In a study of 3½-year-olds Lewis and Wilson (1970) attempted to investigate this problem. The children were presented with four variations of a picture and, after looking at that picture, were presented with a standard— a picture exactly like one of the four variations. The subject's task was to point to the variation which matched the standard. This task is called "matching familiar figures." Figure 8-5 presents the median response over all the trials (20) for six different subjects. Heart rate in bpm is presented along the Y axis and R–R intervals along the X axis. The first R–R interval after presentation is indicated as the first R–R interval after "P." These curves are discontinuous because each of the trials does not contain the same number of R–R intervals. In order to correct for this, the median curves are constructed by using the presentation point (P) and response point (R) as points of reference (see Lewis & Wilson, 1970, for a complete discussion).

The data indicate that HR deceleration was the primary response to presentation of the variations and information processing. Since each subject received 20 trials of different variations and standard, it was possible to compare the HR deceleration for each subject on his correct and incorrect responses. It was hypothesized there would be greater deceleration for the correct responses than the incorrect responses. While the data indicated that this was the case, there was a confounding effect of response time. Subjects took longer to produce correct responses and longer response times were related to greater HR deceleration. When response time was statistically removed, there was no difference between HR deceleration and the accuracy of the cognitive–perceptual task.

Of interest is the fact that when a subject's response was correct, his HR quickly returned to baseline level, whereas, when he was incorrect his HR remained depressed until he guessed again. Figure 8-6 demonstrates this phenomenon for two subjects. Observe that for correct responses the slope after "R" is quite steep and positive, whereas when the same subjects were incorrect, the slope was either zero or continued to be negative. This suggests that while we could not demonstrate a relationship between amount of HR deceleration and accuracy of the cognitive–perceptual task, there is a relationship between the accuracy of the response and the subsequent

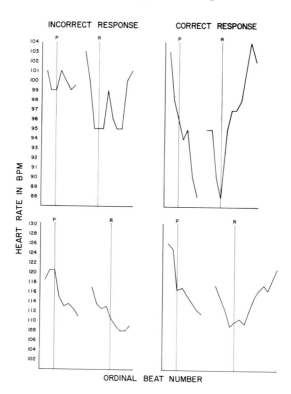

FIG. 8-6. Two individual median cardiac response curves for correct and incorrect trials.

HR response—specifically, the speed with which it returns to base level. It is clear that further investigations using other cognitive tasks are needed before the relationship between HR deceleration and accuracy of cognitive performance can be assessed.

Lacey et al. (1963) not only demonstrated HR deceleration to stimulus situations in which the organism wanted to take in the stimuli, but HR acceleration to stimulus situations in which the organism wanted to exclude the information. In the infancy and early childhood literature, there is little research on this problem. In a recent study by Steele and Lewis (1968), subjects 6–27 years of age were presented with a mental arithmetic problem and their HR was monitored. For all groups there was HR acceleration. Moreover, for the 26 subjects 6–8 years of age there was a significant correlation between degree of HR acceleration to the mental arithmetic task and general intellectual functioning as measured by the WISC ($r = .34$, $p < .05$). Thus, there is some support for HR acceleration accompanying problem-solving and general cognition ability in children, although there are no data of this kind for infants. More work on this aspect of the HR response is needed.

VI. Discussion and Summary

The preceding discussion makes any summary difficult, for it has touched on many different aspects of the heart rate response and research with infants. There are several issues which underlie much of the confusion in the research with infants and we shall touch on some of them in concluding.

Infancy is a broad period; while only 2–3 years in length, it covers a huge developmental period. During this time—short compared to 2–3 years for adults—there is much change, and this requires change in techniques, stimuli, etc. What is appropriate for the newborn may be inappropriate for the 1-year-old. While this holds especially true for general laboratory procedures, it must also be considered for other aspects. There are state differences, if not in quality then at least in distribution, which can significantly affect results. Moreover, a pure tone may have different meanings and thereby elicit different responses for the 1-year-old than for the newborn.

A second issue is the realization that the heart rate response is in the service of different factors. It is necessary to consider that the response is within a vector system and influenced by many functions. Physical activity accompanying orientation may result in no heart rate change, monophasic, or polyphasic responses. Moreover, this vector system may be developmentally linked. Newborn infants may have to learn to suppress activity in order for their heart rate to decelerate.

As a corollary to this is the problem of assigning meaning—interpreting —to heart rate response. Whereas heart rate acceleration during an arithmetic problem may be related to excluding the environment, acceleration to a loud auditory signal may be related to startle. Moreover, heart rate acceleration to threatening pictures may be related to some arousal state such as anxiety. It should not be taken for granted that these cases of acceleration have the same meaning.

Reference has been made to heart rate response as if it is some unitary term. The data are limited, but those that do exist strongly suggest that the different parameters used to describe the heart rate response do not generate the same results. It is possible to observe the presence of a phenomenon like habituation for one parameter and the absence or reverse in another. The parameters of the heart rate response may form a constellation which remains relatively constant while the individual elements rearrange themselves. Multivariate—in this case, multiparameter—analyses would appear in order. Alternatively, we may discover that individual parameters within a response such as heart rate react like the different responses within the ANS. That is, Lacey's notion of autonomic response stereotyping may apply within a response such as heart rate as well as across responses within the ANS.

Throughout the paper we have discussed meaning. It seems clear that any attempt to isolate stimulus and response must result in disarray. In all probability there is no such thing as *a* physiological response to auditory stimuli, for example. Thus, the logic of using "pure" signals of known dimension is unclear. By what reason should one suppose that these signals will reveal anything more about psychophysiology than signals whose dimensions cannot be accurately described? In fact, it appears to be the case that stimuli whose dimensions can be specified are relatively meaningless. Perhaps there is more to be learned from the response to a human voice than a 500 Hz tone of 80 dB; more from rocking motion than air blasts to the abdomen, and more from a human face than 16-square checkerboard pattern. The evidence from a variety of sources would suggest that meaningless and meaningful stimuli result in markedly different consequences in terms of a variety of human responses. Psychophysiology should be no exception.

References

Bartoshuk, A. K. Human neonatal cardiac acceleration to sound: Habituation and dishabituation. *Perceptual and Motor Skills,* 1962, **15,** 15–27. (a)

Bartoshuk, A. K. Response decrement with repeated elicitation of human neonatal cardiac acceleration to sound. *Journal of Comparative and Physiological Psychology,* 1962, **55,** 9–13. (b)

Bartoshuk, A. K. Human neonatal cardiac responses to sound: A power function. *Psychonomic Science*, 1964, **1**, 151–152.

Benjamin, L. S. Statistical treatment of the law of initial values (LIV) in autonomic research: A review and recommendation. *Psychosomatic Medicine*, 1963, **25**, 556–566.

Benjamin, L. S. Facts and artifacts in using analysis of covariance to "undo" the law of initial values. *Psychophysiology*, 1967, **4**, 187–206.

Berg, K. M., Berg, W. K., & Graham, F. K. Infant heart rate response as a function of stimulus and state. Unpublished manuscript, 1968.

Berg, R. L., & Beebe-Center, J. G. Cardiac startle in man. *Journal of Experimental Psychology*, 1941, **28**, 262–279.

Berg, W. K., & Graham, F. K. Cardiac orienting at four months of age. Paper presented at the Society for Psychophysiological Research Meeting, New Orleans, November, 1970.

Boyle, R., Dykman, R., & Ackerman, P. Relationships of resting autonomic activity, motor inpulsivity and EEG tracings in children. *Archives of General Psychology*, 1965, **12**, 314–323.

Bridger, W. H. Sensory discrimination and autonomic function in the newborn. *Journal of American Academy of Child Psychiatry*, 1962, **1**, 67–82.

Clifton, R. K., Graham, F. K., & Hatton, H. M. Newborn heart-rate response and response habituation as a function of stimulus duration. *Journal of Experimental Child Psychology*, 1968, **6**, 265–278.

Dureman, I., & Saaren-Seppälä, P. Electrodermal parameters and sensory–motor response latency. Sixteenth Report, Department of Psychology, University of Uppsala, Sweden, March, 1964.

Eimas, P. D., Siqueland, E. R., Jusczyk, P., & Vigorito, J. Speech perception in early infancy. *Science*, 1971, **171**, 303–306.

Eisenberg, R. B. Auditory behavior in the human neonate: I. Methodologic problems and the logical design of research procedures. *Journal of Auditory Research*, 1965, **5**, 159–177.

Eisenberg, R. B. Stimulus significance as a determinant of newborn responses to sound. Paper presented at the convention of the Society for Research in Child Development, New York, March, 1967.

Ellingson, R. J. The study of brain electrical activity in infants. In L. P. Lipsitt & C. C. Spiker (Eds.), *Advances in child development and behavior*. Vol. 3. New York: Academic Press, 1967. Pp. 53–97.

Graham, F. K., & Clifton, R. K. Heart-rate change as a component of the orienting response. *Psychological Bulletin*, 1966, **65**, 305–320.

Graham, F. K., & Jackson, J. Arousal systems and infant heart-rate responses. In H. W. Reese & L. P. Lipsitt (Eds.), *Advances in child development and behavior*. Vol. 5. New York: Academic Press, 1970. Pp. 59–117.

Gray, M. L., & Crowell, D. H. Heart rate changes to sudden peripheral stimuli in the human during early infancy. *Journal of Pediatrics*, 1968, **72**, 807–814.

Hatton, H. M. Developmental change in infant heart rate response during sleeping and waking states. Unpublished doctoral dissertation, University of Wisconsin, 1969.

Jennings, J. R., Averill, J. R., Opton, E. M., & Lazarus, R. S. Some parameters of heart rate change: Perceptual versus motor task requirements, noxiousness, and uncertainty. *Psychophysiology*, 1970, **7**, 194–212.

Johnson, L. C. A psychophysiology for all states. *Psychophysiology*, 1970, **6**, 501–516.

Kagan, J., & Lewis, M. Studies of attention in the human infant. *Merrill-Palmer Quarterly*, 1965, **11**, 95–127.

Keen, R. K., Chase, H. H., & Graham, F. K. Twenty-four hour retention by neonates of an habituated heart rate response. *Psychonomic Science*, 1965, **2**, 265–266.

Khachaturian, Z. S., Kerr, J., Kruger, R., & Schachter, J. A methodological note: Comparison between period and rate data in studies of cardiac function. *Psychophysiology*, 1972, **9**, 539–545.

Korner, A., & Grobstein, R. Visual alertness as related to soothing in neonates: Implications for maternal stimulation and early deprivation. *Child Development*, 1966, **37**, 867–876.

Lacey, J. I. The evaluation of autonomic responses: Toward a general solution. *Annals of the New York Academy of Sciences*, 1956, **67**, 123–164.

Lacey, J. I. Psychophysiological approaches to the evaluation of psychotherapeutic process and outcome. In E. A. Rubinstein & M. B. Parloff (Eds.), *Research in psychotherapy*. Washington, D. C.: American Psychological Association, 1959. Pp. 160–208.

Lacey, J. I. Of number and syntax in the psychophysiology of the heart. Presidential address, Society for Psychophysiological Research, Denver, October, 1962.

Lacey, J. I., & Lacey, B. C. Verification and extension of the principle of autonomic response stereotypy. *American Journal of Psychology*, 1958, **71**, 50–73.

Lacey, J. I., Kagan, J., Lacey, B. C., & Moss, H. A. The visceral level: Situational determinants and behavioral correlates of autonomic response patterns. In P. H. Knapp (Ed.), *Expressions of the emotions in man*. New York: International Univ. Press, 1963. Pp. 161–196.

Lazarus, R. S., Speisman, J. C., & Mordkoff, A. M. The relationship between autonomic indicators of psychological stress: Heart rate and skin conductance. *Psychosomatic Medicine*, 1963, **25**(1), 19–30.

Lewis, M. A developmental study of the cardiac response to stimulus onset and offset during the first year of life. *Psychophysiology*, 1971, **8**, 689–698.

Lewis, M., & Goldberg, S. The acquisition and violation of expectancy: an experimental paradigm. *Journal of Experimental Child Psychology*, 1969, **7**, 70–80.

Lewis, M., & Spaulding, S. J. Differential cardiac response to visual and auditory stimulation in the young child. *Psychophysiology*, 1967, **3**, 229–237.

Lewis, M., & Wilson, C. D. The cardiac response to a perceptual cognitive task in the young child. *Psychophysiology*, 1970, **6**(4), 411–420.

Lewis, M., Kagan, J., Kalafat, J., & Campbell, H. The cardiac response as a correlate of attention in infants. *Child Development*, 1966, **37**, 63–71.

Lewis, M., Bartels, B., & Goldberg, S. State as a determinant of infants' heart rate response to stimulation. *Science*, 1967, **155**(3761), 486–488.

Lewis, M., Dodd, C., & Harwitz, M. Cardiac responsivity to tactile stimulation in waking and sleeping infants. *Perceptual and Motor Skills*, 1969, **29**, 259–269.

Lewis, M., Wilson, C. D., Ban, P., & Baumel, M. H. An exploratory study of resting cardiac rate and variability from the last trimester of prenatal life through the first year of postnatal life. *Child Development*, 1970, **41**(3), 800–811.

Lewis, M., Wilson, C. D., & Baumel, M. Attention distribution in the 24-month-old child: Variations in complexity and incongruity of the human form. *Child Development*, 1971, **42**, 429–438.

Lipsitt, L. P. Learning in the first year of life. In L. P. Lipsitt & C. C. Spiker (Eds.), *Advances in child development and behavior*. Vol. 1. New York: Academic Press, 1963. Pp. 147–194.

Lipton, E. L., Steinschneider, A., & Richmond, J. B. Autonomic function in the neonate: III. Methodological considerations. *Psychosomatic Medicine*, 1961, **23**, 461–471. (a)

Lipton, E. L., Steinschneider, A., & Richmond, J. B. Autonomic function in the neonate: IV. Individual differences in cardiac reactivity. *Psychosomatic Medicine*, 1961, **23**, 472–484. (b)

Lipton, E. L., Steinschneider, A., & Richmond, J. B. Autonomic function in the neonate: VII. Maturational change in cardiac control. *Child Development*, 1966, **37**, 1–16.

Malcuit, A., & Clifton, R. The pursuit of the OR in the human newborn. Paper presented at the Society for Psychophysiological Research meeting, St. Louis, October, 1971.

McCall, R. B., & Kagan, J. Stimulus–schema discrepancy and attention in the infant. *Journal of Experimental Child Psychology,* 1967, **5,** 381–390.

Moffett, A. R. Speech perception by infants. Unpublished doctoral dissertation, University of Minnesota, 1968.

Moore, P. A. Speech perception in six-week-old infants. Paper presented at Society for Research in Child Development, Minneapolis, April, 1971.

Obrist, P., Hallman, S., & Wood, D. Autonomic levels and lability and performance time on a perceptual task and a sensory–motor task. *Perceptual and Motor Skills,* 1964, **18,** 753–762.

Obrist, P. A., Webb, R. A., & Sutterer, J. R. Heart rate and somatic changes during aversive conditioning and a simple reaction time task. *Psychophysiology,* 1969, **5,** 696–723.

Obrist, P. A., Webb, R. A., Sutterer, J. R., & Howard, J. L. Cardiac deceleration and reaction time: An evaluation of two hypotheses. *Psychophysiology,* 1970, **6,** 695–706.

Parmalee, A. H., Wenner, W. H., & Schulz, H. R. Infant sleep patterns from birth to 16 weeks of age. *Journal of Pediatrics,* 1965, **65,** 576–582.

Raskin, C. D., Hattle, M., Harris, L., & DeYoung, G. The effects of stimulus intensity, duration, and interstimulus interval on evocation and habituation of orienting and defensive reflexes. Paper presented at the Society for Psychophysiological Research Meetings, San Diego, October, 1967.

Schachter, J., Williams, T. A., Khachaturian, Z., Tobin, M., & Druger, R. The multiphasic heart rate response to auditory clicks in neonates. Paper presented at the meeting of the Society for Psychophysiological Research, Washington, D. C., October, 1968.

Schachter, J., Williams, T. A., Khachaturian, Z., Tobin, M., Kruger, R., & Kerr, J. Heart rate responses to auditory clicks in neonates. *Psychophysiology,* 1971, **8**(2), 163–179.

Schulman, C. A. The development of heart rate deceleration in premature infants. Paper presented at the Society for Psychophysiological Research meetings, Washington, D.C., October, 1968. (a)

Schulman, C. A. Differentiation in the neonatal period between infants at high risk and infants at low risk for subsequent severe mental retardation. Paper presented at the Eastern Psychological Association Meetings, Washington, D. C., April, 1968. (b)

Sokolov, E. N. *Perception and the conditioned reflex.* New York: Macmillan, 1963.

Steele, W. G., & Lewis, M. A longitudinal study of the cardiac response during a problem-solving task and its relationship to general cognitive function. *Psychonomic Science,* 1968, **11**(8), 275–276.

Welford, N. T. The SETAR and its uses for recording physiological and behavioral data. *IRE Transactions of Bio-Medical Electronics,* 1962, **9,** 185–189.

Wilder, J. The law of initial values. *Psychosomatic Medicine,* 1950, **12,** 392.

Wilson, C. D., & Lewis, M. A developmental study of attention: A multivariate approach. Paper presented at the Eastern Psychological Association meetings, New York, April, 1971.

Woodcock, J. M. Terminology and methodology related to the use of heart rate responsivity in infancy research. *Journal of Experimental Child Psychology,* 1971, **11,** 76–92.

Chapter 9

Mechanisms of Electrodermal Activity

Don C. Fowles

Department of Psychology
The University of Iowa
Iowa City, Iowa

I. Introduction	232
A. Anatomy of the Skin and Sweat Glands	232
B. Electrodermal Phenomena	235
C. Membrane Potentials	237
II. Skin Conductance Responses	239
A. Duct Filling and Hydration of the Corneum	240
B. Membrane Responses	243
C. Location of the Membrane	244
D. Identification of Duct Filling versus Membrane SCRs	250
III. Skin Conductance Level	251
IV. Skin Potential Responses	253
A. The Monophasic Negative SPR	253
B. The Diphasic or Positive SPR	257
V. Skin Potential Level	259
A. The Sweat Gland Component	259
B. The Epidermal Component	260
VI. Summary and Conclusions	265
References	267

I. Introduction

Electrodermal activity has been of interest to psychologists for many years because of the ease with which electrodermal responses (EDRs) [or "galvanic skin responses" (GSRs)] may be elicited by psychological stimuli. The use of EDRs in psychological research rapidly outpaced the development of theory concerning the underlying mechanisms. In recent years, however, a considerable body of data has developed which offers the hope that in the not-too-distant future the mechanisms of electrodermal activity may be well understood.

The older literature on mechanisms has been well reviewed by Rothman (1954) and by Martin and Venables (1966).[1] Edelberg (1971) has provided an excellent review of the more recent literature. Since limitations of space make it impossible to review all of the relevant literature, an attempt has been made to avoid overlap with the above reviews except where necessary to illustrate crucial evidence for theories being discussed. The emphasis in this review is on an attempt to specify the origin of various electrodermal phenomena and to integrate these conclusions into a consistent model. Although some introductory material is included here, the reader without a strong background in electrophysiology may find it advantageous to read Lykken's (1968) excellent introductory review of basic concepts in this area.

A. Anatomy of the Skin and Sweat Glands

The epidermis may be divided into four strata or layers (Kuno, 1956), which are represented in Fig. 9-1. The inner layer, composed of fully living cells, is known as the stratum germinativum or the mucous layer. The horny layer or stratum corneum is an outer layer composed of dead cells. Separating these layers are the transitional cells of the granular layer (stratum granulosum) and, at least in the palms and soles, the stratum lucidum. On the palms and the soles, which are of most interest because of their more prominent electrodermal activity, the epidermis is very much thicker. The corneum, especially, is much thicker in these areas, and the stratum lucidum can be seen clearly only in these areas (Fig. 9-2).

The major function of skin is to provide a protective barrier between the body and the external environment, and this protection includes the prevention of water loss. At the same time, however, the epidermis must be capable of promoting the evaporation of water in order to regulate body temperature. This dual function appears to have been accomplished by placing the horny layer, which is highly hydrophilic and thus serves as a

---

[1] Readers should also consult reviews by Edelberg (1972) and by Venables and Christie (1973), which had not been published at the time this paper was written.

9. MECHANISMS OF ELECTRODERMAL ACTIVITY

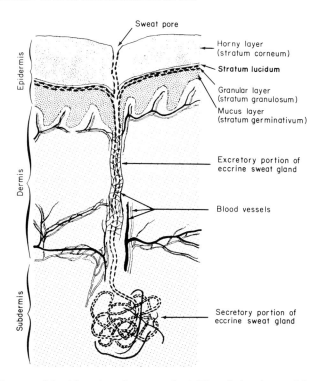

FIG. 9-1. Diagram of the skin and eccrine sweat gland. The spiral pathway of the sweat gland duct in the epidermis is not shown. (Modified from David T. Lykken, Neuropsychology and Psychophysiology in personality research, in Edgar F. Borgatto and William W. Lambert (Eds.), *Handbook of personality theory and research,* © 1968 by Rand McNally & Company, Chicago, Figure 11, p. 466.)

reservoir for water, above a barrier layer which is relatively impermeable to water. The exact location of this barrier is not known, but it is usually believed to lie between the germinating layer and the upper corneum. The potential sites are thus the stratum compactum (the more densely packed lower portion of the corneum), the stratum lucidum, or the granular layer (Edelberg, 1971; Kuno, 1956; Lykken, 1968; Martin & Venables, 1966; Rothman, 1954). Even in the absence of sweating, a small amount of water is lost by diffusion through the barrier layer and subsequent evaporation from the corneum. When additional cooling is necessary, water is delivered to the corneum by the sweat ducts. The corneum may also be hydrated, of course, by exposure to water in the environment.

An impermeability to water carries with it an impermeability to ions and thus the barrier layer is responsible for the major portion of the transepidermal resistance. When dry, the corneum may add considerably to the

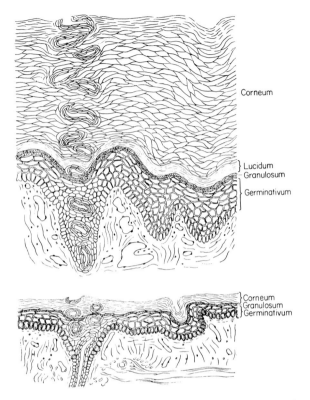

FIG. 9-2. Semidiagrammatic representation of the structure of the epidermis from the palm (upper) and the general body surface (lower). The holes in the epidermis represent cross sections of the sweat gland duct as it follows a corkscrew pathway to the surface. Note the much greater thickness of the epidermis from the palm and the absence of the stratum lucidum in skin from the general body surface. (From Kuno, 1956.)

resistance of the epidermis, but the hydrated corneum appears to be freely permeable to ions down to the level of the stratum compactum (Edelberg, 1971; Rothman, 1954, p. 32). The cells of the germinating layer are surrounded by intercellular clefts filled with tissue fluid (Kuno, 1956, p. 6), making it unlikely that this layer has a high resistance. Although Edelberg (1971) suggested that the basal cells of this layer may form a barrier with some impermeability to ions, the low diffusion resistance (Tregear, 1966, pp. 21–22) and electrical resistance (Lykken, 1968) of the epidermis once the germinating layer has been reached suggests a low resistance to current flow for the basal cells of the germinating layer. Under normal recording conditions the corneum will be at least partially hydrated and it can be assumed that most of the resistance of the epidermis resides in the barrier layer.

The eccrine sweat glands in humans consist of a secretory coil situated at the border of subdermis and dermis and an excretory duct, which begins in the coiled portion of the gland and then ascends to the surface of the skin. In the dermis the duct is relatively straight, but it passes through the epidermis as a spiraling coil reminiscent of tubular nephrons in the kidney (Rothman, 1954, p. 154). Both the secretory and ductal epithelia consist of two layers of cells surrounding the lumen through which the sweat passes. The precursor solution found in the lumen of the secretory coil is slightly hypertonic, suggesting that secretion is accomplished by active solute transport with fluid following passively as a result of the osmotic gradient thus generated (Schulz, 1969). The major function of the subepidermal portion of the duct is to reabsorb sodium chloride (NaCl) from the sweat as it flows to the surface of the skin (Gordon & Cage, 1966; Munger, 1961; Munger & Brusilow, 1961; Schulz, Ullrich, Frömter, Holzgreve, Frick, & Hegel, 1965), thereby preventing the loss of vast quantities of this salt at high environmental temperatures. In view of the importance of hydrating the corneum in order to promote evaporation of water, the corkscrew pathway of the duct in the upper epidermis probably serves to increase diffusion of sweat directly into the corneum.

Far more detailed information concerning the structure and function of the skin and sweat glands may be found in the many reviews on this topic (Elden, 1971; Kuno, 1956; Montagna, 1962; Montagna, Ellis, & Silver, 1962; Montagna & Lobitz, 1964; Rothman, 1954; Weiner & Hellmann, 1960). Gordon and Cage (1966) provide a brief but excellent summary of water and electrolyte secretion in sweat glands.

B. Electrodermal Phenomena

In discussing electrodermal activity it is important to distinguish between tonic or basal levels of potential and resistance and the phasic responses which are superimposed on these levels. The following terminology is useful in discussing electrodermal phenomena (Venables & Martin, 1967b):

SRR = skin resistance response,
SRL = skin resistance level,
SCR = skin conductance response,
SCL = skin conductance level,
SPR = skin potential response,
SPL = skin potential level.

Since conductance is the reciprocal of resistance, SCR and SCL are simply alternative ways of viewing SRR and SRL, respectively. SRR, SRL, SCR, and SCL require the application of an external voltage source and thus are grouped together as *exogenous* phenomena. SPL and SPR, which re-

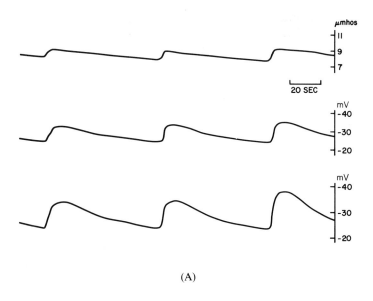

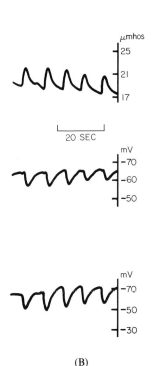

FIG. 9-3. (A) Slow recovery SCRs (upper tracing) and monophasic negative SPRs (middle and lower tracings). The upper and middle tracings were recorded with 67 mEq/liter KCl in Parke-Davis Unibase as recommended by Lykken and Venables (1971). The lower tracing was recorded with an experimental preparation containing the same concentration of KCl in a mixture of equal parts of Unibase and polyethylene glycol 400. The Unibase–glycol mixture produces less epidermal hydration.

(B) Rapid recovery SCRs (upper tracing) and monophasic positive SPRs (middle and lower tracings), recorded as in Fig. 9-2A. Note the difference in scale between the middle and lower tracings.

quire no external voltage, are characterized as *endogenous* phenomena. The term *electrodermal responses* (EDRs) encompasses both exogenous and endogenous responses. The SCRs and SRRs are always monophasic responses, showing a change in the direction of increased conductance or decreased resistance. The SPRs, on the other hand, are often diphasic or even triphasic, showing both positive and negative components. They may also be monophasic negative or monophasic positive. The most frequent SPR is a diphasic response with an initial negative wave followed by a positive wave. A final distinction is that between responses with a slow recovery limb and those with a rapid recovery limb. Slow recovery limb responses are those which require a long time to return to baseline, whereas rapid recovery limb responses show a complete or almost complete recovery within a few seconds. This distinction applies to monophasic negative SPRs as well as to SCRs and SRRs. For convenience, SCR will be used to refer to all exogenous responses except in cases in which it seems important to preserve the units of measurements utilized in the original study. This general use of the term SCR reflects in part the author's agreement with Lykken and Venables' (1971) conclusion that conductance units are more appropriate than resistance units.

The various types of responses are illustrated in Fig. 9-3A and B. The slow recovery, monophasic negative SPRs and the slow recovery SCRs in Fig. 9-3A were recorded during a period of relative inactivity and were elicited by asking the subject to take shallow breaths (if shallow breaths failed to produce responses, deeper breaths were requested). In Fig. 9-3B the SCRs are of the rapid recovery type, while the SPRs are monophasic positive, or else diphasic with a very small initial negative component. These responses were elicited by asking the subject to take very fast, deep breaths just after the subject had blown up a balloon until it burst. This association between positive SPRs and rapid recovery SCRs, on the one hand, and slow recovery, monophasic negative SPRs and slow recovery SCRs, on the other hand, is characteristic (see Section II,D).

The responses in Fig. 9-3A and B were chosen to illustrate pure instances of each type of response. In routine recordings compound responses are often obtained. The type of electrode jelly or cream used will affect the responses obtained, at least in the case of skin potential, as illustrated by the greater amplitude of the SPRs using Unibase–glycol than when using a standard Unibase preparation.

C. Membrane Potentials

A membrane is said to be semipermeable when it is permeable to some ions but not others. When the solutions on either side of the membrane are at different concentrations, those ions which can pass through the membrane tend to do so under the diffusion force generated by their own concentration gradient. If the membrane is permeable to cations but not to anions, there will be an accumulation of cations on the side of the membrane with the lower cation concentration, leaving an excess of negative charges on the other side. This separation of charge, however, generates a potential which restrains the diffusion of cations. An equilibrium condition develops very rapidly in which the potential across the membrane exactly balances the diffusion force, thereby preventing any further net diffusion of cations.

In the simplest case of a membrane with no appreciable anion permeability and a single cation, the potential difference at equilibrium can be predicted from the concentration ratio according to the Nernst equation, which states that at body temperature the potential in millivolts equals 61 times the logarithm of the concentration ratio of the cation (Guyton, 1966, Chapter 5). Thus a rule of thumb is that a perfectly selective membrane should generate a potential of approximately 61 mV for a 10:1 concentration ratio. The potential will be positive on the side with lower cation concentration. If anion permeability is not negligible and/or if additional cations are involved, it is necessary to use an expanded version of the Nernst equation (Guyton, 1966, Chapter 5). For a membrane separating solutions containing sodium (Na), potassium (K), and chloride (Cl) ions, the membrane potential (E_m) would be predicted by

$$E_m = -61 \log \frac{[P_K(K_o^+) + P_{Na}(Na_o^+) + P_{Cl}(Cl_i^-)]}{[P_K(K_i^+) + P_{Na}(Na_i^+) + P_{Cl}(Cl_o^-)]},$$

where P_K, P_{Na}, and P_{Cl} represent the relative permeabilities of the membrane to potassium, sodium, and chloride ions, respectively. The letters in parentheses refer to the concentrations of the respective ions, with the subscripts i and o indicating whether the concentration is that of the inner or outer solution. Thus, each ion contributes to the potential as a function of its ability to penetrate the membrane. When all permeabilities are the same, the potential is zero. It should be clear from this equation that changes in permeability may produce changes in potential. For example, if a membrane initially permeable only to sodium undergoes a sudden increase in chloride permeability, the potential difference will be much reduced as a result of the increased contribution of chloride ions. The membrane is then said to be depolarized because there is a reduced separation of charge across the membrane.

The potentials generated in the skin are believed to be produced by semipermeable membranes such as those just described, and this will be assumed throughout the paper. It should be stressed, however, that one or more of the potentials may be generated in association with the active transport of ions. Although such active transport has been traditionally viewed as not directly producing potentials, in recent years there have been reports of an electrogenic sodium pump (Thomas, 1968; McDonald & MacLeod, 1971). Whether or not potentials of the type described below for the dermal portion of the sweat duct are due to a sodium membrane (that is, a membrane selectively permeable to sodium) as suggested or to an electrogenic sodium pump remains to be determined at some future date. This highly complicated issue has been discussed by Wesson (1969) in an excellent review of membrane potentials in the kidney.

II. Skin Conductance Responses

The eccrine sweat glands are the most obvious contributors to both SCL and to SCRs. The importance of sweat glands for electrodermal activity was suggested initially by the observation that SCR frequency was greatest in areas where sweat glands were densest (Martin & Venables, 1966). Similarly, SCL was greater in areas having a high density of sweat glands, such as the palms and soles, in spite of the much thicker epidermis found in these regions, suggesting that the sweat glands provided a high conductance pathway through the poorly conducting epidermis (Edelberg, 1971).

More impressive evidence has come from reports that procedures which block sweat gland activity also eliminate EDRs, and that stimulation of the nerves innervating the sweat glands produces EDRs. No SCRs or SPRs (other than a slow positive wave) can be obtained in humans who suffer from a congenital absence of sweat glands (Richter, 1927; Wagner, 1952). Similar results have been obtained with peripheral nerve section or sympathetic ganglionectomy (Richter, 1927; Richter & Woodruff, 1941). This treatment eliminates sweat gland activity, reduces SCL, and abolishes both SCRs and SPRs. Parallel findings have been reported in the case of electrodermal activity in the cat footpad: elimination of sweat gland activity by cutting the plantar nerve reduced SCL and eliminated both SCRs and SPRs (Lloyd, 1960; Lloyd, 1961; Wilcott, 1965). Finally, in experiments with humans the iontophoretic application of atropine, a cholinergic blocking agent which prevents the excitation of sweat glands, eliminated SCRs and reduced SCL (Lader & Montagu, 1962; Venables & Martin, 1967a; Wilcott, 1964). SPRs are eliminated as well (Venables & Martin, 1967a). These studies, taken together, constitute impressive evidence for the importance of sweat glands in electrodermal measurements. They do not, however, provide precise information on the exact manner in which sweat glands affect electrodermal measurements, nor do they preclude a contribution to electrodermal measurements by factors other than the sweat glands.

A. Duct Filling and Hydration of the Corneum

The obvious way in which sweat glands may contribute to SC measurements is by providing a low resistance pathway through the epidermis (for example, Darrow, 1964; Kuno, 1956, pp. 13, 380; Lloyd, 1959; Lykken, 1968; Montagu & Coles, 1966). Thus, the sweat glands may be viewed as many resistors in parallel with each other and with the epidermal pathway. Because the cross-sectional area for each sweat gland is very small, it is to be expected that the conductivity of each sweat gland will be relatively small. On the palmar and plantar surfaces, however, the total conductivity of the sweat glands *when filled* is large because of the higher density of sweat glands in these areas. The natural corollary to this assumption is that variations in the extent of duct filling produce variations in skin conductance. SCRs would then reflect increased sweat in the ducts consequent upon sweat gland responses. If sweat is slowly reabsorbed from the ducts as appears to be the case (Gordon & Cage, 1966; Lloyd, 1960; Rothman, 1954, p. 190), the increased conductance of a duct-filling SCR would decline slowly, producing a slow recovery SCR. At the same time, there would be an increase in SCL, since slow recovery SCRs produce tonic rather than phasic increases in conductance.

Evidence consistent with this hypothesis has been obtained in experiments concerned with SR measurements and the latency of visible sweating in the cat (Lloyd, 1960). Starting with ducts filled with sweat as a result of stimulation of the plantar nerve, it was found that SRL increased slowly (to a maximum) over a 1–2 hr period. Similarly, reabsorption, as estimated by the latency of visible sweating (when the nerve was stimulated again), proceeded at a slow, constant rate over a 1–2 hr period. Thus the similarity in the time course of reabsorption and SRL changes supports the duct-filling hypothesis.

The temporal relationship between SCRs and visible sweating is also consistent with the duct-filling hypothesis. Since duct filling begins well before the sweat reaches the surface, SCRs should precede visible sweating if they are the result of duct filling. All available evidence suggests that SCRs begin approximately 1 sec prior to the appearance of sweat (Darrow, 1932; Darrow, 1964; Wilcott, 1962).

This analysis suggests that duct-filling SCRs could occur without any visible sweating if the amount of sweat secreted was enough to only partially fill the ducts. An association between slow-recovery SCRs and a lack of visible sweating, however, led Darrow (1964) to conclude that these responses must be of epidermal origin. Similarly, Wilcott (1967) concluded that monophasic negative SPRs and some SCRs must be of purely epidermal origin because they were not accompanied by visible sweating. Both Darrow and Wilcott could equally well have attributed these responses to duct filling which did not reach the surface, as suggested here and in Edelberg's (1968a) model.[2]

A second way in which the sweat glands may increase conductance is by an effect on the horny layer. When dry, the corneum is a poor conductor, but when wet, the resistance is much reduced (Edelberg, 1968b; Edelberg, 1971). Thus, the sweat will increase the conductance of the skin by hydrating the peritubular corneum. This point has been stressed by Adams (1966; Adams & Vaughan, 1965) in work demonstrating the "hydraulic capacitance" of the skin, a term which encompasses both duct filling and epidermal hydration.

Adams (1966) measured evaporative water loss (EWL) by passing dry air over the surface of the cat footpad and measuring the water absorbed by the dry air from the surface of the skin. Starting with the fresh preparation 25 min were required for the corneum to lose its water and for EWL to reach a minimal level. The hydraulic capacitance of the epidermis was indicated by the fact that there was always a latency for the beginning of increased EWL when starting with a completely unhydrated preparation and stimulating the plantar nerve at a constant frequency. The latency for the beginning of increased EWL was inversely related to frequency of stimulation, as would be expected with a fixed hydraulic capacitance and with the rate of secretion of sweat being proportional to the frequency of stimulation. In contrast, if the frequency of stimulation was increased after first producing a steady-state hydration level by stimulating at a constant frequency over a period of time, there was no latency for the appearance of increased EWL. These results can be explained by assuming that the corneum has a hydraulic capacitance which is unused at the stable-state minimal EWL levels and which must be filled before EWL is visible at the surface of the skin. Under conditions of steady-state EWL above this minimal level, on the other hand, the hydraulic capacitance has already been filled and any change in rate of sweat secretion is reflected in surface EWL measurements.

In experiments with humans Adams and Vaughan (1965) measured both EWL and SR simultaneously. Plots of SR versus time and of EWL versus time superimposed on each other clearly demonstrated substantial correspondence between the two measures. The relationship between SR and EWL depended on whether sweat gland activity was increasing or decreasing. When activity was increasing, both conductance and EWL increased

---

[2] Although the paper by Edelberg (1968a) is not generally available, it was extracted from Edelberg (1972), which is available. The 1968a paper is cited throughout this review because the 1972 paper was not available to the author at the time of writing.

together. When activity was decreasing, SR remained at a low level while EWL decreased. Adams and Vaughan suggested that water remaining in the corneum around the sweat gland tubules was responsible for maintaining the low SR. If reabsorption proceeds more slowly than EWL when the skin is exposed to dry air, on the other hand, it is possible that the effect of duct filling on SRL may account for the slower decline in SRL than in EWL.

Further support for the duct filling and hydration models comes from an experiment (Thomas & Korr, 1957) relating intrasubject variations in SCL measured with a dry electrode to counts of active sweat glands. Product-moment correlations (within subjects) between SCL and number of active glands ranged from 0.44 to 0.96 with a median of 0.91, providing strong support for the hypothesis that filled sweat glands provide parallel pathways through the less conductive epidermis. Estimates of the epidermal contribution to conductance[3] (that is, the conductance of the nonsudorific pathway) were higher when sweat gland activity was decreasing than when it was increasing, and both of these estimates were higher than actual values of total conductance of the skin under conditions of complete rest. These findings were interpreted as reflecting the greater conductance of the epidermis when hydrated by sweat than when completely dry. Thus both duct filling and hydration of the corneum contribute to changes in conductance.

The importance of epidermal hydration in the production of SCRs must be questioned for several reasons. Edelberg (1971) pointed out that the time required for diffusion of the sweat into the corneum is too long for hydration to affect the initial decrease in resistance of the SCR, although it may very well affect the recovery limb. Another consideration is that SCRs are obtained even when the finger is immersed in solutions which may produce complete hydration (Edelberg, 1968b). While it has been argued that even under these conditions hydration may not reach completion (Adams, 1966; Edelberg, 1968b), the case for an important contribution of sweat-produced hydration to SCRs at a well-hydrated site remains to be demonstrated. The conclusion to be reached at present is that hydration, whether from sweat or from an applied jelly, will increase SCL but is probably not a major contributor to SCRs.

Hydration has another effect which makes it important in electrodermal measurements (Fowles & Venables, 1970a). The swelling of the corneum associated with hydration is so great that it produces a simple mechanical obstruction of the duct as indicated by microscopic observations of the skin surface (Sarkany, Shuster, & Stammers, 1965). Several studies have demon-

[3] These estimates were based on an extrapolation to zero active sweat glands using the plot of SCL versus number of active sweat glands.

strated that this hydration-produced poral closure prevents sweat from reaching the surface of the skin and that this effect is obtained with salt solutions as well as with water (Peiss, Randall, & Hertzman, 1956; Randall & Peiss, 1957). The degree of inhibition of sweating was inversely related to the concentration of salt in the solution with a solution of 10% NaCl (approximately 1.7 M) seldom having an effect on sweating (Randall & Peiss, 1957). Sweat continues to be secreted at normal rates, however, suggesting that hydration results in increased reabsorption (Fowles & Venables, 1970a). Since poral closure prevents the ducts from filling completely, it is likely that hydration will reduce the amplitude of many SCRs, but this problem has not been investigated.

In summary, duct filling appears to play a major role in the production of SCRs and also produces changes in SCL. In view of the slowness of reabsorption, duct filling can account only for SCRs which have a slow recovery, as suggested by Edelberg (1968a). Hydration, on the other hand, contributes primarily to SCL, although it is possible that hydration reduces the amplitude of some duct filling SCRs as a result of poral closure.

B. Membrane Responses

Neither the duct filling nor the epidermal hydration model can account for membranelike effects in SCRs, and yet the evidence for such effects is unequivocal. In a classic series of experiments (Edelberg, Greiner, & Burch, 1960) SRRs and SPRs were altered by a number of treatments known to affect membranes. When 1.0 M solutions of different salts were applied at an anodal site, the amplitude of SRRs increased with increasing size of (hydrated) cations (that is, in the order KCl, NaCl, LiCl, NH_4Cl, $CaCl_2$, and $AlCl_3$). Comparisons of SRR amplitudes for each salt at an anodal versus cathodal site indicated that anodal sites, at which the cations are driven into the skin, much more effectively increased SRR amplitude in the case of large cations ($CaCl_2$ and $AlCl_3$). When a salt with a large anion (K_2SO_4) was used, larger responses were obtained at the cathodal site. These results may be understood in terms of an effect of ion size on the change in membrane permeability to the penetrating ions during a response. A simple explanation would be that prior to the response the membrane shows decreasing permeability as ion size increases but that during the response it is equally permeable to all ions tested. Since the permeability would be the same during the response, larger SCRs would be produced for those ions with lower permeability during the resting state. The effects of the various salt solutions were readily reversed by rinsing the site in tap water for 2–3 min, indicating that the membrane is readily accessible to solutions applied to the surface.

A number of methods known to have a disruptive effect on membranes were tried. These included anionic and cationic detergents, basic and acidic solutions, and high current densities. As expected, all of these had the effect of reducing both SRRs and SRL. The accessibility of the membrane to surface solutions was demonstrated by the effectiveness of the detergents regardless of polarity of the applied voltage. Because of this accessibility, the authors concluded that the membrane must be located in the epidermis. Similar results were said to have been obtained for SPRs, suggesting that they originate in the same membrane, but details of the SPR results were not included.

C. Location of the Membrane

In contrast to the conclusion by Edelberg et al. (1960) that the membrane responsible for the SCR is located in the epidermis, a number of reviewers have called attention to the possible contribution of the membrane in the secretory portion of the sweat gland (Darrow, 1932; Darrow & Gullickson, 1970; Lykken, 1968; Martin & Venables, 1966; Rothman, 1954). According to this view, the SCR is the result of an increased permeability of the secretory cells associated with the secretion of sweat. Rothman (1954, p. 18), for example, states the general principle that resting cells have a high polarizing capacity which decreases upon stimulation because of increased permeability of their membranes. The increased permeability would be visible as the SCR. It has also been suggested that the secretory membrane has a lumen positive standing potential and that the depolarization of this membrane produces the negative wave of the SPR (Edelberg, 1963; Martin & Venables, 1966), a view consistent with the close association between the latencies of the SCR and the negative wave of the SPR (Goadby & Goadby, 1936; Wilcott, 1958).

An objection to the secretory theory, however, is that the membrane appears to be located in the epidermis, as suggested above. It appears, moreover, that only a small fraction of the current passing through the skin actually reaches the secretory portion of the gland, making it unlikely that changes in permeability of the secretory membrane are important contributors to SCRs.

Edelberg (1971) has reviewed the evidence that the major pathway for current is down the duct and then through the duct wall at the level of the germinating layer or the dermis or both. Considerable penetration of the duct has been reported for thorium-X (Witten, Ross, Oshry, & Hyman, 1951) and radioactive phosphate (Witten, Brauer, Loevinger, & Holmstrom, 1956). Methylene blue has been observed to travel down the duct, passing through the duct wall into the germinating layer and the uppermost portion of the dermis (Papa & Kligman, 1966b). Suchi (1950) observed the passage of ferrous ions (ferrous sulfate) down the duct and laterally into the intercellular spaces of the stratum germinativum. Once in the stratum germinativum, the ions traveled parallel to the surface of the skin and then turned inward and approached the papillae. The demonstration that sodium is reabsorbed in the dermal portion of the duct (Dobson, 1962; Fowles & Venables, 1970a; Schulz et al., 1965) also suggests a permeability of that portion of the duct to ions. Finally, the rapid dissemination of intradermally injected ferritin in the interspaces all the way up to the level of the junction of the granular layer and the stratum corneum (Nordquist, Olson, & Everett, 1966) demonstrates that the intercellular spaces in the germinating layer are permeable to ions. On the basis of this evidence and the additional observation that secretory events in the cat are not accompanied by any potential change (Shaver, Brusilow, & Cooke, 1965), Edelberg (1968a, 1971) concluded that secretory events are not evident at the surface of the skin.

The secretory membrane may be viewed as a resistor in parallel with the duct wall but in series with the resistance along the length of the duct (Lykken, 1968). Because of the small cross-sectional area of the duct, this series resistance may be appreciable and thus minimize the importance of any possible decreases in resistance at the secretory membrane. In view of the permeability of the duct wall, it seems clear, therefore, that the major portion of the current flows through the duct wall rather than the secretory wall. If the current is strong enough, however, the secretory tube may be stained with methylene blue (Kuno, 1956, pp. 354–355); thus, a small portion of the current does reach the secretory portion of the gland. Although it is not possible to exclude a small contribution of the secretory membrane to electrodermal measurements, the evidence points to another membrane as being more important. It appears, therefore, that Edelberg's (1971) conclusion is valid and that one must look elsewhere for the membrane.

As noted in the discussion of the work of Edelberg, Greiner, and Burch (1960), there were several indications that the membrane demonstrated there was accessible to substances applied to the surface of the skin even in the absence of applied voltages. Because of this and evidence from SP measurements, many authors (Darrow, 1964; Edelberg, 1963; Edelberg, 1968b; Lykken, 1968; Rothman, 1954; Wilcott, 1967) have hypothesized the existence of an epidermal membrane. The location of this membrane has varied, as has the mechanism by which it is depolarized. It has most often been located, however, at the boundary between the living and cornified layers of the epidermis (Lykken, 1968; Martin & Venables, 1966; Rothman, 1954) and several authors have suggested an innervation distinct from the nerves innervating sweat glands (Darrow, 1964; Edelberg, 1968a; Edelberg, 1971; Wilcott, 1967). Since all electrodermal responses are eliminated by cholinergic blocking agents such as atropine, the diffusing mediator would have to be cholinergic (Martin & Venables, 1966; Wilcott, 1967).

The purely epidermal membrane theory encounters some difficulty in explaining the absence of SCRs when the sympathetic nerves innervating the sweat glands are severed and when sweat glands are congenitally absent. There has been, furthermore, no unambiguous identification of nerves which, when stimulated, produce membrane responses but not sweat gland activity. In view of the close coupling between SCRs and sweat gland activity, it would also be necessary to postulate a close association between the central mechanisms activating the two pathways. A simpler approach would make the membrane response in some manner secondary to sweat gland activity. Lykken (1968), for example, suggested that bradykinin produced in association with sweat gland activity produces the epidermal membrane response. A second mechanism by which the membrane response could be secondary to sweat gland activity would be for the sweat itself to activate the membrane, either by increased hydrostatic pressure or by increased NaCl concentration as suggested below. While these arguments do not demolish the epidermal membrane theory, they do point to some of its weaknesses and to the desirability of developing more adequate models of the membrane response.

Edelberg's (1968a, 1971) most recent version of his epidermal membrane theory, on the other hand, has a number of attractive features. After considering the possibility that the epidermal membrane could be in the germinating layer or at the dermoepidermal boundary (possibly the basal cell layer), he suggested that the most attractive possibility is that it is located at the duct wall at the level of the germinating layer. This theory is consistent with the idea that the major pathway for current is down the duct to the germinating layer and then laterally through the duct wall. Increased permeability of the duct wall at this point would certainly be visible as an SCR. The basal cells of the germinating layer were also considered as a possible location of a membrane response but were excluded because they would not be as accessible from the surface of the skin and because SPRs recorded from the epidermal duct in the cat (Shaver, Brusilow, & Cooke, 1965) supported the idea of a membrane response in the duct wall.

Other evidence supports the location of the membrane in the germinating layer and suggests that it is involved in a water reabsorption reflex. Using a number of techniques to record the vapor pressure above the skin and the conductivity of the corneum, Edelberg (1966) demonstrated the existence of a water reabsorption reflex by which sweat is reabsorbed from the skin surface. Because the reabsorption reflex was always associated with an SCR, it seems likely that a major function of the membrane response is the reabsorption of water. The water reabsorption reflex, like the SPRs recorded in the epidermal duct, is suggestive of a living membrane. Since the cells of the corneum are dead and those of the stratum granulosum are transitional, it is unlikely that the membrane could be located in either of these structures (Edelberg, 1971).

Both the assumption of an epidermal membrane responsible for SCRs and the water reabsorption reflex and the location of this membrane in the epidermal duct wall at the level of the germinating layer appear to be well founded. For convenience, the term "epidermal duct" will be used in order to distinguish between this membrane and the dermal duct. It should be stressed, however, that the reference is only to that portion of the epidermal duct in contact with the stratum germinativum.

If one accepts Edelberg's hypothesis that the epidermal duct wall membrane is concerned with the reabsorption of sweat but rejects the hypothesis of neural control of this membrane, activation of the membrane could be accomplished by some aspect of sweat gland functioning. The nature of this reabsorption reflex is suggested by a comparison with the dermal portion of the duct (Fowles & Venables, 1970a). Because of the large numbers of sweat glands in humans, regulation of the salt content of sweat is essential and has been accomplished by the selective reabsorption of NaCl from the sweat as it passes through the dermal duct (Gordon & Cage, 1966). The duct at this point is not very permeable to water, as is necessary to accomplish the reabsorption of NaCl without water. The reabsorption mechanism has a limited capacity which cannot keep up with increased rates of sweating and thus the NaCl content of sweat reaching the surface rises with increased rates of secretion.

Assuming, on the basis of Edelberg's (1966) demonstration of a water reabsorption reflex, that the epidermal portion of the duct is concerned with the reabsorption of water as well as of salts (at least of NaCl), this portion of the duct would be more permeable to water and thus more conductive. It would also have a lower potential because the membrane would be less selective. A similar differentiation of function is found in the kidney tubules in which the proximal tubule reabsorbs NaCl and water while the distal tubule (in the absence of ADH) reabsorbs NaCl but not water (Guyton, 1966, Chapter 33; Koch, 1965; Wesson, 1969). In terms of this parallel the dermal duct resembles the distal tubule of the kidney, while the epidermal duct resembles the proximal tubule. Electrical potentials average 50–60 mV in the distal tubule but may be as high as 120 mV when sodium reabsorption is rapid; in the proximal tubule potentials may rise only slightly above the average of 20 mV (Guyton, 1966, pp. 477–478).

Although many investigators have assumed that sweat is reabsorbed elsewhere, for example, near the base of the gland (Lloyd, 1962), the epidermal duct appears to be a more likely site. Reabsorption at the germinating layer would place the regulation of sweat at the layer of living cells closest to the corneum and thus near the area to be served by the sweat. Support for this hypothesis comes from the histologic examination (following thermal stress) of the sweat glands of patients or experimental subjects suffering from anidrosis, a lack of surface sweating which is produced by an obstruction of the sweat duct in the corneum (Papa & Kligman, 1966a; Shelley & Horvath, 1950). Under these conditions the intraepidermal portion of the duct was routinely widely dilated, while the deeper dermal portion of the duct was only occasionally dilated. The accumulation of sweat in the epidermal portion of the duct suggests that the epidermal duct is more elastic than the dermal duct, providing a reservoir for excess sweat. Thus, a reabsorption mechanism might well be located there. The additional finding that materials believed to have been derived from the secretory cells had formed a granular cast in the epidermal duct (Papa & Kligman, 1966a) suggests that sweat had been flowing from the secretory portion to the epidermal duct where water and NaCl were reabsorbed, leaving behind the substances forming the granular cast.

The most likely mechanism by which water and NaCl are reabsorbed through the duct in the germinating layer is by the active transport of sodium. Direct transport of water has never been demonstrated and thus water reabsorption is secondary to ion transport mechanisms (Cage & Dobson, 1965). Evidence from microelectrodes placed in the duct of the cat footpad (Shaver *et al.*, 1962; Shaver *et al.*, 1965) indicates that only monophasic negative SPRs are found in the epidermal portion of the duct. Since lumen negative potentials may be expected when cations are being transported (the anions following passively down an electrochemical gradient), sodium rather than chloride must be the ion transported. Again, this suggestion is consistent with present theories concerning the mechanism by which water and NaCl are absorbed in the proximal tubule of the kidney (Guyton, 1966, Chapter 33; Koch, 1965). Suchi's (1950) description of the passage of ferrous ions from the duct into the luminal cells and thence into the intercellular spaces (without passing through the basal cells of the duct wall) is also consistent with the pathway found in frog skin (Ussing, 1960; Ussing, 1965), another epithelial tissue with an active sodium transport mechanism. If these assumptions are correct, the mechanisms involved in the epidermal duct would resemble those suggested for the dermal portion of the duct (Fowles & Venables, 1970a) except that the epidermal duct would be less selective and more permeable. This greater permeability would make it the preferred pathway for current, and variations in the permeability of the membrane would produce SCRs.

One possible stimulus which may serve to trigger the response of the epidermal duct is sodium concentration. Such a mechanism would be adaptive since sodium concentration rises with increasing sweat rates, activating the sweat reabsorption mechanism when it is most needed. Thus the epidermal duct would prevent excessive water loss and would serve as a second line of defense against loss of NaCl (being activated when the dermal duct cannot keep up). Such an increased permeability with higher NaCl concentrations has been reported in frog skin (Koefoed-Johnsen & Ussing, 1958; Steinbach, 1937). In the cat footpad, however, in which sweat is isotonic, the epidermal membrane would be activated regardless of the rate of sweating as long as there was sweat to be reabsorbed.

A second stimulus which may trigger the water reabsorption response is an increase in luminal hydrostatic pressure. Nutbourne (1968, 1969) demonstrated that sodium transport in frog skin varied with small changes in hydrostatic pressure gradients between the inside and the outside of the skin. Sodium transport increased when the pressure was higher on the outside and decreased when the gradient was in the opposite direction. Very large changes in sodium transport were obtained with pressure differences of 5 mm H_2O and even transient pressure gradients as low as 0.1–0.5 mm H_2O had "a pronounced effect" on sodium transport. Comparing the outside of the frog skin with the lumen of the proximal tubule of the kidney, Nutbourne suggested that the well-established increase in sodium reabsorption with increased luminal volume may be attributed to increased luminal pressure. Assuming that the same mechanism is operative in sweat glands, the increased luminal pressure associated with the rapid influx of sweat during a sweat gland response would activate the sodium transport mechanism, initiating a phase of increased permeability.

The adaptability of a pressure-sensitive reabsorption mechanism is obvious when the sweat pores are blocked, preventing the sweat from reaching the surface. Since sweat may exert a pressure of up to 500 mm Hg (Schulz, 1969), it is unlikely that the increased pressure which would result from poral closure would prevent sweating. Papa and Kligman (1966a) found that secretory cells showed changes characteristic of sweating during heat stimulation even though surface sweating was prevented. In addition, there is no evidence of a damming up of sweat in the ducts when surface sweating is prevented by pressure applied to the skin (Shuster, 1963) or when an occlusive dressing is painted onto the skin (Fox & Hilton, 1958). Both of these findings can be explained by assuming that reabsorption increases when surface sweating is prevented. The frequency with which blockage of the pores occurs naturally is indicated by the recommendation that when studying the sweat glands of the cat footpad one should use only younger cats because many of the glands are blocked in older cats (Munger & Brusilow, 1961) and by the conclusion that blockage occurs in many people during the summer (Shelley & Horvath, 1950). Immersing the skin in water, moreover, blocks the pores because the corneum swells with hydration, creating a mechanical obstruction to the flow of sweat to the surface (Fowles & Venables, 1970a). Thus a mechanism which would increase reabsorption when pressure increased in the lumen would be essential. Finally, SCRs and positive SPRs may be elicited by applying pressure to the skin or by stretching it (Edelberg, 1971), suggesting that the membrane responsible for SCRs is pressure sensitive.[4]

D. Identification of Duct Filling versus Membrane SCRs

In view of the strong evidence for both a duct filling and a membrane contribution to SCRs, one must conclude that both play an important role in producing SCRs. Edelberg (1968a) suggested that the contribution of these two components may be identified in terms of the recovery limb of the SCRs and the nature of simultaneously recorded SPRs. Specifically, the duct filling component produces SCRs and monophasic negative SPRs, both of which have slow recovery limbs, while the membrane component produces rapid-recovery SCRs and either diphasic SPRs or monophasic negative SPRs with a rapid recovery. Although not emphasized by Edelberg (1968a), it is reasonable to include among the membrane SPRs the monophasic positive SPRs sometimes reported (for example, Darrow, 1964; Fowles & Rosenberry, 1970; Lykken, Miller, & Strahan, 1968).

Although it is not possible to test this hypothesis directly, it is possible to evaluate it indirectly. Once the ducts are filled, there will be a minimal contribution from duct filling, but the membrane response should be readily apparent. According to the model, therefore, rapid-recovery SCRs and diphasic or positive SPRs should be associated with periods of considerable sweat gland activity as indicated by either surface sweating or high (intrasubject) SCL or both. Several reports have supported this prediction. Darrow (1932, 1964) found a strong association between rapid recovery SCRs, high initial SCLs, surface sweating, and the positive component of the SPR. Similarly, Edelberg (1970) found that SCRs accompanied by positive SPRs showed a more rapid recovery than those accompanied by monophasic negative SPRs. Several experimenters have found

---

[4] Edelberg (1972) has independently modified his earlier model (1968a) to include stretching of the duct wall by sweat pressure as one of the possible stimuli responsible for the membrane response.

that positive SPRs are produced by more arousing stimuli (Darrow, 1964; Fowles & Rosenberry, 1970; Wilcott, 1958). Although complete information is not provided, another study (Lykken et al., 1968) appears to have found rapid recovery SCRs and SPRs with a large positive wave under conditions of at least partially filled ducts. Finally, Edelberg (1966) found that the reabsorption reflex, as indicated by decreases in surface moisture, was strongly associated with the occurrence of positive SPRs and that the two responses had a similar latency measured relative to the onset of the SCR. SCRs associated with the reabsorption reflex showed a more rapid recovery than those not accompanied by reabsorption (Edelberg, 1970). Like positive SPRs, rapid recovery SCRs (Edelberg, 1970) and the reabsorption reflex (Edelberg, 1966) are produced by more arousing stimuli.

Under resting conditions, on the other hand, the ducts should be relatively empty and duct filling responses should be prominent. According to Edelberg's model, therefore, one should find slow recovery SCRs and monophasic negative SPRs during rest periods and relatively mild stimulation. Here again the results are consistent with the prediction. Darrow (1932, 1964) found that under resting conditions with low SCL and no surface sweating the SCRs were of the slow recovery type and the SPRs were monophasic negative. Slow recovery SCRs have been reported during rest periods (Edelberg, 1970) and in response to a loud tone presented to sleeping subjects (Lykken et al., 1968).

The evidence to date is consistent with Edelberg's identification of two types of SCR and their relation to the two effectors in his model. Although many electrodermal responses will reflect both components, it is possible to obtain nearly pure instances of each, thus making further investigation of these two effectors relatively easy. Investigators utilizing electrodermal responses for psychological experiments may wish to attempt to work with one or the other type of SCR, or may find the use of measures of recovery time more useful than measures of amplitude. Since the basic shape of the recovery limb is approximated by an exponential decay, it can be quantified in terms of the time constant of the decay (Edelberg, 1970). The time constant may be estimated by manual curve fitting requiring about 7 sec per response (Edelberg, 1967) or it may be automated (Edelberg, personal communication).

III. Skin Conductance Level

As a first approximation there are two paths by which current may pass through the epidermis. The first of these is a nonsudorific pathway in which the current would pass directly through the epidermis. It would include the resistance of both the stratum corneum and the barrier membrane. When the stratum corneum is dry and thus not very conductive, as in the Thomas and Korr (1957) experiment, the contribution of this exclusively epidermal pathway will be small. When the corneum is hydrated and thus more conductive, one might expect the contribution of the epidermal pathway to be substantial. In keeping with this expectation, Edelberg (1971) found that the sweat glands when full can account for less than 50% of the total conductance of a hydrated site. A more accurate estimate of the contribution of the epidermal pathway may be obtained by comparing SC measurements during sweat gland activity with those obtained under conditions of complete rest, when it may be assumed that the ducts are empty and that the sweat gland contribution is negligible. Such data are not systematically available, but Christie and Venables (1971b) present results for a "typical" subject showing a minimum SCL of approximately 25 μmho/cm² with a peak value of 90 μmho/cm². If these results are typical, then the contribution of the epidermal pathway would be around 28%. In this study the subjects were well habituated and there was no attempt to arouse them. Thus, it is quite likely that the purely epidermal pathway contributes even less than 28% of the total conductance when the ducts are filled. The second pathway, in parallel with the first, is the sweat gland pathway. When the ducts are filled, the current will pass down the duct and laterally through the duct wall, probably at both the germinating layer and in the dermis.

A combination of these pathways is also likely (Edelberg, 1971). When the corneum is hydrated, the current may pass laterally from the lower portion of the corneum into the sweat gland duct and then follow the sweat gland pathway into the corium. When the ducts are incompletely filled, this combined pathway would bypass the unfilled upper portion of the duct as well as the poorly conductive barrier layer of the epidermal pathway. The importance of this pathway may also be seen by considering what must happen when electrodermal measurements are obtained with hypotonic solutions or jellies. Such solutions will produce hydration of the corneum and an attendant swelling which mechanically blocks the sweat gland pores. This poral closure precludes the possibility of complete duct filling and forces all current to pass through at least a portion of the corneum. A substantial portion of the current must then pass through the upper duct wall, thereafter following the usual sweat gland pathway.

It is clear from this analysis that intrasubject variations in SCL may be produced by duct filling, corneal hydration, and changes in the permeability of the duct wall. In the study by Edelberg et al. (1960) there was only a small effect on SCL when the polarity of the applied current was reversed, and cation size appeared to have little effect except in the case of aluminum chloride. Both of these findings indicate that the membrane contributes only a small fraction of the total resistance and that under normal conditions the permeability of the epidermal duct is not a major factor in SCL. If duct filling is to be a major variable, as appears to be the case, it is likely that variations in the salt content and thus in the conductivity of sweat will also affect conductance (Kuno, 1956, pp. 379–380). The major factor influencing the salt content is the rate of sweating, NaCl content increasing as sweat rate increases (Cage & Dobson, 1965). Conversely, the decline in SCL following duct filling will involve a decline in NaCl concentration (because of reabsorption by the dermal duct) as well as a decrease in duct filling as a result of reabsorption of sweat by the epidermal duct.

In addition to the factors influencing intrasubject changes in SCL, when comparisons are made across subjects SCL will be affected by variations in the density and power of sweat glands and differences in the conductivity of the epidermis. Since most investigators are not interested in these factors, it is desirable to make corrections for them. While this is not completely possible, Lykken's (1968) suggestion that range-corrected scores be used for this purpose represents a major advance in this direction. In this technique the minimum SC is determined during a prolonged rest period and the maximum during a stressful task. At any other point in the experiment the subject's arousal is quantified in terms of his position relative to his own range of SC measurements. For example, if a subject showed a minimum of 2 μmho and a maximum of 38 μmho, an SCL of 29 μmho would be scored as $(29 - 2)/(38 - 2) = 0.75$ or 75% of his arousal range. This technique has been used with considerable success in studies comparing electrodermal measurements with two flash fusion thresholds, a perceptual measure of arousal (Lykken, Rose, Luther, & Maley, 1966). The correlations between these two measures are remarkable in view of the fact that one is a perceptual measure and one a psychophysiological measure. This approach might well be extended to other psychophysiological systems.

IV. Skin Potential Responses

A. *The Monophasic Negative SPR*

The negative component of the SPR is eliminated by the application of atropine or by sympathectomy; it precedes visible sweating by approximately 1 sec, and is associated with surface sweating (Wilcott, 1962). Other experiments have reported that the latency of the negative SPR and the SRR are the same (Goadby & Goadby, 1936; Wilcott, 1958). On the basis of this evidence, a number of authors have concluded that the negative

SPR has a physiological basis in sweat gland activity. In one version of this theory, the negative SPR is attributed to a secretory potential produced in the coiled portion of the sweat glands (Edelberg, 1963; Lloyd, 1961; Martin & Venables, 1966; Trehub, Tucker, & Cazavelan, 1962). Since it is generally assumed that membranes depolarize when stimulated, the negative response implied a surface positive standing potential across the secretory membrane which depolarized upon stimulation by the cholinergic transmitter (Edelberg, 1963; Martin & Venables, 1966). This theory may be rejected for the same reasons the secretory membrane theory of the SCR was rejected (see Section II,C).

Darrow (1964; Darrow & Gullickson, 1970) attributed the negative SPR to increased permeability of the epidermis, which produces the SPR by virtue of improved contact with the negatively active outer tissue of the sweat gland. It is difficult to see how such a mechanism could produce a negative wave. The reference electrode at an abraded site is effectively in contact with the outer tissue of the sweat gland. Even large changes in epidermal resistance could not be expected to produce changes because of improved contact with the outside of the gland.

Edelberg (1968b) offered an alternative to the membrane depolarization explanation of the negative SPR. He hypothesized four components of the electrodermal effector system as indicated in Fig. 9-4: A surface-negative sweat gland potential (S), a surface-negative epidermal membrane potential (E), the resistance in the sweat gland duct (R_s) between the sweat gland potential and the surface electrode, and the resistance of the epidermis (R_e) between the epidermal membrane and the surface electrode. S is assumed to have a larger magnitude than E. It can be shown that the po-

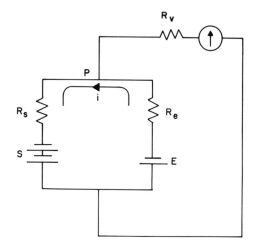

FIG. 9-4. Circuit model for origin of transcutaneous potential in human skin as suggested by Edelberg (1968b), where S and E are potentials across the sweat gland membrane, and the epidermal membrane, respectively; R_s and R_e are combined internal and series resistance of the sweat gland and epidermis, respectively; P is the transcultaneous potential; R_v is the input impedance of the voltmeter.)

tential measured across the skin will be affected by the relative value of the resistors R_s and R_e as well as by the size of the generator potentials, becoming more negative as R_s decreases relative to R_e. With the further assumption that R_s decreases with duct filling, it is possible to relate the negative component of the SPR to duct filling rather than to depolarization of a positive membrane. Thus if the sweat gland membrane is lumen negative as in Edelberg's (1968a) model, the negative SPR could be attributed to duct filling rather than to a membrane response. This duct filling SPR is the slow recovery negative wave discussed above (see Section II,D) in connection with the duct filling SCR in Edelberg's (1968a) model.

A second alternative to the membrane depolarization hypothesis and a possible source of the negative sweat gland potential in Edelberg's model has been suggested by Lykken (1968) and by Fowles and Venables (1970a). These authors cited evidence that there is a "sodium pump" in the sweat duct wall which transports sodium from the sweat back into the body, thereby forming a sweat with low salt concentration and reducing the amount of salt lost in sweating. Such a mechanism is known to produce potentials which would make the lumen of the sweat duct negative relative to the interstitial fluid and would be of the same order as the negative potentials reported in humans. In studies with human sweat glands, moreover, Schulz and her colleagues (Schulz *et al.*, 1965) provided convincing evidence that sodium reabsorption in the duct is intricately associated with the sweat gland potential. Potentials of approximately 50 mV negative relative to interstitial fluid were found in the subepidermal portion of the duct and decreased as the electrode was pulled out of the duct through the epidermis. Since the potentials were obtained in the duct even when the secretory coil and the lower part of the duct were filled with oil, the possibility that the potential comes from the secretory coil is ruled out. Subcutaneous injections of g-strophanthin, an inhibitor of the sodium pump, reduced the surface potential and increased the concentration of NaCl in the surface sweat, supporting the hypothesis that the large negative potentials found in humans are related to sodium reabsorption. Other experiments indicated that samples of sweat taken from the dermal duct at the point at which it enters the epidermis had NaCl concentrations similar to sweat on the surface of the skin. Thus it seems that all or most of the sodium reabsorption process has been completed by the time the duct enters the epidermis, suggesting that the potential is generated in the dermal portion of the duct.

Arguing by analogy from other epithelial tissues such as frog skin, toad bladders, and salivary glands, all of which resemble sweat glands in that sodium is actively transported across the multicellular membrane, Fowles and Venables (1970a) suggested that the inner border of the surface cells (in contact with the lumen) of the sweat gland duct are permeable to both sodium and chloride ions but much more so to sodium and thus act as a sodium electrode. An active sodium pump transports sodium out of these surface cells, maintaining low sodium and high potassium inside the cells. Thus the sodium diffuses into the surface cells, generating a potential which is a function of the relative concentrations in the duct and in the surface cells. Since the sodium concentration is always higher (or the same) in the duct, the potential across the duct wall is lumen negative. There will be a greater concentration gradient across the duct wall at high rates of sweating than at low rates of sweating because of the higher lumen concentration of NaCl and this will produce larger negative potentials. As the rate of sweating declines (or stops), the sodium reabsorption mechanism would slowly lower the concentration of NaCl in the sweat with a consequent slow decline in the potential across the duct wall. Thus this explanation can account for most features of negative SPRs which have a slow recovery limb. The initial increase in negativity results from the increase in lumen sodium while the slow recovery limb reflects the gradual return to resting sodium concentrations. It seems likely, however, that these changes operate in conjunction with duct filling as suggested in Edelberg's model. The slow-recovery negative SPR, therefore, probably reflects changes in both the generator potential and duct filling.

The sodium reabsorption through the dermal duct wall demonstrated in human sweat glands is not found in the cat. The sweat from the foot and toe pads is isotonic and histological examination reveals that the dermal portion of the duct is shortened and lacks other specialized features believed to contribute to sodium reabsorption (Munger, 1961; Munger & Brusilow, 1961). Thus the negative SPRs recorded in the cat cannot be attributed to the mechanism just described. A comparison of electrodermal activity in the cat versus the human, however, provides a natural experiment which clarifies the effects of the dermal and epidermal portions of the duct.

Using microelectrodes placed in the sweat gland duct, Shaver *et al.* (1962) were able to obtain only monophasic negative SPRs from the cat. This negative SPR could be measured in the epidermal portion of the duct but not in the dermal portion (Shaver *et al.*, 1965). The only difference between the typical SPRs recorded with the usual surface electrodes and those obtained with microelectrodes in the duct was in the recovery limb. A slow positive epidermal potential observed in the same study appeared to combine with the fast negative sweat gland potential to produce a more rapid recovery (and at times a positive overshoot) for the responses recorded with a macroelectrode than for those recorded from the duct itself. It seems clear, therefore, that the sweat glands in the cat produce only monophasic negative responses and that these are produced somewhere in the epidermal portion of the duct.

All examples of these negative SPRs in the cat showed a rapid recovery limb and thus may be taken as examples of Edelberg's (1968a) membrane SPR. Along with the membrane SCR then, the negative SPR in the cat may be attributed to the reabsorption of water and NaCl in the epidermal duct by means of a sodium transport mechanism. In contrast to SPRs in humans, the membrane SPRs in the cat do not have a rapid positive wave (Shaver *et al.,* 1962; Shaver *et al.,* 1965). The positive SPR, therefore, appears to be dependent in some way on sodium reabsorption in the dermal duct in humans.

B. The Diphasic or Positive SPR

The positive component of the SPR is eliminated by the application of atropine or by sympathectomy, has a latency indistinguishable from that of the water reabsorption reflex (Edelberg, 1966) or of visible sweating (Wilcott, 1962), and is even more closely related to surface sweating than is the negative SPR (Wilcott, 1962). Most theories have attributed the positive SPR to depolarization of a membrane with a consequent reduction of the generator potential, either in the secretory coil (Darrow, 1964; Darrow & Guillickson, 1970) or in the epidermis (Edelberg, 1963; Martin & Venables, 1966; Wilcott, 1967).

As in the case of SCRs and negative SPRs, the secretory membrane theory may be eliminated because it is unlikely that secretory potentials would be visible at the surface of the skin. Placing the depolarizing membrane in the epidermis is consistent with the evidence from many sources that the epidermis produces a surface-negative potential under the usual conditions of recording (Martin & Venables, 1966; Rothman, 1954), since depolarization of a negative potential would produce a positive-going wave. The location of this membrane is problematical, however. When the base potential is increased by increasing the concentration of the applied electrolyte, there is no effect on positive SPRs (Edelberg, 1963; Fowles & Johnson, in press) although an increased amplitude would be expected if the same membrane were involved in both basal potential and positive SPRs. Moreover, this theory faces the same difficulties in finding a mechanism by which depolarization is effected as does the purely epidermal theory of the membrane-produced SCR—that is, the complications attendant upon postulating a separate neural pathway (see Section II,C).

A second approach to understanding the positive SPR attributes it to a change in resistance rather than to a decrease in the generator potential. Edelberg (1968b) argued that just as the negative wave could be produced by duct filling (see Section IV,A) positive waves could be produced by decreased resistance in the epidermal pathway (R_e in Fig. 9-4) as a consequence of hydration of the corneum. Thus, duct filling would at first produce a negative wave by decreasing the resistance along the duct and then would produce a positive wave as the sweat diffused into the corneum and lowered corneal resistance. The significance of this factor when measurements are obtained with a wet electrode is unclear since much of the corneum would be hydrated by the electrolyte. Edelberg (1968b), however, argues that hydration is not likely to reach completion during the hour required for most experiments. More recently, Edelberg (1971) noted that while this phenomenon might affect the recovery limb of SPRs, the diffusion process would be too slow to produce the characteristic positive components of the diphasic response.

The close association between the positive SPR and both the water reabsorption reflex and the rapid recovery SCR suggest a common basis for all three and thus point to the epidermal duct wall as the probable location of the positive SPR (Edelberg, 1968a; Edelberg, 1971). In this more recent version of Edelberg's epidermal membrane hypothesis the epidermal duct is viewed as a cation-permeable membrane which, in response to acetylcholine shows first an increased permeability to cations and then an increased permeability to anions and thus a loss of selectivity. The initial increase in cation permeability is short lived and produces a brief negative SPR. The loss of selectivity (that is, increased permeability to anions) depolarizes the membrane, causing the negativity to decrease and thereby producing a positive wave. Since both phases involve increased permeability, both would contribute to the SCR and to water reabsorption. Thus the SCR would be monophasic while the SPR would be diphasic with an initial negative component.

In addition to the simplicity of making the positive wave an integral part of the membrane-produced SCR, this theory has the advantage of being able to explain the linkage between sweat gland activity and the membrane responses: if the ducts are not filled, changes in the duct wall permeability would have little effect because most of the current would be flowing directly through the epidermis. The theory does not, on the other hand, explain why positive waves have not been recorded from the epidermal duct of the cat (Shaver *et al.*, 1962; Shaver *et al.*, 1965) and only rarely from surface electrodes (Lykken, 1968; Wilcott, 1967).

An alternative explanation for the positive SPR in terms of a shunting of the *dermal* duct potential by the increased permeability of the epidermal duct would explain the absence of SPRs in the cat, since the cat has no dermal duct potential. The dual assumptions that a larger negative potential is produced at the dermal duct than at the epidermal duct and that during the membrane SCR the epidermal duct shows increased permeability make it highly likely that the membrane SCR would produce a positive wave. The argument here is identical to that of Edelberg's (1968b) model presented in Fig. 9-4 except that R_e, rather than representing the transverse resistance of the corneum, now represents the resistance of the epidermal duct wall plus the pathway through the germinating layer and thence to the corium, while E represents the less negative epidermal duct potential seen in cats. It is clear from the previous discussion of this circuit that the positive response could be produced by increased permeability of the epidermal duct without any change in the potential at that site. It is unlikely that the epidermal duct potential does necessarily change, moreover, since a sustained negativity can be produced in the cat by stimulation at the rate of twice per second (Lloyd, 1961) and since potentials in the kidney and other epithelial tissues are tonic rather than phasic as long as crucial parameters such as concentration and pressure remain constant.

Both the potential and the conductance at the epidermal duct wall will change, however, as a function of stimulation by the sweat. As suggested above, it is likely that increased hydrostatic pressure or increased sodium concentration as a result of sweating will activate the reabsorption reflex. The occurrence of positive SPRs, then, would be a function of rate of sweating as well as of the relative magnitudes of the dermal and epidermal duct potentials.

A possible explanation of the initial sharp negative wave of the diphasic SPR is suggested by Wilcott's (1965) finding that in cats the negative SPR begins approximately .2 sec before the SCR. The latencies were based only on responses having a sharp onset and were obtained mostly from waking animals. Thus, the SCRs probably were membrane SCRs, as is also suggested by the finding that large negative SPRs can only be obtained from cats when there has been a considerable degree of duct filling (Lloyd, 1961). If Wilcott's comparison was based on membrane SCRs, his results indicate that the conductance change (that is, the SCR) does not begin until .2 sec after the onset of the negative wave. The initial negative component of many diphasic responses in humans, then, could represent the first .2 sec of the membrane-produced negative SPR in the cat. After this delay the positive wave would develop simultaneously with the membrane SCR because of the shunting effect of the SCR on the dermal duct potential. The amplitude of the initial negative component in humans would be considerably smaller than in the cat because of the dominance of the dermal duct potential.

V. Skin Potential Level

A. *The Sweat Gland Component*

The slow-recovery negative SPRs produce a change in SPL which may be attributed to the same mechanism which produces the SPRs. Thus,

SPLs above a complete resting level may be attributed to the sodium reabsorption mechanism in the dermal duct (Fowles & Venables, 1970a). Support for this conclusion again comes from a comparison with recordings from the cat. Although most investigators do not report SPL when discussing electrodermal activity in the cat, Wilcott (1965) did record SPLs and his reference to an SPL of -8 mV as a "relatively high basal potential" is in striking contrast to the much higher SPLs found in humans during sweat gland activity. For example, using a polyethylene glycol solution to prevent epidermal hydration Fowles and Rosenberry (1970) found an average SPL of 71 mV during periods of sweat gland activity. Similarly, inducing sweat gland activity in humans can raise the potentials as much as 63 mV above resting SPL with the glycol solution (Fowles & Venables, 1970b), whereas in the cat sweat gland activity produces a decrease in SPL (Lloyd, 1961). These comparisons with the cat plus the general finding of large potentials in connection with the selective reabsorption of NaCl across epithelial tissue (Fowles & Venables, 1970a) make it clear that the higher SPLs produced by sweat gland activity in humans can be attributed to a potential across the dermal duct.

B. The Epidermal Component

When all sweat gland activity is blocked by atropine, there is still a sizable potential across the palm (Venables & Martin, 1967a), and a similar potential is obtained during prolonged rest periods in which the absence of sweat gland activity is indicated by an absence of electrodermal responses. There are many indications that the skin behaves like a membrane with a fixed negative charge which makes it more permeable to cations than to anions (Edelberg, 1971; Martin & Venables, 1966; Rothman, 1954). The evidence for this conclusion includes the results of studies of electroosmosis in which water moved through isolated skin or epidermis (used as a diaphragm) in the direction expected for a negatively charged membrane (Rothman, 1954), findings of larger surface negative potentials with higher concentrations of applied sodium or potassium chloride (Christie & Venables, 1971c; Edelberg, 1963; Floyd & Keele, 1937; Shimizu, Tajimi, Watanabe, & Niimi, 1969; Venables & Sayer, 1963), and the greater permeability of the skin to basic dyes (having a positive charge) than to acid dyes (having a negative charge) (Rothman, 1954).[5] Since the outer corneum is highly permeable to salt solutions, it is probably too porous to produce these results (Edelberg, 1971; Rothman, 1954). The boundary between the

[5] The failure to find an effect on SP of variation in NaCl (Niimi, Yamazaki, & Watanabe, 1968) has since been attributed to "methodological defects" by the same authors (Shimizu et al., 1969).

cornified and noncornified epidermis is not very permeable, on the other hand, and thus is the most likely site for this purely epidermal membrane (Edelberg, 1971; Martin & Venables, 1966; Rothman, 1954). This boundary includes the stratum compactum and, in the palms and the soles, the stratum lucidum. That is, it is the structure suggested as the barrier layer of the epidermis (see Section I,A).

Christie and Venables have recently reported a series of experiments which provide important additional information about this nonsudorific potential. In these experiments subjects, often well habituated to the laboratory, either lie or sit quietly for periods of up to 45 min until SPL changes from the usual slow drift in a positive direction to a slow drift in the negative direction (Christie & Venables, 1971b). Simultaneous measurements of SCL show a continued decline and thus preclude the possibility that the negative drift is caused by sweat gland activity. A reading is taken of the potential at the point at which the drift changes (that is, the minimum value of SPL) and this is designated as the basal skin potential level (BSPL). Christie and Venables (1971a) reported a correlation of -0.70 between BSPL and the amplitude of the T wave of the EKG, a finding which was replicated in a second study ($r = -0.61$). No relationship was found between BSPL and other components of the EKG. Citing evidence that the T wave is strongly dependent on plasma potassium concentration, the authors concluded that BSPL reflects potassium concentration in the interstitial fluid, which is itself an ultrafiltrate of plasma. Since interstitial fluid contains many times more sodium than potassium, this membrane must be much more permeable to potassium than to sodium. It is possible that BSPL is an extremely accurate measure of plasma potassium, since the correlations obtained with T wave amplitude are very close to the value of 0.68 for the correlation between plasma potassium and T wave amplitude (Papadimitriou, Roy, & Varkarakis, 1970).

If the slow decline in SPL during rest and prior to BSPL is to be attributed to a decreasing contribution of the dermal duct potential, then BSPL represents the disappearance of this potential, leaving only the potassium potential as the determinant of SPL (Fowles & Venables, 1970a). One would expect, therefore, that SPL prior to the BSPL point would not correlate well with the EKG T wave because of the contamination of the potassium potential by the sodium potential from the duct. When SPL and T wave amplitude were measured at a point 4 min before BSPL, the correlation was not significant, confirming this expectation (Christie & Venables, 1971a). At a time 4 min after BSPL, which should still reflect the potassium potential, the correlation was still very high ($-.64$). These results further support the conclusion that the resting potential is a potassium potential, whereas the activated potential primarily reflects a sodium potential.

The expectation that the resting potential should reflect a potassium potential is strengthened by a comparison with frog skin (Ussing, 1960; Ussing, 1965) and with kidney tubules (Koch, 1965), both of which have an inner membrane responsive to potassium.

The electrolyte used to record BSPL in the studies above was 67 mEq/liter KCl. When the concentration of this outer electrolyte was varied from 2 mEq/liter to 67 mEq/liter the potentials obtained were larger with higher KCl concentrations as would be expected if the outer solution were in contact with a membrane more permeable to potassium than to chloride ions (Christie & Venables, 1971c). The variation in potential with concentration in this study and in others (Edelberg, 1963; Floyd & Keele, 1937; Venables & Sayer, 1963), however, is less than would be expected if the membrane were perfectly selective for potassium over chloride. This result has traditionally led to the suggestion that the membrane shows an appreciable permeability to chloride ions (Martin & Venables, 1966). It is tempting, therefore, to suggest that there is a single membrane or barrier layer in the epidermis which is somewhat more permeable to potassium than to chloride and which generates the BSPL by virtue of the concentration gradients of K and Cl between the interstitial fluid and the outer electrolyte. Given the very low concentration of potassium and the much higher concentration of chloride in interstitial fluid, however, this simple model will not fit the results obtained. Using the Goldman equation (Guyton, 1966, pp. 59–60) and assuming a single membrane, there are no values for the permeabilities to K^+ and Cl^- which will predict the potentials obtained by Christie and Venables (1971c).

An alternative explanation can be derived from the circuit in Fig. 9-4. This time E would represent the value of a single membrane (or of a structure which acts like a single membrane) which is highly permeable to potassium but not very permeable to Cl^- or Na^+. R_e represents the internal series resistance of this membrane plus the resistance of the corneum and R_s represents the combined resistance of the ductal pathway. S will be the potential at the epidermal duct and is assumed to be low or negligible when BSPL is measured with KCl electrolyte. It is to be expected that some of the potential generated by the potassium membrane will be lost across R_e, thus reducing the variation in potential with concentration and giving the appearance of a lack of potassium selectivity at the membrane. The surface *positive* BSPL found with 2 mEq/liter (Christie & Venables, 1971c) is consistent with this model but cannot be explained by a single membrane with appreciable chloride permeability because the high chloride concentration in interstitial fluid would generate surface-negative potentials.

An implication of this model is that variations in NaCl concentration would have little effect on the potassium membrane. As the applied electro-

9. MECHANISMS OF ELECTRODERMAL ACTIVITY

lyte, NaCl would, however, probably produce a potential at the epidermal duct wall, possibly activating the sodium transport system to some degree. Thus variations in NaCl concentration would produce changes in SPL, as has been reported many times (Edelberg, 1963; Floyd & Keele, 1937; Shimizu et al., 1969). A comparison of the potentials recorded with NaCl and with KCl reveals that there is no correlation between the two sites (Edelberg, 1963), as would be expected if separate membranes were involved.

It should be emphasized that the characteristics of the epidermal membrane have not been determined. Whereas it is possible that there is a single membrane located somewhere in the barrier layer, or perhaps even in the basal cells of the stratum germinativum, it is equally possible that potentials are generated at several points in the epidermis (Edelberg, 1971). A single membrane is suggested here in the interest of parsimony and because it appears to handle the Christie and Venables results very nicley. A single membrane permeable to potassium is also an attractive hypothesis because it would provide a pathway for the diffusion of potassium back into the body. Since potassium accumulates on the skin as sweat evaporates, some mechanism for its return to the body is highly desirable.

The same model can be extended to explain the positive drift of SPL and the increased amplitude of SPRs with repeated stimulation reported by Lloyd (1961) in the cat. In this experiment the slow positive drift was associated with the activation of the sweat glands, followed by a slow negative drift once stimulation of sweat gland activity was terminated. The negative drift of the recovery phase bears a striking resemblance to the BSPL. In humans, however, the positive drift during sweat gland activity is largely masked by the potential from the dermal duct, which becomes more negative with sweating.

An important clue to the origin of the slow positive drift in Lloyd's study comes from his finding of a perfect parallelism between the development of this positive drift and changes in impedance. Since these changes in impedance had elsewhere (Lloyd, 1960) been related to duct filling, the positive drift is most likely to be related to changes in the value of R_s in the model just described. Assuming that between responses the epidermal duct potential (*S*) is *lower* than the potassium potential (*E*), decreases in R_s would decrease the measured potential. One of the noteworthy aspects of the SPRs was that the peak value increased only slightly with repeated stimulation, but the amplitude increased because of the lower initial level. It appears, therefore, that during each response the epidermal duct potential became only slightly more negative than the potassium membrane potential. Prior to duct filling, the value of R_s would be high, causing the measured SPL to be almost as high as the potassium membrane potential. Under these conditions the response could produce only a small increase above baseline. With duct filling and the positive drift, however, the peak of the response is further and further above the base line. The small increase in the peak value with repeated stimulation is probably also the result of duct filling, which would result in a larger measured potential when S is greater than E, as would be the case at the peak of a response.

Although rare, diphasic SPRs have been reported in the cat (Lykken, 1968; Wilcott, 1965). Wilcott (1965) found that such responses could be obtained only during the first 5–10 min of the recording period and that they were associated with higher than usual basal potentials. If these basal potentials reflect a value of E which is higher than the peak value of S, then the increased permeability associated with the epidermal duct response would produce a positive SPR. That is, whether the epidermal duct response produces a negative or a diphasic SPR depends on whether there is another potential which is larger than the peak of the negative wave of the epidermal duct response. If there is a more negative potential, either in the dermal duct as in humans or from the epidermis as appears to be the case in Wilcott's study, the increased conductance of the epidermal duct response will act as a shunt conductance, producing a positive wave.

The higher basal potentials found during the first few minutes of recording in Wilcott's study probably reflect a transient potential associated in some way with hydration. A dramatic decline in SPL during epidermal hydration has been reported in humans (Fowles & Venables, 1970b; Fowles & Rosenberry, 1970). Although in humans much of this decline may be attributed to a reduced effect of the sweat gland potential as a result of poral closure, it is possible that some of it may reflect a decreased epidermal potential. When unhydrated, the upper layers of the corneum are possibly even less permeable than the lower layers (Tregear, 1966, p. 64). During the early stages of hydration, therefore, it is possible that the upper corneum acts as a semipermeable membrane which produces a surface-negative potential. Since this potential would be very near the surface of the skin, there would be little resistance between it and the recording electrode, and thus this membrane would be subject to virtually none of the attenuation characteristic of the deeper potassium potential. As hydration progresses, the upper layers of the corneum lose their selectivity, possibly as a result of a weakening of the structure of the corneum (Tregear, 1966, pp. 35–36), and would no longer produce a potential.

The circuit of Fig. 9-4 appears to be surprisingly satisfactory as an explanation of potentials recorded from the cat and of the basal potentials recorded in humans. With the addition of a third potential and resistance representing the dermal portion of the sweat duct, the model appears to be able to encompass the results of activated as well as basal potential measurements in humans.

VI. Summary and Conclusions

The major conclusions of this review may be summarized by reference to Fig. 9-5, which shows three sources of potential and three major pathways for current. The lumen negative potential generated across the duct wall in the dermis, E_1, is primarily a function of the sodium concentration in the lumen of the duct. The lumen negative potential generated across the epidermal duct wall at the level of the germinating layer is E_2. It is probably a function of both sodium and chloride concentration in the lumen but, because of its lesser selectivity, will be lower than E_1 during sweat responses. The potential generated by the barrier membrane in the stratum compactum, E_3, is a function of potassium concentration in the interstitial fluid and in the applied electrolyte. It will be surface negative as long as the outer potassium concentration is greater than that of interstitial fluid. The most important pathway for current is along R_1 and R_3, representing the resistances of the epidermal duct and the duct wall in the germinating layer, respectively. Here, R_1 and R_3 are both variable resistors, R_1 changing with duct filling and R_3 with membrane responses. The second pathway for current is along R_1, R_2, and R_4, where R_2 is the dermal duct resistance and R_4 is the resistance of the dermal duct wall. The variable resistors in this pathway are R_1 and R_2, both of them decreasing with duct filling. The final pathway is along R_5 and R_6, which represent the resistance of the upper corneum and the barrier layer, respectively; R_5 varies with hydration of the corneum, whereas R_6 remains relatively constant. A fourth pathway in which current flows into the duct from the corneum and then along the two sweat gland pathways has been omitted in order to avoid complicating the circuit further. Similarly, the resistance of the germinating layer and possibly of the basal cell layer, which would be found between the reference electrode and the two potentials in the epidermis (E_2 and E_3), have been omitted for simplicity. Whereas these resistances may not be negligible,

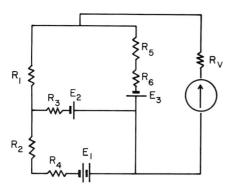

FIG. 9-5. A simplified model of the electrodermal effector system. R_v is the input impedance of the voltmeter. See text for identification of the other components of the circuit.

they appear to be relatively constant and contribute only a small portion of the total resistance.

Under completely resting conditions, reabsorption outpaces secretion, maintaining high values of R_1, R_2, and R_3 and minimal potentials at E_1 and E_2. The potential at E_3 is the major factor in the measured potential and thus reflects interstitial potassium. A small or moderate sweat response decreases the resistance of R_2 and probably of R_1, producing a slow-recovery SCR. At the same time, lumen sodium concentration increases with a resultant increase in E_1. The increased potential plus the decreased ductal resistance produce the slow-recovery negative SPR. These responses cause an increase in both SCL and SPL. Larger sweat responses or those occurring with the ducts partially full will further decrease R_1 and R_2. If the hydrostatic pressure (or the sodium concentration) reaches a high enough level, a response will be initiated at the epidermal duct, decreasing the value of R_3 and generating a small lumen negative potential at E_2. This membrane response produces the rapid recovery SCR and, at the same time, produces a positive SPR by virtue of a shunting effect on E_1. Most of the time, however, there will be a small initial negative component to the SPR as a result of the earlier onset of the negative wave than of the increase in permeability. Once the ducts are maximally full, further sweat gland responses will produce only membrane responses.

Most of the essential features of Edelberg's (1968a) model appear to be well supported. These include the identification of two effectors (duct filling and the epidermal duct membrane) and the characteristics of the responses they produce. That is, slow-recovery negative SPRs and SCRs are associated with duct filling, while the membrane response produces rapid-recovery SCRs, the positive SPR and possibly the sharp negative SPR which may be seen either alone or as the initial portion of many diphasic SPRs. Edelberg's model has been modified by attributing the positive SPR to a shunting of the dermal duct potential rather than to the generation of a positive potential at the epidermal duct wall and by suggesting that the membrane response is triggered by either the hydrostatic pressure or the NaCl concentration of the sweat (or both) rather than by a cholinergically mediated neural pathway. Edelberg's model has been extended by relating the origin of the duct potentials to the sodium transport mechanism and by attributing the nonsudorific epidermal membrane potential to the concentration of potassium in interstitial fluid and the applied electrolyte.

There are several implications of these conclusions for future research with electrodermal measurements. Investigators concerned with SCRs would do well to distinguish between membrane and duct filling responses if amplitude is to be scored, and they should try to minimize intrasubject changes in SCL. Where the assessment of arousal is concerned, range-corrected measures of SCL (Lykken, 1968; Lykken *et al.*, 1966) and quantification of the recovery limb of SCRs (Edelberg, 1970) are the most promising approaches. An entirely new area has been opened up by the finding that BSPL reflects the concentration of potassium in interstitial fluid. If the suggestions (Christie & Venables, 1971a; Venables, 1970) that potassium concentration may be expected to vary with stress are true, this measure may become an extremely important tool for research into stress. Whether or not a response to stress is involved, interstitial potassium concentration can be expected to affect resting potentials in nerve cells and thus may be an important variable in its own right.

The long tradition of research into the effector system of electrodermal activity has begun to bear fruit in terms of more precise and testable hypotheses. Although the complexity of this effector system may dismay some investigators wishing to use electrodermal measurements as a tool in a particular content area, it is likely that advances in understanding of electrodermal activity will increase its value as a tool for research. Lykken's (1968) comment is apropos here: the continued interest in electrodermal activity in spite of inadequate technique and lack of understanding implies that these phenomena must be unusually rich in psychological significance.

Acknowledgments

The preparation of this manuscript was supported in part by Biomedical Sciences Support Grant FR-07035 from the National Institutes of Health. The money was made available in the form of a small grant for health-related research awarded by the Graduate College of the University of Iowa.

The author wishes to thank P. H. Venables, Robert Edelberg, and David Lykken for their critical comments.

References

Adams, T. Characteristics of eccrine sweat gland activity in the footpad of the cat. *Journal of Applied Physiology*, 1966, **21**, 1004–1012.

Adams, T., & Vaughan, J. A. Human eccrine sweat gland activity and palmar electrical skin resistance. *Journal of Applied Physiology*, 1965, **20**, 980–983.

Cage, G. W., & Dobson, R. L. Sodium secretion and reabsorption in the human eccrine sweat gland. *Journal of Clinical Investigation*, 1965, **44**, 1270–1276.

Christie, M. J., & Venables, P. H. Basal palmar skin potential and the electrocardiogram T-wave. *Psychophysiology*, 1971, **8**, 779–786. (a)

Christie, M. J., & Venables, P. H. Characteristics of palmar skin potential and conductance in relaxed human subjects. *Psychophysiology*, 1971, **8**, 525–532. (b)

Christie, M. J., & Venables, P. H. Effects on "basal" skin potential level of varying the concentration of an external electrolyte. *Journal of Psychosomatic Research*, 1971, **15**, 343–348. (c)

Darrow, C. W. The relation of the galvanic skin reflex recovery curve to reactivity, resistance level, and perspiration. *Journal of General Psychology*, 1932, **7**, 261–272.
Darrow, C. W. The rationale for treating the change in galvanic skin response as a change in conductance. *Psychophysiology*, 1964, **1**, 31–38.
Darrow, C. W., & Gullickson, G. R. The peripheral mechanism of the galvanic skin response. *Psychophysiology*, 1970, **6**, 597–600.
Dobson, R. L. The correlation of structure and function in the human eccrine sweat gland. In W. Montagna, R. A. Ellis, & A. S. Silver (Eds.), *Advances in biology of skin. Vol. III: Eccrine sweat glands and eccrine sweating*. New York: Macmillan, 1962. Pp. 54–75.
Edelberg, R. *Electrophysiologic characteristics and interpretation of skin potentials*. (Tech. Documentary Rep. No. SAM-tdr-63-95) Brooks Air Force Base, Texas: United States Air Force School of Aerospace Medicine, 1963.
Edelberg, R. Response of cutaneous water barrier to ideational stimulation: A GSR component. *Journal of Comparative and Physiological Psychology*, 1966, **61**, 28–33.
Edelberg, R. A proposed physiological model of the electrodermal effector system. Progress Report MH-08656 (National Institute of Mental Health), December, 1968. (See footnote 2.) (a)
Edelberg, R. Biopotentials from the skin surface: The hydration effect. *Annals of the New York Academy of Sciences*, 1968, **148**, 252–262. (b)
Edelberg, R. The information content of the recovery limb of the electrodermal response. *Psychophysiology*, 1970, **6**, 527–539.
Edelberg, R. Electrical properties of skin. In H. R. Elden (Ed.), *Biophysical properties of the skin*. Vol. I. New York: Wiley, 1971. Pp. 513–550.
Edelberg, R. Electrical activity of the skin: Its measurement and uses in psychophysiology. In N. S. Greenfield & R. A. Sternbach (Eds.), *Handbook of psychophysiology*. New York: Holt, 1972.
Edelberg, R., Greiner, T., & Burch, N. R. Some membrane properties of the effector in the galvanic skin response. *Journal of Applied Physiology*, 1960, **15**, 691–696.
Elden, H. R. (Ed.) *Biophysical properties of the skin*. Vol. I. New York: Wiley, 1971.
Floyd, W. F., & Keele, C. A. Some observations on skin potentials in human subjects. *Transactions of the Faraday Society*, 1937, **33**, 1046–1049.
Fowles, D. C., & Johnson, G. The influence of variations in electrolyte concentration on skin potential level and response amplitude. *Biological Psychology*, in press.
Fowles, D. C., & Rosenberry, R. The effects of epidermal hydration on the positive skin potential response and on prestimulus skin potential level. Presented at the 42nd Annual Meeting of the Midwestern Psychological Association, May, 1970.
Fowles, D. C., & Venables, P. H. The effects of epidermal hydration and sodium reabsorption on palmar skin potential. *Psychological Bulletin*, 1970, **73**, 363–378. (a)
Fowles, D. C., & Venables, P. H. The reduction of palmar skin potential by epidermal hydration. *Psychophysiology*, 1970, **7**, 254–261. (b)
Fox, R. H., & Hilton, S. M. Bradykinin formation in human skin as a factor in heat vasodilatation. *Journal of Physiology*, 1958, **142**, 219–232.
Goadby, K. W., & Goadby, H. K. Simultaneous photographic records of the potential and resistance effects of the psycho-emotive response. *Journal of Physiology*, 1936, **86**, 11P–13P.
Gordon, R. S., Jr., & Cage, G. W. Mechanism of water and electrolyte secretion by the eccrine sweat gland. *Lancet*, 1966, **1**, 1246–1250.
Guyton, A. C. *Textbook of medical physiology*. (3rd ed.) Philadelphia, Pennsylvania: Saunders, 1966.
Koch, A. The kidney. In T. C. Ruch & H. B. Patton (Eds.), *Physiology and biophysics*. (19th ed.) Philadelphia, Pennsylvania: Saunders, 1965. Pp. 843–870.

Koefoed-Johnsen, V., & Ussing, H. H. The nature of the frog skin potential. *Acta Physiologica Scandinavica*, 1958, **42**, 298–308.

Kuno, Y. *Human Perspiration*. Springfield, Ill.: Thomas, 1956.

Lader, M. H., & Montagu, J. D. The psycho-galvanic reflex: A pharmacological study of the peripheral mechanism. *Journal of Neurology, Neurosurgery and Psychiatry*, 1962, **25**, 126–133.

Lloyd, D. P. C. Average behavior of sweat glands as indicated by impedance changes. *Proceedings of the National Academy of Science*, 1959, **45**, 410–413.

Lloyd, D. P. C. Electrical impedance changes of the cat's foot pad in relation to sweat secretion and reabsorption. *Journal of General Physiology*, 1960, **43**, 713–722.

Lloyd, D. P. C. Action potential and secretory potential of sweat glands. *Proceedings of the National Academy of Science*, 1961, **47**, 351–358.

Lloyd, D. P. C. Secretion and reabsorption in eccrine sweat glands. In W. Montagna, R. A. Ellis & A. S. Silver (Eds.), *Advances in biology of skin. Vol. III: Eccrine sweat glands and eccrine sweating*. New York: Macmillan, 1962. Pp. 127–151.

Lykken, D. T. Neuropsychology and psychophysiology in personality research. Part II. Psychophysiological techniques and personality research. In E. F. Borgatta & W. W. Lambert (Eds.), *Handbook of personality theory and research*. Chicago, Illinois: Rand McNally, 1968. Pp. 454–509.

Lykken, D. T., & Venables, P. H. Direct measurement of skin conductance: A proposal for standardization. *Psychophysiology*, 1971, **8**, 656–672.

Lykken, D. T., Miller, R. D., & Strahan, R. F. Some properties of skin conductance and potential. *Psychophysiology*, 1968, **5**, 253–268.

Lykken, D. T., Rose, R., Luther, B., & Maley, M. Correcting psychophysiological measures for individual differences in range. *Psychological Bulletin*, 1966, **66**, 481–484.

Martin, I., & Venables, P. H. Mechanisms of palmar skin resistance and skin potential. *Psychological Bulletin*, 1966, **65**, 347–357.

McDonald, T. F., & MacLeod, D. P. Maintenance of resting potential in anoxic guinea pig ventricular muscle: Electrogenic sodium pumping. *Science*, 1971, **172**, 570–572.

Montagna, W. *The structure and function of the skin*. (2nd Ed.) New York: Academic Press, 1962.

Montagna' W., & Lobitz, W. C., Jr. (Eds.) *The epidermis*. New York: Academic Press, 1964.

Montagna, W., Ellis, R. A., & Silver, A. S. (Eds.) *Advances in biology of skin. Vol. III: Eccrine sweat glands and eccrine sweating*. New York: Macmillan, 1962.

Montagu, J. D., & Coles, E. M. Mechanism and measurement of the galvanic skin response. *Psychological Bulletin*, 1966, **65**, 261–279.

Munger, B. L. The ultrastructure and histophysiology of human eccrine sweat glands. *Journal of Biophysical and Biochemical Cytology*, 1961, **11**, 385–402.

Munger, B. L., & Brusilow, S. W. An electron microscopic study of eccrine sweat glands of the cat foot and toe pads—evidence for ductal reabsorption in the human. *Journal of Biophysical and Biochemical Cytology*, 1961, **11**, 403–417.

Niimi, Y., Yamazaki, K., & Watanabe, T. Pseudo-effects of external electrolyte concentration on measured skin potential levels. *Psychophysiology*, 1968, **5**, 188–191.

Nordquist, R. E., Olson, R. L., & Everett, M. A. The transport, uptake, and storage of ferritin in human epidermis. *Archives of Dermatology*, 1966, **94**, 482–490.

Nutbourne, D. M. The effect of small hydrostatic pressure gradients on the rate of active sodium transport across isolated living frog-skin membranes. *Journal of Physiology*, 1968, **195**, 1–18.

Nutbourne, D. M. Effect of small hydrostatic pressure gradients on the rate of active sodium transport across isolated living frog-skin membranes. *Proceedings of the Royal Society of Medicine*, 1969, **62**, 1122.

Papa, C. M., & Kligman, A. M. Mechanisms of eccrine anidrosis: I. High level blockade. *Journal of Investigative Dermatology*, 1966, **47**, 1–9. (a)

Papa, C. M., & Kligman, A. M. Sweat pore patterns. *Journal of Investigative Dermatology*, 1966, **46**, 193–197. (b)

Papadimitriou, M., Roy, R. R., & Varkarakis, M. Electrocardiographic changes and plasma potassium levels in patients on regular haemodialysis. *British Medical Journal*, 1970, **2**, 268–269.

Peiss, C., Randall, W. D., & Hertzman, A. B. Hydration of the skin and its effect on sweating and evaporative water loss. *Journal of Investigative Dermatology*, 1956, **26**, 459–470.

Randall, W. C., & Peiss, C. N. The relationship between skin hydration and the suppression of sweating. *Journal of Investigative Dermatology*, 1957, **28**, 435–441.

Richter, C. P. A study of the electrical skin resistance and the psychogalvanic reflex in a case of unilateral sweating. *Brain*, 1927, **50**, 216–235.

Richter, C. P., & Woodruff, B. G. Changes produced by sympathectomy in the electrical resistance of the skin. *Surgery*, 1941, **10**, 957–970.

Rothman, S. *Physiology and biochemistry of the skin*. Chicago, Ill.: Univ. of Chicago Press, 1954.

Sarkany, I., Shuster, S., & Stammers, M. C. Occlusion of the sweat pore by hydration. *British Journal of Dermatology*, 1965, **77**, 101–104.

Schulz, I. J. Micropuncture studies of the sweat formation in cystic fibrosis patients. *Journal of Clinical Investigation*, 1969, **48**, 1470–1477.

Schulz, I., Ullrich, K. J., Frömter, E., Holzgreve, H., Frick, A., & Hegel, U. Mikropunktion und elektrische potentialmessung an schweißdrüsen des menschen. *Pflügers Archiv*, 1965, **284**, 360–372.

Shaver, B. A., Brusilow, S. W., & Cooke, R. E. Origin of the galvanic skin response. *Proceedings of the Society of Experimental Biology and Medicine*, 1962, **110**, 559–564.

Shaver, B. A., Brusilow, S. W., & Cooke, R. E. Electrophysiology of the sweat gland: Intraductal potential changes during secretion. *Bulletin of the Johns Hopkins Hospital*, 1965, **116**, 100–109.

Shelley, W. B., & Horvath, P. N. Experimental miliaria in man. III. Production of miliaria rubra (prickly heat). *Journal of Investigative Dermatology*, 1950, **14**, 193–204.

Shimizu, K., Tajimi, T., Watanabe, T., & Niimi, Y. Effects of external electrolyte concentration on both skin potential-level and -reflex: Successful observation. *Japanese Psychological Research*, 1969, **11**, 32–36.

Shuster, S. Graded sweat-duct occlusion: A technique for studying sweat-gland function. *Clinical Science*, 1963, **25**, 89–95.

Steinbach, H. B. Potassium in frog skin. *Journal of Cellular and Comparative Physiology*, 1937, **10**, 51–60.

Suchi, T. (1950) Cited in Kuno, Y., *Human perspiration*. Springfield, Illinois: Thomas, 1956. Pp. 311.

Thomas, R. C. Measurement of current produced by the sodium pump in a snail neurone. *Journal of Physiology*, 1968, **195**, 23–24 P.

Thomas, P. E., & Korr, I. M. Relationship between sweat gland activity and electrical resistance of the skin. *Journal of Applied Physiology*, 1957, **10**, 505–510.

Tregear, R. T. *Physical functions of skin*. New York: Academic Press, 1966.

Trehub, A., Tucker, I., & Cazavelan, J. Epidermal b-waves and changes in basal potentials of the skin. *American Journal of Psychology*, 1962, **75**, 140–143.

Ussing, H. H. Transport through epithelial membranes. In H. H. Ussing, P. Kruhøffer, J. H. Thaysen & N. A. Thorn (Eds.), *The alkali metal ions in biology*. Berlin and New York: Springer-Verlag, 1960.
Ussing, H. H. Transport of electrolytes and water across epithelia. In *Harvey Lectures: 1963–64*. New York: Academic Press, 1965.
Venables, P. H. Electrolytes and behaviour in man. In R. Porter & J. Birch (Eds.), *Chemical influences on behaviour*, London: Churchill, 1970.
Venables, P. H., & Christie, M. J. Mechanisms, instrumentation, recording techniques, and quantification of responses. In W. Prokasy & D. Raskin (Eds.), *Electrodermal activity in psychological research*. Chapter 1. New York: Academic Press, 1973.
Venables, P. H., & Martin, I. The relation of palmar sweat gland activity to level of skin potential and conductance. *Psychophysiology*, 1967, **3**, 302–311. (a)
Venables, P. H., & Martin, I. Skin resistance and skin potential. In P. H. Venables & I. Martin (Eds.), *A manual of psychophysiological methods*. Amsterdam: North-Holland Publ., 1967. Pp. 53–102. (b)
Venables, P. H., & Sayer, E. On the measurement of the level of skin potential. *British Journal of Psychology*, 1963, **54**, 251–260.
Wagner, H. N. Electrical skin resistance studies in two persons with congenital absence of sweat glands. *Archives of Dermatology and Syphilology*, 1952, **65**, 543–548.
Weiner, J. S., & Hellman, K. The sweat glands. *Biological Review*, 1960, **35**, 141–186.
Wesson, L. G. *Physiology of the human kidney*. New York: Grune & Stratton, 1969.
Wilcott, R. C. Correlation of skin resistance and potential. *Journal of Comparative and Physiological Psychology*, 1958, **51**, 691–696.
Wilcott, R. C. Palmar skin sweating vs. palmar skin resistance and skin potential. *Journal of Comparative and Physiological Psychology*, 1962, **55**, 327–331.
Wilcott, R. C. The partial independence of skin potential and skin resistance from sweating. *Psychophysiology*, 1964, **1**, 55–66.
Wilcott, R. C. A comparative study of the skin potential, skin resistance and sweating of the cat's foot pad. *Psychophysiology*, 1965, **2**, 62–71.
Wilcott, R. C. Arousal sweating and electrodermal phenomena. *Psychological Bulletin*, 1967, **67**, 58–72.
Witten, V. H., Ross, M. S., Oshry, E., & Hyman, A. B. Studies of thorium X applied to human skin. I. Routes and degree of penetration and sights of deposition of thorium X applied in selected vehicles. *Journal of Investigative Dermatology*, 1951, **17**, 311–322.
Witten, V. H., Brauer, E. W., Loevinger, R., & Holmstrom, V. Studies of radioactive phosphorus (P^{32}) applied to human skin. I. Erythema and autoradiographic findings following applications in various forms. *Journal of Investigative Dermatology*, 1956, **26**, 437–447.

Chapter 10

Recording of Electrodermal Phenomena

William W. Grings[1]

Department of Psychology
University of Southern California
Los Angeles, California

I. Nature of the Physical Measurement 273
II. The Electrode 277
III. Input Signal Conditioning, Amplification, and Registration 281
IV. Measurement Units 286
    References 292

I. Nature of the Physical Measurement

Electrodermal recording methods differ with the phenomena of interest. The major subdivision is between measurement of skin potential (termed *endodermal phenomena* and referred to historically as the Tarchanoff effect) and skin resistance, conductance, or impedance (termed *exodermal phenomena* and identified historically with Feré). The endodermal measure is an expression of voltage between locations on the skin. Resistance record-

---

[1] This chapter was written while the author was an NIMH research fellow (MH 46955) at the Institute of Psychiatry, University of London.

ings designate opposition to passage of a small direct current through the skin, whereas conductance measures express the quantity of current passage per unit of surface area per unit of time. Skin impedance is recorded when reactance quantities as well as d.c. resistance are of interest. Within these classes of phenomena, a major distinction is made between measurements of tonic levels (states) and phasic reactions (responses).

During recent years there has been increased argument for standard terminology and abbreviation. One dominant version is as follows: skin conductance (SC), skin resistance (SR), skin potential (SP), and skin impedance (SZ). Abbreviations are commonly extended to add an L to indicate a tonic level measurement or an R to indicate a phasic or response measurement, as in SCL for skin conductance level, SCR for skin conductance response, SPL for skin potential level, etc. The abbreviation which has been used most widely in the past is GSR, for galvanic skin response. As of this date, the term galvanic skin response is still the most general heading for electrodermal phenomena in *Psychological Abstracts*, although it is expected that GSR will eventually be replaced by the more specific designations SRR, SCR, SPR, or SZR. Another useful expression is electrodermal response (EDR), recommended as a substitute for GSR and PGR (psychogalvanic reflex) over 35 years ago (Ruckmick, 1937). Readers interested in the history of electrodermal terminology may note also Landis and DeWick (1929) and Landis (1932).

In referring to responses, a further distinction is of importance. Sometimes the response can be said to be elicited by a stimulus. Such instances are designated as specific, elicited, or evoked responses to contrast them with changes that are not related to specifiable external stimuli and which are thus termed spontaneous or nonspecific. Most common measures of responses are in terms of magnitude or amplitude, usually from onset to peak, although with spontaneous responses a count of their number in a specified period of time is common. The latency (time from stimulus onset to response onset) may be of interest, as may the recruitment time (time from response onset to peak, which ranges from .5 to 5 sec). Recovery time is usually slower than rise time and varies with many factors. Freeman and Katzoff (1942) recommended the use of a recovery quotient. More recently, the slope of the recovery limb of the response has been proposed as a useful measure (Edelberg, 1970).

Skin potential levels (SPLs) vary with a number of factors (for example, electrode location) over a range extending to 60 mV or more. Quantification is in millivolts with polarity specified in terms of a designated reference. Palmar potential is frequently expressed with reference to an indifferent electrode on the forearm or some more distant point (an earlobe). Skin potential responses (SPRs) are also expressed in millivolts and may be

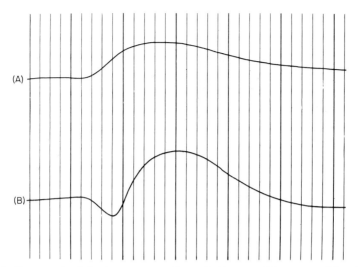

FIG. 10-1. Photograph of simultaneous recordings of skin potential response (B) and skin conductance response (A). The phasic activity resulted from a stimulus introduced at the left-hand edge of the graph. Curve A shows an increase in conductance, whereas curve B shows an initial negative swing followed by a positive potential.

biphasic with an initial negative "a" swing followed shortly by a positive "b" potential, or monophasic. The total change is in the order of a few millivolts, generally less than 10. The latency of the SPR also varies with factors like stimulus intensity over a range from 1–3 sec with a mean value around 1.5–2.0 sec. The general shape of an SPR is shown in Fig. 10-1, where it is compared to an SCR. Edelberg (1972) speaks of a third "c" wave in SPR and characterizes it as a slow return past the initial point.

When a voltage is impressed between the electrodes upon the skin, and the passage of current is the item of interest, several alternative measures may be used. The three most common are resistance in ohms, conductance in micromhos (reciprocal megohms), and impedance in ohms. All are expressed in terms of area of electrode surface (for example, micromhos per square centimeter). The equivalent resistance to the skin is often described as a series-parallel circuit of resistance and capacitance components. General usage favors d.c. measures over a.c., although arguments have been advanced both ways and correlational studies have been reported.

The d.c. resistance offered by 1 cm^2 of skin ranges from a few thousand ohms to several hundred thousand ohms with most palmar values between 5000 and 50,000. Most common skin conductance levels are between 10 and 100 μmhos/cm^2. When impedance is measured, the frequency of the applied voltage is specified and the phase angle is typically given. Impedance is generally cited as dropping to 25% or less (of d.c. resistance) by 1000 Hz, and to 5% or less by 10,000 Hz. Alternating current measures corresponding to conductance are those of admittance (reciprocal impedance), also expressed in micromhos per square centimeter.

There are many physical variables which affect electrodermal measurements. Some of these are associated with the electrode–skin junctions, such as the electrolyte used, the pressure and mechanical stability of the placement, electrode polarization, and preparation of the skin. Separate consideration of these factors will be made in the next section on electrodes. Another physical measurement question will be given further discussion in connection with input circuitry. It concerns the choice between controlling the current through the subject and reflecting differences as changes in resistance or controlling the voltage applied to the subject and reflecting differences as changes in conductance.

The skin does not act as a pure ohmic resistance in many of its properties, like the time course of its reaction to applied voltage; and it does not obey Ohm's law throughout a very wide range of values. Therefore, efforts have been made to find the optimum range of current densites, that is, the range of current where the relation between current and voltage will be linear. In an early study Edelberg, Greiner, and Burch (1960) demonstrated linearity up to a mean current density of 11 $\mu A/cm^2$ and recommended the use of 8 $\mu A/cm^2$ as the maximum current density, a specification of importance if a constant current system is utilized. A lower limit is also set on current density by the fact that if current is too small the voltage changes produced by impressed sources and those resulting from SPRs will be in the same range and lead to measurement complications. It is thus recommended that currents close to the maximum of the linearity range be used.

A later study (Edelberg, 1967) demonstrated that the limiting factor for linearity is the voltage developed rather than the current passing between the electrodes. With high resistances only small current densities (for example, 4 $\mu A/cm^2$) could be tolerated without nonlinearity, whereas with low resistances relatively high currents (for example, 75 $\mu A/cm^2$) were reached before nonlinearity occurred. This led to the recommendation for specification (in the constant voltage system) of a value of .5 V across each site as a limit.

Another variable important to both the physical measurement and the underlying response mechanisms is temperature. Skin resistance increases with decreases in temperature at the rate of about 3%/°C. Whereas the amplitude of the SRR may increase originally with a drop in temperature, the long-term effect of low temperature (for example, 20°C) is to decrease the response (Maulsby & Edelberg, 1960). The effects of temperature in SPL measurements vary with type of electrode and electrolyte concentration.

An error will occur if both electrodes are not at the same temperature (Venables & Sayer, 1963). The positive and negative components of the potential responses have been found to be differently affected by differences in room temperature. The positive component is less for lower temperatures at which the negative component may be more prominent (Yokota, Takahashi, Kondo, & Jujimori, 1959). Latencies of electrodermal changes are markedly increased by decreases in skin temperature. Negative correlations have been reported between SCR and relative humidity (Venables, 1955; Wenger & Cullen, 1962).

The choice between endodermal and exodermal methods will depend upon the particular needs of a given research problem. There are many factors, other than the methodological issues reviewed here, which underlie such a choice. For a comprehensive discussion of such factors see Edelberg (1972). A sample of more specific comparisons of the two major types of phenomena may be obtained from Wilcott (1958), Lykken, Miller, and Strahan (1968), Gavira, Coyne, and Thetford (1964), Hupka and Levinger (1967), and Burstein, Fenz, Bergeron, and Epstein (1965).

Throughout all of the material to follow, excellent overall guides to more detail will be found in Edelberg (1967, 1972), Venables and Martin (1967), and Venables and Christie (1973). In addition to references already cited, the interested reader will find further detail on the physical nature of the response as follows: (1) on the range of values for SPLs, Venables and Sayer (1963) and O'Connell, Tursky, and Evans (1967); (2) on description of equivalent resistance of the skin, Montagu (1958, 1964), Montagu and Coles (1966); Lykken (1970), Lloyd (1960), Brazier (1933), Lykken, Miller, and Strahan (1968); (3) on current and current density, Edelberg and Burch (1962), Lykken (1959, 1971), Wenger and Gustafson (1962), Grings (1953); Floyd and Keele (1936), and Davis (1930); (4) on impedance, frequency effects, and a.c.-d.c. relations, Lykken (1971), Forbes and Landis (1935), Montagu (1958, 1964), Plutchik and Hirsch (1963), Grings (1953), Barnett (1938), Floyd and Keele (1936), Brazier (1933); (5) on comparative response latencies, Uno and Grings (1965); (6) on temperature effects, Neumann (1968), Smith (1937), Wilcott (1963), Wenger and Gustafson (1962), Wenger and Cullen (1962), Venables (1955), and Floyd and Keele (1936).

II. The Electrode

The initial element in the electrodermal recording chain is the pickup electrode. Its role in tonic measurement is particularly crucial, for small changes in properties of the electrode-skin junction lead to large changes in "apparent resistance" of the subject. This applies with added emphasis where exodermal procedures are used, that is, where a voltage is applied to the skin.

Arrangements of two electrodes are most common, although three-electrode pickups have been discussed (for example, Montagu & Coles, 1966). Electrodes may be placed over active areas (that is, over areas with active sweat glands) providing a bipolar input, or an indifferent or inactive electrode may be paired with an active one for a monopolar input. The skin under the indifferent electrode should be prepared by some abrasive method (such as mild sandpapering) to achieve very low resistance contact with the fluid levels of the skin, (Venables & Sayer, 1963; Shackel, 1959). The most frequently used site for the indifferent electrode is the forearm. Most common active sites are the palm of the hand, middle and distal phalanges of the fingers, and the sole of the foot. Data surveying skin conductance and skin potential for a variety of anatomical locations are given by Edelberg (1967).

Material used for the metallic portion of the electrode has varied more with SR (or SC) measurements than with SP. In both cases, however, the generally preferred substance is silver, coated electrolytically with silver chloride. The important feature, from the standpoint of reducing artifacts (for example, bias potentials and polarization) at the skin-electrode boundary in SC and SR recording is to have the metal communicate with one of its salts, in this case the junction of Ag and AgCl. For conductance and resistance pickup a zinc–zinc sulfate junction finds some use.

Structure and shape of electrodes may vary with the measurement type. Area is of critical importance in conductance work. Partly for this reason disk or plate electrodes are rather universal for SC or SR, whereas sponge Ag–AgCl electrodes are also recommended for potentials (for example, O'Connell, Tursky, & Orne, 1960). Disk electrodes commonly have the metal mounted in a shallow chamber or plastic cup, into which the electrode paste is inserted to communicate between the metal and the surface of the skin. Care is taken to avoid sources of artifacts through location and soldering of the lead-in wire to the back of the disk. Area of exposed skin is controlled through the use of an adhesive mask in conjunction with the electrode. Quite satisfactory adhesive masks and fully fabricated disk electrodes are available commercially (for example, Beckman Corp). Alternatives in use range from "corn pads" (Lykken, 1959) to rubber grommets (Venables & Martin, 1967). Details for construction of both sponge and disk electrodes are given by Venables and Martin (1967). For long periods of measurement where movement of the hand or foot is involved, a metal cloth electrode has been used effectively (Edelberg, 1967). A number of electrode configurations are illustrated in Fig. 10-2.

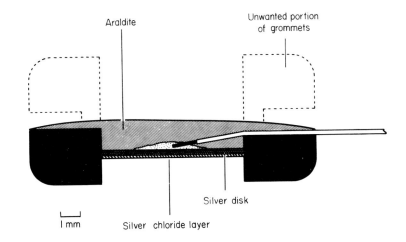

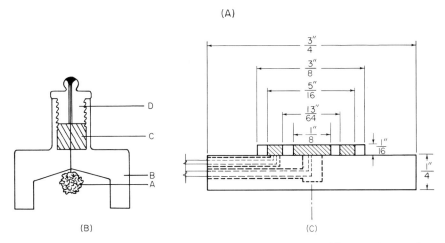

Fig. 10-2. Common electrode configurations: (A) disk; (B) sponge; (C) two-element electrode. (Illustrations courtesy of (A) Venables and Martin, 1967; (B) O'Connell and Tursky, 1960; and (C) Lykken, 1959.)

Much has been written about the problem of polarization with d.c. resistance or conductance recording, and the need to use "nonpolarizing" electrodes. This refers to the fact that passage of current through the electrolyte–electrode boundary will lead to a buildup of counter electromotive force. Further complication may arise from differences in half-cell potential. The half-cell potentials result from the battery effect at the junction of metal and electrolyte. Bias potential can be prevented by careful preparation of electrodes. The general procedure for producing AgCl electrodes involves the use of pure silver, carefully cleaned of dirt and grease and with insulating cement placed over areas where lead-in wires have been soldered. Various additional forms of preliminary cleaning may be recommended—such as immersion in dilute HCl or brief treatment as the cathode in an NaCl solution with Ag anode and allowing current to pass long enough for bubbles to form on the surface. The chloriding process is then carried out by placing the electrode disk as the anode in a 1 M NaCl solution while a second silver strip is used as the negative terminal and a current from a small battery is permitted to pass until the surface of the disk is a uniform brownish purple color. Use of a series resistor is desirable to obtain a current of from .5 mA to 2.5 mA/cm^2 through the electrode. When the coating is completed, the electrodes are washed in distilled water and placed in fresh NaCl solution to age for at least 24 hr.

For production of matched pairs, several electrodes may be chlorided at once (increasing the applied voltage so that the desired current passes through each electrode). Electrodes may be stored wet or dry and care should be taken to clean the electrodes of paste after each use. This can be done conveniently with a jet of water, like that provided from a dental "water pick." Any form of abrasive cleaning will destroy the anodization. In selecting electrode pairs and for periodic checks during use, the electrodes can be placed in a saline bridge for measurement of bias potential and apparent resistance and rechlorided as soon as values reach unacceptable levels (for example, 1 mV or 300–500 Ω). Details for construction and care of electrodes are given by Venables and Martin (1967), Edelberg (1967), Miller (1968), O'Connell and Tursky (1960), Lykken (1959), Quilter and Surwillo (1966), and Malmo and Davis (1961).

Polarization with current passage cannot be prevented completely. It can only be reduced by keeping current density low and providing a good electrode–electrolyte junction. This is provided by a reversible junction of the metal and a solution of its own ions, such as occurs with chloridized silver in a sodium chloride (or potassium chloride) electrolyte. Among other methods for reducing polarization, a two-element electrode has been recommended (Lykken, 1959). The two elements, arranged as a disk and annulus are connected to each other by electrolyte in such a way that apparent resistance caused by polarization is reflected across the annulus while voltage across the subject (apparent resistance of the subject) is reflected across the disk. One major difficulty with this arrangement is that it is not adaptable to constant voltage measurement systems (Montagu & Coles, 1966). As another approach, periodic polarity reversals are recommended by some (for example, Montagu & Coles, 1968).

Electrode attachment needs to be stable in order to keep exposed areas constant and to retain an even moisture level (prevent loss or change of concentration of electrolyte). Pressure is an important variable to hold constant and care needs to be taken to ensure that adhesive tapes or other fasteners are not so tight as to affect circulation and local skin conditions.

For all forms of electrodermal measurement the junction between electrode and skin is electrolytic, mediated by a liquid medium or an electrode paste. The substance employed should have ions compatible with those of skin fluids (sweat) to permit recording of current passage and surface voltage without distortion during ionic transmission. Moisture content should remain stable over time. A certain amount of stability (viscosity) is needed and achieved by mixing the electrolyte with an inert base.

The most commonly used electrolyte for SC and SR measures is a .05 M solution of NaCl uniformly distributed in an inert agar or methyl cellulose jelly. Among substances previously used but found to be insufficiently inert are gum tragacanth and bentonite pastes. Commercial EKG pastes may be undesirable because they are typically made of hypertonic solutions. When electrodes other than Ag–AgCl are used, different electrolytes are required. The most common is $ZnSO_4$ used when zinc electrodes are used (for example, Lykken, 1959). Because $ZnSO_4$ (or $ZnCl_2$) has a tendency to interact with skin ions, a double boundary is sometimes used, a Zn salt next to the electrode and an NaCl paste next to the skin.

There are some differences in practice where skin potentials are involved. Many workers use the NaCl paste suggested above. Others (for example, Venables & Sayer, 1963) have demonstrated that the skin potential response is best recorded with a potassium chloride rather than a sodium chloride electrolyte. They recommend a particular KCl formula and have recently extended their arguments for special chemical contributors to SP measurements (Fowles & Venables, 1970; Christie & Venables, 1971, 1972).

In addition to the references cited above, the reader will find data on current, polarity, and electrolyte effects in Edelberg, Greiner, and Burch (1960), Edelberg and Burch (1962), Blank and Finesinger (1946), and O'Connell *et al.* (1960). Data on preparation of indifferent electrode sites are given by Shackel (1959) and Venables and Sayer (1963). Electrode paste formulas are indicated by Edelberg (1967), Venables and Martin (1967), Miller (1968), and Montagu and Coles (1966).

III. Input Signal Conditioning, Amplification, and Registration

For SP measurements the critical requirements occur at the electrode–skin junction and in signal amplification. Both the tonic levels and phasic changes are fairly large (in the order of millivolts) so that medium gain amplification is sufficient. The period of response is quite long (seconds), making d.c. amplification desirable. Under some circumstances (for example, where the incidence of phasic changes is of primary interest and the effects of slow potential level shifts are to be minimized), a.c. amplification with long time constants may be used. Because of the direct pickup from the skin electrodes, it is important that the input impedance of the amplifier be high with reference to that of the skin. The only other signal conditioning that is required is a bucking voltage at the input to quantify initial levels and to balance or offset the zero point, plus conventional step-calibrated attentuation and gain control.

As was indicated earlier, there is some diversity of practice in measuring SC and SR, depending upon whether a constant voltage or a constant current system is used. Before turning to that central question, however, the nature of the measurement task will be reviewed. For exodermal measurement the essential requirement is that a known voltage be applied to the electrodes and a known current passed through the skin in order that either the resistance or conductance of the skin may be determined by Ohm's law.

Until recently the most commonly used circuit for this purpose was some form of bridge [for example, Wheatstone bridge, Darrow (1932); Wynn–Williams bridge, Haggard and Gerbrands (1947); a.c. impedance bridge, Edelberg (1967]. One advantage of the bridge circuit is ease in separating measurements of initial level and change by relating tonic level to the resistance (or impedance) quantities required to initially balance the bridge, and relating phasic changes to the voltage of imbalance occurring across the bridge with changes in resistance in the subject (electrodes) who occupies one arm of the bridge. In other types of arrangements the balancing feature is provided by a bucking voltage produced by a voltage divider. The bridge remains a useful input arrangement, and excellent reviews of its basic operation in this context are given by Edelberg (1967) and Venables and Martin (1967).

Returning to the simplest requirements of the measurement problem, the task is to impress a voltage across the skin electrodes in such a way that requirements are met for either maintaining a relatively constant current through the electrodes so that an initial resistance level can be read and momentary changes will be reflected by changes in apparent resistance of the skin; or maintaining a relatively constant voltage across the electrodes so that initial skin conductance can be read and momentary changes can be reflected which are proportional to phasic changes in conductance. The former (constant current) system is usually achieved by having the subject in series with a resistance that is very large compared to the resistance of the subject. Voltage across the series is set to the value desired for passage of an appropriate current through the electrodes (for example, 8 $\mu A/cm^2$). Momentary changes in skin resistance will thus produce only small changes

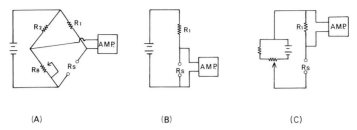

FIG. 10-3. Some basic circuits for measuring skin resistance and skin conductance. See text for description of components.

in the current through the subject (because of the large total resistance). This small change can then be reflected to the amplifier as a signal proportional to the change in resistance between the electrodes. Two forms of arrangements for this type of pickup are shown in Fig. 10-3A and B.

When serving as constant current inputs, the circuits in Fig. 10-3A and B share the feature that a voltage is applied across a series circuit involving the subject (R_s) and a series resistor (R_1) which is very large with reference to R_s. In the bridge circuit (Fig. 10-3A) a parallel branch is introduced in which R_2 is most likely to equal R_1 and R_b is a variable resistor to balance the bridge for differences in R_s. The applied voltage is set at a value appropriate to produce the prescribed current through R_s. When the bridge is balanced, no voltage is reflected to the amplifier and the initial resistance can be read from R_b. Changes in the subject's resistance will be reflected as voltages of imbalance in the bridge.

Simple modifications of the above bridge arrangement would introduce a small calibration resistor (say 1000 Ω) in the arm with R_s, which can be switched out to register a known change. A microammeter might be introduced to monitor current through R_s. Still another form of this circuit (often termed a Darrow bridge, see Hicks, Giesige and Fick, 1968) places R_b in the same arm of the bridge as R_s and infers initial resistance by the amount of R_b which must be taken out in order for the bridge to balance.

The other, simple series circuit (Fig. 10-3B), suffices with only R_1, the large resistance in series with R_s. Again current through R_s is set by adjusting the applied voltage. As drawn, the circuit would lack a method for allowing for differences in initial values of R_s. This can be added conveniently in either of two ways: inserting a voltage divider circuit between R_s and the amplifier to produce a small bucking voltage proportional to the base level, or putting a variable resistor (R_b) in the input circuit with R_s and subtracting from R_b the amount of R_s to keep the sum (of R_b and R_s) constant over events. Thus, R_b should have a maximum value slightly above the largest anticipated value of R_s. A calibration resistor and current monitoring meter can be added as before (that is, in immediate series with R_s).

To record by the second method (constant voltage) the electrodes are placed in series with a resistor which is small in comparison to the subject resistance and a known constant voltage is applied across them both. Change in current through the electrodes will then be reflected by changes of voltage across the small series resistor which serves as a signal resistor for calibration in units of conductance. Circuits for constant voltage pickup are also illustrated in Fig. 10-3. The circuit in Fig. 10-3A can be used to serve as a constant voltage input if appropriate changes in component sizes are made. To function as a constant voltage bridge, the value of R_1 would need to be small (for example, .01 to .05 of the lowest value of R_s) and R_b would again serve to balance the bridge for initial conductance. The ratio of R_2 to R_b would need to match the ratio of R_1 to R_s at the null condition for the bridge and a voltage divider rather than a single battery would be employed to set the voltage levels. Such an arrangement is included in the voltage supply of the simplified constant voltage circuit in Fig. 10-3C. Provision for different voltage levels is necessary because no single source voltage will suffice for all initial levels (if one is to operate with some limits to the range of current values which might occur with differences in basal skin conditions). One procedure for coping with this problem is to select levels of constant voltage to limit current to some starting value (like 10 μA, Lykken & Roth, 1961). Another procedure provides a means for balancing conductance through the signal resistor (R_1) at the outset for different starting levels, then measures changes from that level (Hagfors, 1964).

Both systems (constant current and constant voltage) have advantages and disadvantages which have been discussed at some length (Edelberg, 1967; Montagu & Coles, 1966, 1968; Wilcott & Hammond, 1965; Lykken & Venables, 1971). The central argument for current control is the previously mentioned nonlinearity between current and apparent resistance which leads to the recommendation of a standardized current density of 8 μA/cm^2. Control of current and use of the same current density for all circumstances thus minimizes differences between persons and measurement periods attributable to current-resistance nonlinearity. The central argument for voltage control rests upon recent evidence that voltage is a critical determiner of nonlinearity, plus an assumption about the nature of the source of the skin impedance. This assumption is that the primary determiner is the number of sweat glands that are active, and that the sweat glands act as conductance "units" or as resistances in parallel (Thomas & Korr, 1957; Montagu & Coles, 1966, 1968). It is then asserted that if only a few "units" are active and constant current is used, a disproportionately large current is forced through each gland, or unit, perhaps sufficient to alter its function. When constant voltage is used, the current through individual units will not change greatly but the total current through the electrodes will be proportional to the number of active skin (conductance) units. This last fact creates the disadvantage that it is impossible to set a constant voltage which will operate within tolerable current levels throughout the range of observable conductances produced by different conditions of skin hydration and initial resistance. Some of these difficulties and methods for dealing with them have been mentioned previously (Hagfors, 1964; Lykken & Roth, 1961).

Both systems are now in use. At least one manufacturer (Beckman) provides an input coupler with both constant voltage and constant current capabilities. A strong plea has been made recently (Lykken & Venables, 1971) for standardized use of the constant voltage system with direct readout of conductance values. It is worth noting in passing that in both systems described, the preferred initial unit on physiological grounds is conductance in micromhos per square centimeter. In the constant current system this is typically accomplished through point-by-point conversion of resistance values into reciprocal resistances, whereas in the constant voltage system, conductance values are read out directly. The conversion process has the disadvantage that it creates the possibility of conversion error (for example, as when one reciprocalizes a resistance change quantity rather than expressing it as a change in reciprocalized quantities) and in certain situations where equal quantities would not be obtained by the two methods (Lykken, 1968).

Previously mentioned characteristics of amplifiers for SP measures are generally appropriate for receipt and handling of the signal from the SC or SR input segment, although amplifier gain requirements will vary widely with the type of input used. Another major consideration will be the impedance matching of the input segment to the amplifier. Examples of circuits to meet these requirements appear periodically. Using the Grass Model 7 and Beckman Type R as examples, Fig. 10-4 presents a constant current input device designed for use with Grass equipment. Figure 10-5 presents a constant voltage system adapted to the Grass amplifier and Fig. 10-6 a constant voltage circuit adapted to the Beckman Dynagraph amplifiers. A description of instrumentation for electrodermal measurement prior to 1954 is given by Grings (1954); comprehensive accounts of currently used circuits are given by Edelberg (1967) and Venables and Martin (1967). Some recently published circuits other than those previously mentioned include Davis, Siddons, and Stout (1954), Malmo and Davis (1961), Simons and Perez (1965), Welford (1969), and Kuechenmeister (1970). Special-purpose devices provide for measuring total conductance of a group of subjects (Edelberg, 1967), for counting occurrence of SRRs (Burch & Childers, 1963), and provide a mixture of true and false GSR feedback to the subject (Borrey, Longstreet, & Grings, 1969).

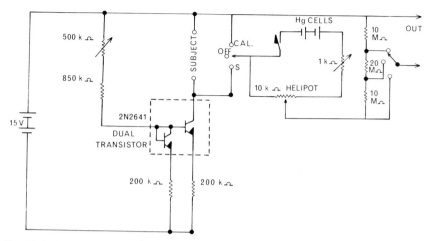

Fig. 10-4. A constant current input circuit, designed originally for use with the Grass Model 7 polygraph. (Courtesy of L. N. Law, Institute of Psychiatry, University of London.)

Final registration of the EDR information for inspection, analysis, and storage may be variously accomplished. Graphic chart readouts are most common, usually as a basis for scoring "by hand" or as a backup visual record for automatic processing. The period of the response is sufficiently long that inexpensive ink-writing milliammeters may be used, although usage requires an awareness of the high inertia in such systems. More sophisticated devices may provide a choice of features like rectilinear or curvilinear registering oscillographs. A few manufacturers have marketed special purpose digital writeout systems, typically providing visual displays (for example, Nixie tubes) and binary coded decimal outputs of base and change values for operating printers, tape or card punches, or digital magnetic recorders.

Continued expansion in the use of small hybrid computers (like the PDP-12) as general laboratory devices has greatly increased their use in processing electrodermal data. The computer may operate from an intermediate analog magnetic tape recording for convenience. For analog to digital conversion a slower rate of sampling is required than for most other phenomena (values between 5 and 20 per second cover most situations). Typical analysis follows what Ax (1967) terms a "points of interest" computer program.

IV. Measurement Units

Closely tied to questions of physical measurement are those concerning the units in terms of which the results are expressed. With endodermal mea-

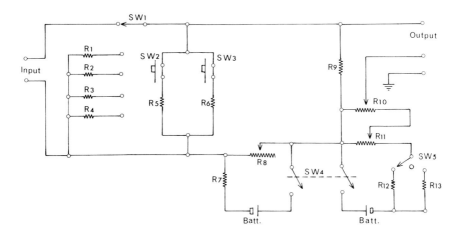

FIG. 10-5. Constant voltage circuit for use with Grass low-level dc preamplifier or equivalent. The output is 100 μV/μmho. Using values in parentheses for R_9, R_{12}, and R_{13} will increase output to 1 mV per micromho. R_8 is adjusted for .5 V across the input terminals. Component values are as follows:

R_1	500 kΩ, 1%	R_9	200 Ω, 1% (2000 Ω)	SW_1	5-position rotary "input switch"
R_2	200 kΩ, 1%	R_{10}	500 Ω, 10-turn	SW_2	pushbutton "1 μV"
R_3	100 kΩ, 1%	R_{11}	200 Ω trim pot	SW_3	pushbutton "5 μmho"
R_4	50 kΩ, 1%	R_{12}	10 kΩ, 1% (1 kΩ)	SW_4	DPST toggle "on–off"
R_5	1 MΩ, 1%	R_{13}	100 kΩ, 1% (10 kΩ)	SW_5	SPDT center off, toggle "suppression range" "1–off–100 μmho"
R_6	200 kΩ, 1%	Batt.	1.35 V Hg		
R_7	150 Ω				
R_8	100 Ω trim pot				

(From Lykken & Venables, 1971, by permission.)

surements the voltage scale is appropriate and convenient. Thus, it is used quite universally and the principal considerations are matters of polarity and properties of response dynamics rather than units or scales. Exodermal measurements, on the other hand, have had a long history of debate about what units to use and what transformations, if any, to perform on the obtained scores.

The issues about scale units stem from at least three major sources. One is theory, and represents the desire to choose a unit that best integrates the electrical measures with other known properties of the response, partic-

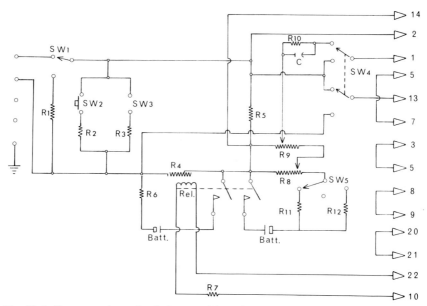

Fig. 10-6. Constant voltage circuit for use with Beckman Type R Dynograph. With SW_4 in down position, adjust R_4 for .5 V to subject. With dummy subject of 10 μmho (that is, SW_1 in up position), adjust R_8 for zero output when zero suppression (R_9) is full on and suppression range switch (SW_5) is set at 10 (that is, on R_{11}). The output of the preamplifier will be 100 μV/ μmhos. With the power amplifier attenuator set at "X.1," the preamplifier attenuator will read directly in μmhos per centimeter. Component values are as follows:

R_1	100 kΩ, 1%	R_9	500 Ω, 10-turn	SW_1	SPDT miniature toggle input switch
R_2	2 MΩ, 1%	R_{10}	10 Ω, 1%	SW_2	"Subj., 10 μmho"
R_3	200 kΩ, 1%	R_{11}	10 kΩ, 1%	SW_2	Miniature push-button ".5 μmho"
R_4	200 Ω trim pot	R_{12}	100 kΩ, 1%	SW_3	miniature push-button "5 μmho"
R_5	200 Ω, 1%	Batt.	1.35 V Hg	SW_4	DPDT miniature toggle "cal.—operate"
R_6	200 Ω	C_1	500 mfd, 3 V	SW_5	SPDT center off, min. toggle. Suppression range switch. "10–100 μmho"
R_7	200 Ω, ½ watt	Rel.	DPST Reed Relay (Hathaway J-2A)		
R_8	200 Ω trim pot				

(From Lykken & Venables, 1971, by permission.)

ularly its underlying physiology. A second source is statistical and evaluates electrodermal units in terms of frequency distribution characteristics (for example, normality) and the extent to which assumptions are met for application of test of significance (for example, independence of means and variances). The third source of concern deals with the fact that magnitudes of phasic reaction may be determined, in part, by initial level, making it sometimes desirable to seek a measure of change which is independent of initial level (that is, to seek a "base-free" unit).

The most commonly used physical units, those of conductance, find support from the argument that the principal resistive units in the skin are sweat glands and that those units operate as resistances in parallel. This suggests that their effects will add according to the reciprocal rule of parallel resistance. Since conductance is the reciprocal of resistance, it becomes the natural unit.[2] The "mho," which is the reciprocal "ohm," is too large for convenience in the present context so that a multiplier of 10^{-6} is used to express the number in micromhos. One of the most recent arguments for use of SCR rather than SRR was presented by Lader (1970), who introduced atropine into the skin and observed both specific and spontaneous GSRs as the drug gradually took effect, inhibiting sweat gland activity. He found SCRs to present an orderly picture of change, whereas SRRs were erratic. Darrow (1964) was led to similar conclusions about the theoretical basis of the response (namely, the preference for conductance over resistance units) from a somewhat different premise, that the response has a dual origin. Also, because of the reactive nature of the sweat glands and the tendency of biological tissue to react proportionately rather than absolutely, a logarithmic transformation is recommended (Darrow, 1937; Lader & Wing, 1966). Such a transformation has the advantage of normalizing SCL values (for example, Lacey, 1956).

Initial levels and measure of change in skin conductance, resistance, impedance, and potential all typically yield frequency distributions which are positively skewed when expressed in "raw" units. As a result, caution is required when summaries are made using descriptive statistics which are sensitive to or weighted by extreme values (for example, means and standard deviations). Where significance tests are run, transformations are customarily made to increase distribution symmetry and to achieve other desired properties, such as independence of means and variances as required by analysis of variance, or linearity of regression and homoscedasticity if regression adjustments are to be made. One recommended procedure is to employ a table like that provided by Bartlett (1947) as an aid in selecting the most appropriate transformation for a given set of data.

[2] At the time of this writing the journal *Psychophysiology* is adopting the editorial policy of specifying exodermal data as conductance in micromhos per square centimeter, with current density specified.

The literature contains many empirical checks of electrodermal units and resulting recommendations. Such empirical checks have not provided an answer applicable to all situations, and unfortunately descriptive accounts of research still appear in diverse units. Modal usage for description would probably be for expression of SCL as the logarithm of conductance. Measures of change may be expressed as change in log conductance (that is, $\Delta \log C$, or $\log C_{peak} - \log C_{onset}$) which will be recognized as being algebraically equivalent to the log of the ratio of resistance at outset and resistance at peak response, or $\log (R_{onset}/R_{peak})$. On predominately empirical grounds (increasing distribution symmetry), some workers have consistently measured change in the conductance scale, then transforming those change values into logs or square roots ($\log \Delta C$ or $\sqrt{\Delta C}$). The defense of such usage rests upon the effectiveness of the transformation in improving symmetry of distribution and the maintenance of data continuity within a particular context (e.g., discrimination conditioning where the last-mentioned units have been employed in about 80% of published studies in the last decade, thus providing comparability among studies). While such idiosyncratic decisions about units for expressing change may be defensible at the present time for descriptive purposes and may always be necessary to accommodate different statistical models in significance testing, there is strong argument for adoption of standard practice rather than the current range of unit scales in use.

The fact that the magnitude of electrodermal change may depend upon initial level of excitation (the so-called law of initial values) has led to efforts by some workers to produce base-free units. One of these was introduced by Lacey (1956) as an attempt to differentiate between measures of autonomic "tension" and autonomic "lability." His autonomic lability score is an adjusted poststimulus level (of log conductance) obtained by removing that part of the poststimulus value which is predictable by linear regression methods from the initial level. This value is then put into standard score form by expressing the adjusted value in units of the standard error of estimate and putting it on an arbitrary scale with mean of 50 and standard deviation of 10. This form of adjustment is essentially the same as those applying analysis of covariance to measures of change (for example, Benjamin, 1963, 1967).

Issues involved in such adjustments of change scores for initial level vary with the research use to be made of the electrodermal data. At least two general classes of situations occur. In one case the initial level and phasic values may be seen as two correlated indices of individual behavior, both of which contribute unique variation to the observational event. For optimum individual description it may be best to combine the information from the measures of level and change to provide a score which maximizes predictions of some other class of variables or external criteria (as, for example, when the canonical correlation between "electrodermal behavior" and some personality measures or clinical symptoms is of interest). At other times, emphasis will be upon one measure with an effort to exclude the effects of the other. An example would be the use of SCRs as an index of orienting behavior or of conditioned responding when it is desired to have differences between classes of an independent variable reflect measures of change, not differences in level. One method for achieving this control is to employ groups matched on initial level. Another is to use an adjustment process like analysis of covariance. The requirement most likely to create difficulty in this last situation is the need to keep the covariate (initial level) independent of the experimental manipulations (Evans & Anastasio, 1968).

One recent series of discussions argues that both statistical and excitation level problems can be solved by using an index based on response range (Lykken, Rose, Luther, & Maley, 1966; Lykken, 1968; Lykken & Venables, 1971). Tonic SC values are expressed as a ratio defined by the difference in momentary conductance level and minimum conductance level divided by the total range of conductance values (maximum minus minimum). Phasic responses, or measures of change, are expressed as the ratio of change in conductance to a prescribed stimulus and the maximum change elicited by a standard (strong) stimulus. This last measure is basically similar to a previously proposed "Paintal" index (Elliott & Singer, 1953; Paintal, 1951).

In the opinion of the writer, there are two general limitations to the range ratio procedure. One relates to the emphasis given to the range, which as a statistical quantity is notoriously unreliable with small samples. It will vary with a host of environmental and stimulus conditions from one situation to another, as well as with the number of observations in the sample from which it is estimated. This leads to the second problem, that the range score as described will always be a relative unit in which the reference conditions will be difficult to replicate from situation to situation and population to population.

The preceding statements do not argue against the use of range score in many cases. In fact, it may prove highly useful in reducing error and maximizing correlations within individuals and within specific situations. They do argue, however, against the adoption of such a unit as standard for all purposes as seems to have been advocated (Lykken & Venables, 1971).

As the preceding discussion should suggest, there is no single or universal solution to the "units" problem. Choice depends upon the considerations relevant for the particular application (statistical, theoretical, definitional). For any given research purpose, enough is known to make possible a suitable decision among alternatives.

For a review of early empirical checks of measurement units, see Haggard

(1949), Haggard and Gerbrands (1947), Lacey and Siegel (1949), and Schlosberg and Stanley (1953). Descriptions of distribution characteristics for various measures are given by O'Connell, Tursky and Evans (1967), Grings (1953), Shapiro and Leiderman (1964), and Surwillo (1969). Further discussion relevant to relations of initial (tonic) levels and measures of change will be found in Hord, Johnson, and Lubin (1964), Block and Bridger (1962), Wilder (1962), and Heath and Oken (1965).

References

Ax, A. F. Electronic storage and computer analysis. In P. Venables & I. Martin (Eds.), *Manual of psychophysiological methods*, Chapter 14. New York: Wiley, 1967.
Barnett, A. The phase angle of normal human skin. *Journal of Physiology* (*London*), 1938, **93**, 349–366.
Bartlett, M. S. The use of transformations. *Biometrics*, 1947, **3**, 39–52.
Benjamin, L. S. Statistical treatment of the law of initial values in autonomic research: A review and recommendation. *Psychosomatic Medicine*, 1963, **25**, 556–566.
Benjamin, L. S. Facts and artifacts in using analysis of variance to "undo" the law of initial values. *Psychophysiology*, 1967, **4**, 187–206.
Blank, I. H., & Finesinger, J. E. Electrical resistance of the skin. *Archives of Neurology and Psychiatry*, 1946, **56**, 544–557.
Block, J. D., & Bridger, W. H. The law of initial values in psychophysiology: A reformulation in terms of experimental and theoretical considerations. *Annals of the New York Academy of Sciences*, 1962, **98**, 1229–1241.
Borrey, R., Longstreet, B., & Grings, W. A system for registering both true and false feedback of the galvanic skin response. *Psychophysiology*, 1969, **6**, 366–370.
Brazier, M. A. B. The impedance of the skin. *Journal of the Institute of Electrical Engineers*, 1933, **73**, 204–209.
Burch, N. R., & Childers, H. E. Automatic GSR analyzer. *USAF School of Aerospace Medicine*, Tech. Report 63-74, 1963.
Burstein, K. R., Fenz, W. D., Bergeron, J., & Epstein, S. A comparison of skin potential and skin resistance responses as measures of emotional responsivity. *Psychophysiology*, 1965, **2**, 14–24.
Christie, M. J., & Venables, P. H. Effects on "basal" skin potential of varying the concentration of an external electrolyte. *Journal of Psychosomatic Research*, 1971, **15**, 343–348.
Christie, M. J., & Venables, P. H. Site, state and subject characteristics of palmar skin potential levels. *Psychophysiology*, 1972, **9**, 545–549.
Darrow, C. W. Uniform current for continuous standard unit resistance records. *Journal of General Psychology*, 1932, **6**, 471–473.
Darrow, C. W. The equation of the galvanic skin reflex curve. I. The dynamics of reaction in relation to excitation background. *Journal of General Psychology*, 1937, **16**, 285–309.
Darrow, C. W. The rationale for treating the change in galvanic skin response as a change in conductance. *Psychophysiology*, 1964, **1**, 31–38.

Davis, R. C. Factors affecting the galvanic skin reflex. *Archives of Psychology*, 1930, **18**, No. 115, 1–64.
Davis, R. C., Siddons, G. F., & Stout, G. L. Apparatus for recording autonomic states and changes. *American Journal of Psychology*, 1954, **67**, 343–352.
Edelberg, R. Electrical properties of the skin. In C. C. Brown (Ed.), *Methods in psychophysiology*. Baltimore, Maryland: Williams & Wilkins, 1967.
Edelberg, R. Biopotentials from the skin surface: The hydration effect. *Annals of the New York Academy of Sciences*, 1968, **148**, 252–262.
Edelberg, R. The Information content of the recovery limb of the electrodermal response. *Psychophysiology*, 1970, **6**, 527–539.
Edelberg, R. Electrical activity of the skin: Its measurement and uses in psychophysiology. In N. S. Greenfield and R. A. Sternbach (Eds.), *Handbook of psychophysiology*. New York: Holt, 1972.
Edelberg, R., & Burch, N. R. Skin resistance and galvanic skin response: Influence of surface variables and methodological implications. *Archives of General Psychiatry*, 1962, **7**, 163–169.
Edelberg, R., Greiner, T., & Burch, N. R. Some membrane properties of the effector in the galvanic skin response. *Journal of Applied Physiology*, 1960, **15**, 691–696.
Elliott, D. N., & Singer, E. G. The Paintal index as an indicator of skin resistance change to emotional stimuli. *Journal of Experimental Psychology*, 1953, **45**, 429–430.
Evans, S. H., & Anastasio, E. J. Misuse of analysis of covariance when treatment effect and covariate are confounded. *Psychological Bulletin*, 1968, **69**, 225–234.
Floyd, W. F., & Keele, C. A. Electrolytic phenomena in electrodes used for physiological purposes. *Journal of Physiology, London*, 1936, **86**, 25–27P.
Forbes, T. W., & Landis, C. The limiting AC frequency for the exhibition of the galvanic skin (psychogalvanic) response. *Journal of General Psychology*, 1935, **13**, 188–193.
Fowles, D. C., & Venables, P. H. The effects of epidermal hydration and sodium reabsorption on palmar skin potential. *Psychological Bulletin*, 1970, **73**, 363–378.
Freeman, G. L., & Katzoff, E. T. Methodological evaluation of the galvanic skin response, with special reference to the formula for the recovery quotient. *Journal of Experimental Psychology*, 1942, **31**, 239–248.
Gavira, B., Coyne, L., & Thetford, P. E. Correlation of skin potential and skin resistance measures. *Psychophysiology*, 1964, **5**, 465–477.
Grings, W. W. Methodological considerations underlying electrodermal measurement. *Journal of Psychology*, 1953, **35**, 271–282.
Grings, W. W. *Laboratory instrumentation in psychology*. Palo Alto, California: National Press, 1954.
Hagfors, C. Two conductance bridges for galvanic response measurements. Report No. 60 from Department of Psychology, University of Jyvaskyla, Finland, 1964.
Haggard, E. A. On the application of analysis of variance to GSR data. I. The selection of an appropriate measure. *Journal of Experimental Psychology*, 1949, **39**, 378–392.
Haggard, E. A., & Gerbrands, R. An instrument for the measurement of continuous changes in palmar skin resistance. *Journal of Experimental Psychology*, 1947, **37**, 92–98.
Heath, H. A., & Oken, D. The quantification of "response" to experimental stimuli. *Psychosomatic Medicine*, 1965, **27**, 457–471.
Hicks, R. G., Giesige, R., & Fick, A. An absolute ohmic readout of the galvanic skin response. *Psychophysiology*, 1968, **4**, 349–351.
Hord, D. J., Johnson, L. C., & Lubin, A. Differential effect of the law of initial values on autonomic variables. *Psychophysiology*, 1964, **1**, 79–87.

Hupka, R. B., & Levinger, G. Within subject correspondence between skin conductance and skin potential under conditions of activity and passivity. *Psychophysiology*, 1967, **4**, 161–167.

Kuechenmeister, C. Instrument for direct readout of log conductance measures of the galvanic skin response. *Psychophysiology*, 1970, **7**, 128–134.

Lacey, J. I. The evaluation of autonomic responses: Toward a general solution. *Annals of the New York Academy of Sciences*, 1956, **67**, 123–163.

Lacey, O. L., & Siegel, P. S. An analysis of the unit of measurement of the galvanic skin response. *Journal of Experimental Psychology*, 1949, **39**, 122–127.

Lader, M. H. The unit of quantification of the GSR. *Journal of Psychosomatic Research*, 1970, **14**, 109–110.

Lader, M. H., & Wing, N. *Physiological measures, sedative drugs and morbid anxiety*. London and New York: Oxford Univ. Press, 1966.

Landis, C. Electrical phenomena of the skin. *Psychological Bulletin*, 1932, **29**, 693–752.

Landis, C., & DeWick, H. N. The electrical phenomena of the skin (psychogalvanic reflex). *Psychological Bulletin*, 1929, **26**, 64–119.

Lloyd, D. C. Electrical impedance changes of the cat's footpad in relation to sweat secretion and reabsorption. *Journal of General Physiology*, 1960, **43**, 713–722.

Lykken, D. T. Properties of electrodes used in electrodermal measurements. *Journal of Comparative and Physiological Psychology*, 1959, **52**, 629–634.

Lykken, D. T. Neuropsychology and psychophysiology in personality research. In Borgotta & Lambert (Eds.), *Handbook of personality theory and research*. Chicago, Illinois: Rand McNally, 1968.

Lykken, D. T. & Roth, N. Continuous direct measurement of apparent skin conductance. *American Journal of Psychology*, 1961, **74**, 293–297.

Lykken, D. T., & Venables, P. H. Direct measurement of skin conductance: A proposal for standardization. *Psychophysiology*, 1971, **8**, 656–672.

Lykken, D. T., Miller, R. D., & Strahan, R. F. Some properties of skin conductance and potential. *Psychophysiology*, 1968, **5**, 253–268.

Lykken, D. T., Rose, R., Luther, B., & Maley, M. Correcting psychophysiological measures for individual differences in range. *Psychological Bulletin*, 1966, **66**, 481–484.

Malmo, R. B., & Davis, J. F. A monopolar method of measuring palmar conductance. *American Journal of Psychology*, 1961, **74**, 106–113.

Maulsby, R. L., & Edelberg, R. The interrelationship between the galvanic skin response, basal resistance and temperature. *Journal of Comparative and Physiological Psychology*, 1960, **53**, 475–479.

Miller, R. D. Silver–silver chloride electrodermal electrodes. *Psychophysiology*, 1968, **5**, 92–96.

Montagu, J. D. The psychogalvanic reflex: A comparison of AC skin resistance and skin potential changes. *Journal of Neurology, Neurosurgery, and Psychiatry*, 1958, **21**, 119–128.

Montagu, J. D. The psychogalvanic reflex: A comparison of DC and AC methods of measurement. *Journal of Psychosomatic Research*, 1964, **8**, 49–65.

Montagu, J. D., & Coles, E. M. Mechanism and measurement of the galvanic skin response. *Psychological Bulletin*, 1966, **65**, 261–279.

Montagu, J. D., & Coles, E. M. Mechanism and measurement of the galvanic skin response: An addendum. *Psycholegical Bulletin*, 1968, **69**, 74–96.

Neumann, E. Thermal changes in palmar skin resistance patterns. *Psychophysiology*, 1968, **5**, 103–111.

O'Connell, D. N., & Tursky, B. Silver–silver chloride sponge electrodes for skin potential recording. *American Journal of Psychology*, 1960, **73**, 302–304.

O'Connell, D. N., Tursky, B., & Evans, F. J. Normality of distribution of resting palmar potential. *Psychophysiology*, 1967, **4**, 151–155.

O'Conneli, D. N., Tursky, B., & Orne, M. T. Electrodes for the recording of skin potential. *Archives of General Psychiatry*, 1960, **3**, 252–258.

Paintal, A. S. A comparison of the galvanic skin responses of normals and psychotics. *Journal of Experimental Psychology*, 1951, **41**, 425–428.

Plutchik, R., & Hirsch, H. R. Skin impedance and phase angle as a function of frequency and current. *Science*, 1963, **141**, 927–928.

Quilter, R. E., & Surwillo, W. W. A simple method for preparing silver–silver chloride electrodes for recording skin potential. *American Journal of Psychology*, 1966, **79**, 309–313.

Ruckmick, C. A. *Psychology of feeling and emotion.* New York: McGraw-Hill, 1937.

Schlosberg, H., & Stanley, W. S. A simple test of the normality of twenty-four distributions of electrical skin conductance. *Science*, 1953, **117**, 35–37.

Shackel, B. Skin-drilling: A method of diminishing galvanic skin potential. *American Journal of Psychology*, 1959, **72**, 114–121.

Shapiro, D., & Leiderman, P. H. Studies on the galvanic skin potential level: Some statistical properties. *Journal of Psychosomatic Research*, 1964, **7**, 269–275.

Simons, D. G., & Perez, R. E. The B/GSR module: A combined recording to present base skin resistance and galvanic skin reflex activity patterns. *Psychophysiology*, 1965, **2**, 116–124.

Smith, C. E. The effect of changes in temperature and humidity of the air on the apparent skin resistance. *Journal of Psychology*, 1937, **3**, 325–331.

Surwillo, W. W. Statistical distribution of volar skin potential level in attention and the effects of age. *Psychophysiology*, 1969, **6**, 13–16.

Thomas, P. E., & Korr, I. M. Relationship between sweat gland activity and electrical resistance of the skin. *Journal of Applied Physiology*, 1957, **10**, 505–510.

Uno, T., & Grings, W. Autonomic components of orienting behaviour. *Psychophysiology*, 1965, **1**, 311–321.

Venables, P. H. The relationship between PGR scores and temperature and humidity. *Quarterly Journal of Experimental Psychology*, 1955, **7**, 12–18.

Venables, P. H., & Christie, M. J. Mechanisms, instrumentation, recording techniques, and quantification of responses. In W. Prokasy and D. Raskin (Eds.), *Electrodermal activity in psychological research,* Chapter 1. New York: Academic Press, 1973.

Venables, P. H., & Martin, I. Skin resistance and skin potential. In P. Venables and I. Martin (Eds.), *Manual of psychophysiological methods.* New York: Wiley, 1967.

Venables, P. H., & Sayer, E. On the measurement of the level of skin potential. *British Journal of Psychology*, 1963, **54**, 251–260.

Welford, N. T. A constant current skin resistance coupler. *Psychophysiology*, 1969, **5**, 724–726.

Wenger, M. A., & Cullen, T. D. Some problems in psychophysiological research. Part III. The effects of uncontrolled variables. In R. Roessler and N. S. Greenfield (Eds.), *Physiological correlates of psychological disorder.* Madison: Univ. of Wisconsin Press, 1962.

Wenger, M. A., & Gustafson, L. A. Some problems of psychophysiological research. Part II. Effects of continuous current on measurements of electrical skin resistance. In R. Roessler and N. S. Greenfield (Eds.), *Physiological correlates of psychological disorder.* Madison: Univ. of Wisconsin Press, 1962.

Wilcott, R. C. Correlation of skin resistance and potential. *Journal of Comparative and Physiological Psychology*, 1958, **51,** 691–696.

Wilcott, R. C. Effects of high environmental temperature on sweating and skin resistance. *Journal of Comparative and Physiological Psychology*, 1963, **56,** 778–782.

Wilcott, R. C., & Hammond, L. J. On the constant-current error in skin resistance measurement. *Psychophysiology*, 1965, **2,** 39–41.

Wilder, J. Basimetric approach (law of initial values) to biological rhythms. *Annals of the New York Academy of Sciences*, 1962, **98,** 1211–1228.

Yokota, T., Takahashi, T., Kondo, M., & Jujimori, B. Studies on the diphasic wave form of the galvanic skin reflex. *Electroencephalography and Clinical Neurophysiology*, 1959, **11,** 687–696.

Chapter 11

The Electrogastrogram[1]

Daniel Lilie

*Southern Wisconsin Colony
and Training School
Union Grove, Wisconsin*

I. Introduction 297
II. Recording Technique 298
III. Data Analysis 299
IV. The Data 300
V. Clinical Application of the EGG 301
VI. Problems and Caveats 301
VII. Concluding Statement 304
   References 304

I. Introduction

The electrogastrogram (EGG) is a recording of the bioelectrical potentials associated with gastrointestinal (GI) activity in intact organisms. EGGs may be obtained from electrodes that are introduced into the stomach via intubation or by external electrodes that are placed on the abdomen.

---

[1] In the following paper, I am not using the term "we" in the editorial sense. Rather, I am referring to my colleagues who collaborated with me in this work, in particular, Bernard J. O'Loughlin, Roger W. Russell, and John K. Stevens.

For the purposes of this paper, I will be discussing EGGs obtained from external electrodes, unless otherwise noted.

Although many other techniques have been developed for studying GI activity, including fluoroscopy, telemetry, and various mechanical methods for detecting changes in intragastric pressure, they all require the introduction of foreign substances into the GI tract and thus raise the possibility of interfering with the activity that is to be studied or recorded. One of the great advantages of the EGG over other techniques is that it allows one to record GI activity without directly interfering with it.

Alvarez (1922) was the first to report obtaining an EGG from external electrodes. Very little further work was done with the EGG until Davis, Garafolo, and Gault (1957) reported an exploratory study with human subjects. Subsequently, Davis and his colleagues performed a series of EGG studies in which they investigated the effects on GI motility of food, rest, visual stimulation, and a gastric balloon (Davis, Garafolo, & Kveim, 1959), avoidance conditioning (Davis & Berry, 1963) and body position (Paden, Garrison, & Davis, 1959). Work in Davis' laboratory continued, after his untimely death, along two main lines: (1) interactions between conditioning (of other responses) and the EGG (Fedor & Russell, 1965; Stern, 1966) and (2) methodological refinement (Stevens & Worrall, 1968).

While these studies deal with the EGG as a tool in experimental psychophysiology, Davis *et al.* (1957) were very much aware of its potential use in clinical assessment and expressed the hope that it would be as useful as the EKG. Morton, Davis, and Lipowski (1959) reported using the EGG as an adjunct in the psychiatric interview. However, they placed electrodes into the patients' stomachs. Soviet workers have reported using the EGG to diagnose ulcers and cancer (Krasil'nikov, 1960), but certain methodological questions are left unanswered in their reports.

A fairly comprehensive account of the laboratory EGG is given by Russell and Stern (1966), so that I will recapitulate what is presented there only insofar as it is necessary to discuss the work that has been done since then. Much of our work has been an attempt to assess the EGG as a clinical diagnostic tool. There are some encouraging preliminary results, but much work remains to be done. I will describe our methods, data, and some of the problems that we have encountered.

II. Recording Technique

Two silver cup electrodes (Grass Instrument Co., model E5S) are chlorided and filled with electrode paste; one is placed on the abdomen, approximately over the antrum and the other (reference) electrode is fastened to the right leg, 4–5 inches above the ankle. The electrodes are held firmly in place with adhesive tape. We have also found it useful to strap an EKG belt around the subject and over the electrode when recording from humans. Although the electrodes could probably be used several times without being rechlorided, we rechloride electrodes before each recording session. Without this precaution, the EGG might be very unstable, owing to electrode polarization. The problem of "drift" in the EGG due to polarization of the electrodes can be quite serious. It is very important, for example, to minutely inspect electrodes after they have been chlorided to ensure that there are no cracks or holes in the chloride coating. It is also desirable to test the electrodes before using them by putting two of them in a saline solution and then checking for any potential difference between them.

The areas where the electrodes are placed are first shaved, when necessary, and then rubbed vigorously with alcohol and with a small amount of electrode paste. Human subjects are instructed to lie as quietly as possible in a supine position and to breathe normally.

Animal subjects are physically restrained in such a manner that they cannot pull off the electrodes. In our work with squirrel monkeys, we generally "break in" new subjects by fastening the electrodes to them for several hourly intervals before we do any recording. The subjects habituate to having the electrodes taped to them, thus reducing the amount of movement artifacts in the record. Even though the subjects are restrained, they can engage in a certain amount of movement. We have also found that the amount of physical restraint may be reduced for squirrel monkeys after they become accustomed to having the electrodes taped to them. They generally will not try to remove them at that stage.

The output of these electrodes is fed into an amplifier that has the following characteristics: (1) high input impedance, (2) sensitivity to fractions of a millivolt, (3) capability of registering d.c. potentials, but (4) not necessarily responding to frequencies greater than 1 per sec (Russell & Stern, 1966, p. 231). For example, in a portion of our work, we used a Beckman Type R Dynograph recorder (d.c. coupled) with a low-pass filter to reduce EKG artifact. The recorder is calibrated before and after each recording session. Also, it is necessary to have a device for creating a bucking voltage to deal with the relatively large and fairly constant potential between the leg and the stomach.

The record is traced out on curvilinear paper; some of our data were also stored on analog tape for later analog-to-digital conversion.

III. Data Analysis

The EGG may be analyzed with respect to frequency, period, peak amplitude, total motility, displacement, and peak response time (Russell &

Stern, 1966, pp. 234–235). All of these measures must be obtained by hand, and this procedure is often quite tedious and prone to error. Another method of analysis, first applied to EGG data by Stevens and Worrall (1968), involves autocorrelation and Fourier analysis (Rosenblith, 1962).

As mentioned above, the data were stored on analog tape. They were then fed into an analog-to-digital converter, which punched the digitized data on paper tape at a sampling rate of one per second. We generally digitized two or three 10 minute samples of our analog data from a given session. Very much more than that would have overtaxed both our time and our computer. The data were then analyzed by computer, using an adaptation of the U.C.L.A. Biomedical Program BMD02T. Prior to this analysis, we removed the "resets" in our record (the bucking voltage that we introduced to counteract drift in the record) by means of another program. The results of this analysis are to eliminate the random noise in the data and to enable us to identify the frequency components of the EGG in the range of 0.7 to 12 cycles per minute.

IV. The Data

Figure 11-1 shows an EGG record from a human subject with an empty stomach. Figure 11-2 shows an EGG record of the same subject approximately 20 min later, after he had eaten a slice of bread. Characteristically, there is 3–4 per minute cyclical activity in the record which became more marked after the subject swallowed some food. The power spectra corresponding to these records are shown in Figs. 11-3 and 11-4. Note that there is an increase in the power spectrum (Fig. 11-4) for the postingestion EGG at 3.141 cycles per minute as compared with the prefood power spectrum (Fig. 11-3). This activity in the 3–4 per minute range was also observed by virtually all previous investigators of the EGG. The frequencies in the 1 to 2 per minute range that are shown in the power spectra are shown in the power spectra are something of a puzzle to us. We are not sure whether these frequencies are "phantoms" that are introduced by our method of analysis or whether they actually exist in the data. Visual inspection of the unanalyzed EGGs would lead me to conclude that these early peaks in the power spectra are caused by artifacts, but, of course, simple visual inspection could be misleading here. There is no such problem of interpretation with the activity in the 3–4 per minute range. The difference that is shown between the prefood and the postfood power spectra is a reliable one.

V. Clinical Application of the EGG

Since there are normal patterns of stomach activity that are detectable with the EGG, there may also be characteristic abnormal patterns which are associated with the onset and development of GI pathology. We were hopeful that we could find these patterns which, according to the Soviet literature (Krasil'nikov, 1960), do exist and are different from each other, depending upon the nature of the underlying pathology.

A pilot study which we conducted at Indiana University did not reveal any obvious (or even any subtle) differences between subjects with a history of GI disorders and a matched control group with no such history. The control group was matched with the experimental group on the basis of age, sex, and weight. However, one problem with this study was that we were unable to recruit very many subjects who had current GI disorders; most of the experimental subjects had only a history of such disorders. Also, many subjects were receiving various types of medication; we could not very well ask them to stop taking their medicine while we recorded their EGGs.

Obviously, we needed access to a hospital population. Fortunately, this was possible and we obtained EGG records from 40 patients at the Orange County Medical Center in Orange, California. All of these were patients who had come in for upper GI radiological examinations. In order to minimize "experimenter bias" (Rosenthal, 1966), the experimenter was not told about the nature and extent of pathology in any patient prior to the recording session. After the data were collected, the medical records of all of the subjects were examined.

Figure 11-5 shows a portion of the EGG from a patient with an esophageal hiatal hernia and duodenal ulcer. When compared with a "normal" EGG (as in Figs. 11-1 and 11-2), the record appears to be more irregular and includes episodes of relatively rapid, low-amplitude activity. This can also be seen in Fig. 11-6, which shows a portion of the EGG from another patient with a posterior wall duodenal ulcer. Again, note the relatively rapid, low amplitude activity. This was never observed in EGGs from patients who had no ulceration. This may be an important clinical sign, but further studies must be done to confirm the significance of this activity in the EGG.

VI. Problems and Caveats

One possible problem in interpreting the EGG is determining the locus of the recorded activity. How can we be reasonably certain that we are recording primarily stomach activity? Russell and Stern (1966, pp. 235–241) have discussed several lines of evidence to suggest that we are indeed picking up bioelectric potentials primarily from the stomach in the EGG. Also, further work since then (Stevens & Worrall, 1968) has demonstrated a high correlation between the EGG and mechanogram records obtained from strain gauges sutured directly to the smooth muscle layers of cats' stomachs.

Certain assumptions are made when an autocorrelation is performed. One is that the data are periodic. Another is that they constitute a stationary, stochastic time series. We believe that these assumptions are met closely enough in our data for the purposes of this type of analysis.

Another problem with autocorrelation and Fourier analysis is that it eliminates the aperiodic, "irregular" disturbances in the EGG which may

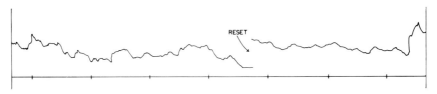

FIG. 11-1. EGG prior to food ingestion. Vertical marks indicate 30 sec intervals. Amplitude of activity is in the order of several hundred microvolts. Reset is shown where the activity drifted out of range.

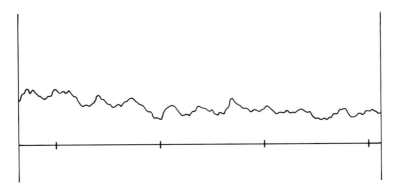

FIG. 11-2. Postingestion EGG. Again, vertical marks indicate 30 sec intervals. Note that the scale is not the same as for Fig 11-1

FIG. 11-3. Power spectrum of autocorrelation of EGG shown in Figure 1. Numbers along the top and bottom indicate frequencies in cycles per minute. Numbers along ordinate are arbitrary; the units are relative. Arrow points to peak in 3–4 per minute range.

FIG. 11-4. Power spectrum of autocorrelation of EGG shown in Fig. 11-2. The scale is not the same as for Fig. 11-3.

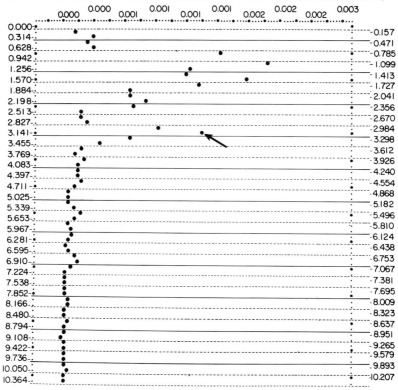

FIG. 11-3.

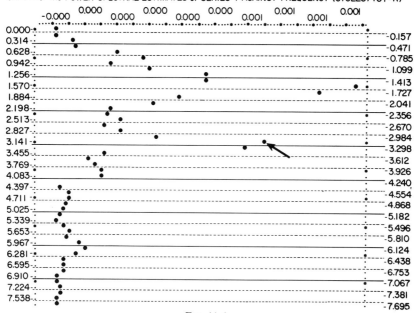

FIG. 11-4.

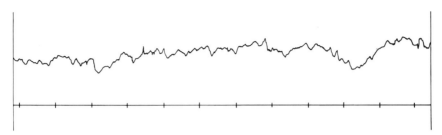

FIG. 11-5. EGG for patient with esophageal hiatal hernia and duodenal ulcer.

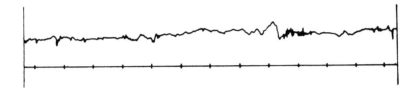

FIG. 11-6. EGG for patient with posterior wall duodenal ulcer.

be important clinical signs. So far, it does not appear that the methods of time series analysis will add importantly to the clinical usefulness of the EGG.

Finally, I believe that we were overconcerned with the "experimenter effect" in our clinical study. As it turned out, very few of our 40 subjects had currently active GI disorders, hence the extreme tentativeness of our conclusions to date. One solution to this problem is to record fairly extensively from a small number of patients who are known to have currently active GI pathology, and to record from these patients under a variety of conditions. This is what we plan to do in future work.

VII. Concluding Statement

It is our hope that more workers in the areas of experimental psychophysiology and clinical diagnosis will investigate the EGG as a tool for their work. We believe that we have made only a very modest beginning and that the technique will prove to be of great value, both in basic and applied research.

References

Alvarez, W. C. The electrogastrogram and what it shows. *Journal of the American Medical Association*, 1922, **78**, 1116–1119.

Davis, R. C., & Berry, F. Gastrointestinal reactions during a noise avoidance task. *Psychological Reports*, 1963, **12,** 135–137.

Davis, R. C., Garafolo, L., & Gault, F. P. An exploration of abdominal potentials. *Journal of Comparative and Physiological Psychology*, 1957, **50,** 519–523.

Davis, R. C., Garafolo, L., & Kveim, K. Conditions associated with gastrointestinal activity. *Journal of Comparative and Physiological Psychology*, 1959, **52,** 466–475.

Fedor, J. H., & Russell, R. W. Gastrointestinal reactions to response-contingent stimulation. *Psychological Reports* (Monograph Supplement 1-V16), 1965, **16,** 95–113.

Krasil'nikov, L. G. Klinicheskoye znocheniye elektrogastrografii. (Clinical importance of the electrogastrography.) *Soviet Medicine*, 1960, **3,** 107–114.

Morton, H. S., Davis, J. F., & Lipowski, Z. J. The use of the electrogastrograph in problem identification in psychoneurotic patients. *Psychiatric Research Reports*, 1959, **11,** 70–82.

Paden, A. B., Garrison, P., & Davis, R. C. The effect of body position on gastro-intestinal activity. Technical Report No. 30, Indiana University, 1959, Contract Nonr-908(03), Office of Naval Research.

Rosenblith, W. A. *Processing neuroelectric data.* Cambridge, Massachusetts: M.I.T. Press, 1962.

Rosenthal, R. *Experimenter effects in behavioral research.* New York: Appleton, 1966.

Russell, R. W., & Stern, R. M. Gastric motility: The electrogastrogram. In I. Martin and P. Venables (Eds.), *Manual of psychophysiological methods.* Amsterdam: North-Holland Publ., 1966. Pp. 221–243.

Stern, R. M. A re-examination of the effects of response-contingent aversive tones on gastrointestinal activity. *Psychophysiology*, 1966, **2,** 217–223.

Stevens, J. K., & Worrall, N. Abdominal surface potentials and their relation to gastric motility. Technical Report No. 23, Indiana University, 1968, Contract Nonr 908-15, Office of Naval Research.

Bibliography: Methods and Techniques

This bibliography was provided us by the extremely kind, efficient, and capable offices of the UCLA Brain Information Service and the Information Services, Pacific Southwest Regional Medical Library Service. It is not meant to be exhaustive or even systematic; rather it provides a sampling of recent and current articles and books relevant to methods of bioelectric recording.

The references given here follow the volume categories exactly. Methods related more to recording of cellular processes and to analytic recording of brain activity in animals are given in the bibliography at the end of Part A, whereas references related to human EEG recording techniques and human evoked potentials and the CNV are in the bibliography at the end of Part B.

Cochlear Potential Methods

Aran, J. M., Portmann, C., Delaunay, J., Pelerin, J., & Lenoir, J. L'electro-cochleogramme — methodes et premiers resultats chez l'enfant. (The electro-cochleogram—methods and results in children. *Revue Laryngologie (Bordeaux)*, 1969, **90**(11), 615–634. (French)

Daigneault, E. A., Brown, R. D., & Blanton, J. F. Alterations of round window recorded N1 by selective sections of the cochlear and vestibular nerves. *Acta Otolaryngologica* (Stockholm), 1970, **70**(4), 254–259.

Dallos, P. Combination tone 2FL-FH in microphonic potentials. *Journal of the Acoustical Society of America,* 1969, **46**(6 Pt.2), 1437–1444.

Dallos, P., Schoeny, Z. G., & Cheatham, M. A. On the limitations of cochlear-microphonic measurements. *Journal of the Acoustical Society of America,* 1971, **49**(4 Pt.2), 1144–1154.

Legouix, J. P. Experimental study of the influence of stimulus duration on cochlear responses to transient signals. (Etude experimentale de l'influence de la durée du stimulus sur les responses cochléaires aux signaux transitoires). *International Audiology,* 1969, **8**(4), 591–594. (French)

Mainen, M. W., Chiong, V. T., Glackin, R. N., Elder, J. C., & Warfield, D. Magnitude of cochlear microphonics elicited by components of broad-band noise. *Journal of the Acoustical Society of America,* 1970, **47**(4 Pt.2), 1139–1143.

Peake, W. T., Sohmer, H. S., & Weiss, T. F. Microelectrode recordings of intracochlear potentials. *Quarterly Progress Report of MIT Research Laboratory Electronics*, 1969, **94**, 293–304.

Portmann, M. Cochleogramme et reflexe psychogalvanique, methodes d'audiometrie objective. (Cochleogram and psychogalvanic reflex. Methods for objective audiometry). *Journal Francais Otorhinolaryngology*, 1970, **19**(1), 39–43. (French)

Portmann, M., & Aran, J. M. Electro-cochleography. *Laryngoscope*, 1971, **81**(6), 899–910.

Praxhma, I. Experiences in recording cochlear potentials with the aid of glass microelectrodes. *Vestnik Otorinolaringologii*, 1969, **31**(4), 71–76. (Russian)

Salomon, G., & Elberling, C. Cochlear nerve potentials recorded from the ear canal in man. *Acta Otolaryngolica* (Stockholm), 1971, **71**(4), 319–325.

Sohmer, H., & Feinmesser, M. Cochlear and cortical audiometry conveniently recorded in the same subject. *Israel Journal of Medical Sciences*, 1970, **7**(2), 219–223.

Yoshie, N., & Yamaura, K. Cochlear microphonic responses to pure tones in man recorded by a non-surgical method. *Acta Otolaryngolica Supplement* (Stockholm), 1969, **252**, 37–69.

Electroretinography

Gouras, P. Electroretinography: Some basic principles. *Investigative Ophthalmology*, 1970, **9**, 557–569.

Ikeda, H., & Pringle, J. Instrument note: A microelectrode advancer for intraretinal recording from the cat. *Vision Research*, 1971, **11**, 1169–1173.

Nagata, M., & Honda, Y. Studies on the local electric response of the human retina. I: An instrument for stimulating local retinal areas in various photic conditions. *Acta Society of Ophthalmology of Japan*, 1970, **74**, 388–394. (Japanese)

Nagayma, R. Studies on the electro-retinogram of the isolated rabbit retina. I: Method for maintaining light-evoked responses *in vitro*. *Acta Society of Ophthalmology of Japan*, 1969, **73**, 1900–1908.

Peckham, R. H. Practical ERG in clinical optometry. *American Journal of Optometry*, 1969, **46**, 725–730.

Peckham, R. H. The AC amplifier artifact in electroretinography. *American Journal of Optometry and Archives of the American Academy of Optometry*, 1971, **48**, 932–935.

Ronchi, L. Effect of tissue–metal contact variation on ERG response. *Atti Fondazione Giorgio Ronchi*, 1969, **24**, 486–491. (Italian)

Satoh, Y. Studies on ERG electrodes with reference to oscillatory potentiation. *Acta Society of Ophthalmology of Japan*, 1970, **74**, 949–950.

Schmidt, B. Contributions to the recording technique for clinical ERG. *Berichte Deutsche Ophthalmologische Gesellschaft (München)*, 1970, **70**, 575–580. (German)

Stangos, N., Rey, P. & Gonzalo-Platero, J. Essai de standardisation d'une méthode pour l'étude des potentiels oscillatoires chez l'homme. *Ophthalmologica*, 1970, **161**, 202–209. (French)

Tamura, O. A system for the measurement of active potential distribution in the retina. *Folio Ophthalmology of Japan*, 1971, **22**, 367–370.

Yonemura, D. Review of Electroretinography. *Ophthalmology (Tokyo)*, (*Japanese Journal of Ophthalmology*), 1971, **13**, 1057–1073.

Yonemura, D., & Hasui, I. Introduction to the ERG. ERG recordings and some problems of artifacts. *Ophthalmology*, 1971, **13**, 1043–1051.

Electro-Olfactogram Methods

Caruso, V., Hagan, J., & Manning, H. Quantitative olfactometry in the measurement of post-traumatic hyposmia. A simple method. *Archives of Otolaryngology,* 1969, **90,** 500–503.

Eyferth, K. A simple olfactometer. *Hals-, Nasen-, Ohrenaerzte,* 1969, **17,** 14–15. (German)

Herberhold, C. Absorption of odor particles on thermistors. *Archiv für klinische und experimentelle Ohren-Nasen-und Kehlkopfheilkunde,* 1969, **194,** 435–440. (German)

Herberhold, C. NTC thermistors as odor sensible elements in a so-called artificial nose. *Biomedizinische Technik (Stuttgart),* 1971, **16,** 127–130. (German)

Kitsera, A. E. New model of a clinical impulse olfactometer. *Zhurnal Ushnykh, Nosovykh i Gorlovyk Boleznei,* 1971, **31,** 95–97. (Russian)

Nishida, H., & Maeda, T. Olfactometry by means of pupillary reaction. *Otolaryngology (Tokyo),* 1971, **43,** 425–431. (Japanese)

Rous, J., & Synek, V. Objektivni olfaktometrie U laryngektomovanych. (Objective method of olfactometry in patients after laryngectomy). *Ceskoslovenska Otolaryngologie,* 1971, **20**(1), 13–22. (Czechoslovakian)

Takagi, S. F. EOG problems. In *Olfaction and taste. Proceedings of the third international symposium.* New York: Rockefeller Univ. Press, 1969. Pp. 71–91.

Eye Movements

Burgeat, M., Fontelle, P., Danon, J., Coevoet, A., & Lecam, D. Automatic calibration of eye movements. *Annals of Otolaryngology (Paris),* 1969, **86,** 101–106. (French)

Dorne, P. A., & Espiard, J. F. The Electroculogram (EOG) principle and technic. Its value in the study of maculopathies. *Archives of Ophthalmology (Paris),* 1971, **31,** 217–224. (French)

Edwards, D. C., Antes, J. R., Adams, R. W., & Trumm, G. A. Comparison of first eye movement detection methods. *Perception and Motor Skills,* 1971, **32,** 435–441.

Gabersek, V., Carre, F., & Augay, C. New electrode-supporting ocular device for the recording of eye movements. *Revue Neurologique (Paris),* 1969, **121,** 266–267. (French)

Gabersek, V., Carre, F., & Augay, C. New electrode carrying spectacles for the recording of ocular movements. *Electroencephalography and Clinical Neurophysiology,* 1970, **28,** 646.

Giulling, E. V. Procedure for implanting electrodes for electronystagmography in chronic experiments on rabbits. *Zhurnal Ushnykh, Nosovykh i Gorlovyk Boleznei,* 1969, **29,** 112–113. (Russian)

Jungmayr, H. Diagnostic possibilities with binocular lead of electronystagmogram. *Zeitschrift für Larngologie, Rhinologie, Otologie und Ihre Grenzgebiete (Stuttgart),* 1971, **50,** 286–292. (German)

Kozaki, M. Studies on objective convergence test by means of EOG. I. Recording instrument and normal convergence. *Acta Societatis Ophthalmologicae Japonicae,* 1970, **74,** 876–881. (Japanese)

Laurent, J. G., & Pialoux, P. Choice of amplifier for better recording of vestibular responses. *Annales d'Oto-Laryngologie et de Chirurgie Cervico-Faciale (Paris),* 1971, **88,** 135–141. (French)

Lawwill, T. Electrodiagnostic techniques in ophthalmology. *Journal of the Kentucky State Medical Association,* 1971, **69,** 97–99.

Minard, J. G., & Krausman, D. Rapid eye movement definition and count: An on-line detector. *Electroencephalography and Clinical Neurophysiology,* 1971, **31**(1), 99–102.

Minowada, H., Niki, T., & Kitahara, M. DC-electronystagmograph. *Journal of Otolaryngology (Japan)*, 1970, **73** (Suppl.), 1092–1093. (Japanese)

Pietruski, J., Danon, J., & Grall, Y. Use of a caloric stimulator with output and temperature controls in electronystagmography. *Annales d'Oto-laryngologie et de Chirurgie Cervico-Faciale (Paris)*, 1971, **88,** 95–98. (French)

Pignataro, O., & Sambataro, G. Recording with amplifiers using continuous and alternating current in electronystagmographic practice. *Archivio Italiano di Otologia*, 1970, **81,** 136–142. (Italian)

Smith, J. R., Cronin, M. J., & Karacan, I. A multichannel hybrid system for rapid eye movement detection (REM detection). *Computers and Biomedical Research*, 1971, **4**(3), 275–290.

Strzyzewski, K., & Barej, A. Simplified method of EOG recording. *Klinika Oczna*, 1971, **41,** 185–189. (Polish)

Takagi, S. F. EOG problems. In *Olfaction and taste. Proceedings of the third international symposium.* New York: Rockefeller Univ. Press, 1969. Pp. 71–91.

Toth, M. F. A new method for detecting eye movement in sleep. *Psychophysiology*, 1970, **7**(3), 516–523.

Woillez, M., Dhedin, G., Constantinides, G., Tilloy, R., & Malbrel, P. Recording of rotatory nystagmus by an opticelectronic (E.Y.E.) method. *Bulletins et Memoires de la Société Française d'Ophtalmologie*, 1970, **83,** 451–465. (French)

Yarbus, A. L. *Eye movements and vision.* New York: Plenum, 1967.

Electromyography

Arutiunov, A. I., Kadin, A. L., Konovalov, A. N., Raeva, S. N., Mershchikova, T. A., & Seliverstov, V. V. Experience in applying the microelectrode method in neurosurgical practice during one-stage stereotaxic operations for Parkinsonism. *Voprosy Neirokhirurgii (Moskva)*, 1970, **34**(6), 12–18. (Russian)

Asgian, B., Drasoveanu, C., & Popoviciu, L. Contributii la metodologia examenului electromiografic al musculaturii velo-palatine, faringiene si laringiene. (Methodologic contributions to the electromyographic examination of velopalatine, pharyngeal and laryngeal muscles.) *Studii si Cercetaride Neurologie*, 1970, **15**(5), 319–326. (Rumanian)

Ball, R. L. Amplifier for electromyographic signals. *American Journal of Physical Medicine*, 1969, **48**(3), 116–118.

Bashkirova, L. S., & Zhizhina, Z. H. I. Tremography method. *Gigiena Truda i Professional'nye Zabolevaniya*, 1970, **14,** 59–61. (Russian)

Berger, W., & Barr, L. Use of rubber membranes to improve sucrose gap and other electrical recording techniques. *Journal of Applied Physiology*, 1969, **26**(3), 378–382.

Blanton, P. L., Biggs, N. L., & Perkin, R. C. Apparatus facilitating the preparation of fine wire electrodes for electromyography. *Electromyography*, 1969, **9**(2), 213–214.

Boyanov, B. Functional methods of investigation in prosthodontics. *International Dental Journal*, 1969, **19,** 273–284.

Bradley, G. W., & Mustafa, K. Y. A digital recording system for respiratory variables. *Respiration Physiology*, 1971, **11,** 390–398.

Buskirk, E. R., & Komi, P. V. Reproducibility of electromyographic measurements with inserted wire electrodes and surface electrodes. *Acta Physiologica Scandinavica*, 1970, **79,** 29A.

Butler, S. R., & Giaquinto, S. Stimulation triggered automatically by electrophysiological events. *Medical and Biological Engineering*, 1969, **7**(3), 329–331.

Chobotas. M. A., & Saplinskas. I. S. An attachment to the electromyography for the investigation of dynamic and static efforts of muscles. *Biulleten Eksperimental'noi Biologii i Meditsiny (Moskva)*, 1970, **70**(7), 121–122.

Cohen, H. L., & Brumlik, J. *A manual of electroneuromyography*. New York: Hoeber Medical Division, Harper, 1968.

Coombes, R. J., Lange, G. W., & Priban, I. P. Experimental and data processing methods for investigating in man the relations between mechanical and electrical activity of the arm. *Journal of Physiology (London)*, 1969, **201**, 88P–90P.

Dedo, H. H., & Hall, W. N. Electrodes in laryngeal electromyography, Reliability comparison. *Annals of Otolaryngology*, 1969, **78**(1), 172–180.

Delhez, L., Deroanne, R., & Petit, J. M. Electromyographie respiratoire autonome par memorisation simple ou double. (Single or double retained recording of autonomic respiratory electromyography). *Electromyography*, 1970, **10**(1), 69–84. (French)

Demus, H. G., & Skurczynski, W. Prerequisite for surface electromyography. Preliminary report. *Zeitschrift für Aerztliche Fortbildung (Jena)*, 1971, **65**, 88–89. (German)

Dietrichson, P., & Sorbye, R. Clinical method for electrical and mechanical recording of the mechanically and electrically elicited ankle reflex. *Acta Neurologica Scandinavica*, 1971, **47**, 1–21.

Dixon, C., Toledo, L. De., & Black, A. H. A technique for recording electromyographic activity in freely moving rats using an all purpose slip ring commutator. *Journal of the Experimental Analysis of Behavior*, 1969, **12**(3), 507–509.

Ekstedt, J., Haggqvist, P., P., & Staalberg, E. The construction of needle multi-electrodes for single fiber electromyography. *Electroencephalography and Clinical Neurophysiology*, 1969, **27**(5), 540–543.

Ekstedt, J., & Staalberg, E. The effect of non-paralytic doses of D-tubocurarine on individual motor end plates in man, studied with a new electrophysiological method. *Electroencephalography and Clinical Neurophysiology*, 1969, **27**(6), 557–562.

Gawronska, I., Mempel, E., Stadnicki, R., & Dziduszko, J. Methods of recording Parkinsonian tremor. *Neurologia i Neurochirurgia Polska*, 1971, **5**, 63–68. (Polish)

Giannazzo, E. Caratteristiche tecniche di uno stimolatore acusticomecanico per ricerche di elettromiografia e di neurofisiologia clinica. (Technical characteristics of an acoustical-mechanical stimulator for research in electromyography and clinical neurophysiology.) *Acta Neurologica (Napoli)*, 1970, **25**(3), 411–413. (Italian)

Gottlieb, G. L., & Agarwal, G. C. Filtering of electromyographic signals. *American Journal of Physical Medicine*, 1970, **49**, 142–146.

Green, E. E., Walters, E. D., Green, A. M., & Murphy, G. Feedback technique for deep relaxation. *Psychophysiology*, 1969, **6**(3), 371–377.

Guld, C., Rosenfalck, A., & Willison, R. G. Technical factors in recording electrical activity of muscle and nerve in man. Report of the committee on EMG instrumentation. *Electroencephalography and Clinical Neurophysiology*, 1970, **28**(4), 399–413.

Gundarov, V. P., & Solodkov, F. B. The electromyogram and methods of its analysis. *Novosti Meditsinskogo Priborastroeniia (Moskva)*, 1969, **1**, 100–104. (Russian)

Hall, E. A. Electromyographic techniques: A review. *Physical Therapy*, 1970, **50**(5), 651–659.

Hamonet, C., Opsomer, G., Aubriot, J. H., & Dubousset, J. F. Electromyographie par fils souples intra-musculaires—par application aux etudes bio-mechaniques. (Electromyography by intramuscular flexible wires: Application to biomechanical studies.) *Revue Neurologique (Paris)*, 1970, **122**(6), 424–426. (French)

Herberts, P. Myoelectric signals in control of prostheses. Studies on arm amputees and normal individuals. *Acta Orthopaedica Scandinavica Supplement*, 1969, **124**, 18.

Hill, D. W. Recording systems for physiological signals. In D. W. Hill (Ed.), *Electronic measurement techniques in anaesthesia and surgery*. London: Butterworth, 1970. Pp. 1–53.

Hill, D. W. Recording electrodes. In D. W. Hill (Ed.), *Electronic measurement techniques in anaesthesia and surgery*. London: Butterworth, 1970. Pp. 147–171.

Himwich, W. A. Physiological measurements in neonatal animals. In W. I. Gay (Ed.), *Methods of animal experimentation*, Vol. 3. New York, Academic Press, 1969. Pp. 241–261.

Hirano, M., & Ohala, J. Use of hooked wire electrodes for electromyography of the intrinsic laryngeal muscles. *Journal of Speech and Hearing Research*, 1969, **12**, 362–373.

Hixon, T. J., Siebens, A. A., & Minifie, F. D. An EMG electrode for the diaphragm. *Journal of the Acoustical Society of America*, 1969, **46**(6 Pt.2), 1588–1590.

Hollis, L. I., & Harrison, E. An improved surface electrode for monitoring myopotentials. *American Journal of Occupational Therapy*, 1970, **24**, 28–30.

Iakimenko, M. A. Method for studying the thermal effect of muscle contraction. *Fiziologicheskii Zhurnal SSSR imeni I. M. Sechenova (Moskva)*, 1970, **56**(12), 1861–1864. (Russian)

Iusevich, I. U. S. Concerning the use of integrators in clinical electromyography. *Zhurnal Neuropatologii i Psikhiatrii imeni S. S. Korsakova (Moskva)*, 1970, **70**(4), 537–539. (Russian)

Jonsson, B. Topography of the lumbar part of the erector spinae muscle. An analysis of the morphologic conditions precedent for insertion of EMG electrodes into individual muscles of the lumbar part of the erector spinae muscle. *Zeitschrift für Anatomie und Entwicklungsgeschichte (Berlin)*, 1970, **130**(3), 177–191.

Jonsson, B., & Reichmann, S. Radiographic control in the insertion of EMG electrodes in the lumbar part of the erector spinae muscle. *Zeitschrift für Anatomie und Entwicklungsgeschichte (Berlin)*, 1970, **130**(3), 192–206.

Jonsson, B., Reichmann, S., & Bagge, U. Synchronization of film camera and EMG recorder in kinesiologic investigations. *Acta Morphologica Neerlando-Scandinavica (Utrecht)*, 1970, **7**, 247–251.

Kadefors, R., Reswick, J. B., & Martin, R. L. A percutaneous electrode for long-term monitoring of bio-electrical signals in humans. *Medical and Biological Engineering*, 1970, **8**(2), 129–135.

Kiwerski, J., Zmyslowski, W., Gawronski, R., & Wirski, J. Use of digital myoestimator in electromyography. *Polski Tygodnik Lekarski*, 1970, **25**, 41–42. (Polish)

Krajca, K., & Orolin, D. Hemitetanicky syndrom u sclerosis multiplex. (Hemitetanic syndrome in the course of disseminated sclerosis). *Ceskoslovenska Neurologie*, 1969, **32**(4), 193–196. (Czechoslovakian)

Kreifeldt, J. G. Signal versus noise characteristics of filtered EMG used as a control source. *IEEE Transactions on Bio-Medical Engineering*, 1971, **18**(1), 16–22.

Likovsky, Z., & Votava, Z. Technical report: Bipolar electrode for the use in electromyography of laboratory animals. *Ceskoslovenska Fysiologie*, 1971, **20**, 45–53. (Czechoslovakian)

McPhee, M. D., & Koch, R. D. Placement device for surface electrodes on the hand. *Archives of Physical Medicine*, 1970, **51**, 307.

Monges, H., & Salducci, J. A method of recording the gastric electrical activity in man. *American Journal of Digestive Diseases*, 1970, **15**, 271–276.

Nagaki, J. Evaluation of electromyographic examination by the concentric needle electrode. *Japanese Journal of Clinical Pathology*, 1970, **18**, 848–851.

Potter, A., & Menke, L. Capacitive type of biomedical electrode. *IEEE Transactions on Bio-Medical Engineering*, 1970, **17**(4), 350–351.

Rogalewski, R., & Laudanski, T. Differential D.C. preamplifier for biopotentials registration. *Acta Physiologica Polonica*, 1970, **21**(1), 124–129. (English)

Rogoff, J. B. Logarithmic amplification: A new method of electromyographic presentation. *Electroencephalography and Clinical Neurophysiology*, 1969, **27**, 727.

Rubow, R. T., & Smith, K. U. Feedback parameters of electromyographic learning. *American Journal of Physical Medicine*, 1971, **50**(3), 115–131.

Sato, G. Automated Electromyogram. *Japanese Journal of Clinical Medicine,* 1969, **27,** 2924–2936. (Japanese)

Scott, R. N., & Thompson, G. B. An improved bipolar wire electrode for electromyography. *Medical and Biological Engineering,* 1969, **7**(6), 677–678.

Shipp, T., Fishman, B. V., Morrissey, P., & McGlone, R. E. Method and control of laryngeal EMG electrode placement in man. *Journal of the Acoustical Society of America,* 1970, **48**(2 Pt.1), 429–430.

Simons, D. G., & Lamonte, R. J. Automated system for the measurement of reflex responses to patellar tendon tap in man. *American Journal of Physical Medicine,* 1971, **50,** 72–79.

Stepita-Kiauco, M., & Ujec, E. Physical principles of new technique for measuring cable properties and myoplasmic resistance of muscle fibres. *Physiologica Bohemoslovenica,* 1971, **20,** 43–52.

Vaughan, W., & Locke, S. A circuit for stimulating and recording through a single capillary micropipette. *IEEE Transactions on Bio-Medical Engineering,* 1971, **18**(1), 71–72.

Vincent, S. A. Portable EMG monitoring and recording equipment. *Journal of Anatomy,* 1970, **106,** 187.

Vovk, M. I. A method of registering muscle contractions using a piezo electric transducer. *Fiziologicheskii Zhurnal SSSR imeni I. M. Sechenova (Moskva),* 1970, **56,** 132–133. (Russian)

Walker, A., Talbot, J. M., Walker, D. W., & Newman, W. Technique for physiological measurement in the human fetus. *Lancet,* 1969, **2,** 31–32.

Walter, W. G. Telemetry of electrophysiological data in human subjects. *Proceedings of the Royal Society of Medicine,* 1969, **62**(5), 449–450.

Zuniga, E. N., Truong, X. T., & Simons, D. G. Effects of skin electrodes position on averaged electromyographic potentials. *Archives of Physical Medicine,* 1970, **51**(5), 264–272.

Electrocardiogram

Alexander, D. C., & Wortzman, D. Computer diagnosis of electrocardiograms. I. Equipment. *Computers and Bio-Medical Research,* 1968, **1,** 348–365.

Arthur, R. M., Geselowitz, D. B., Briller, S. A., et al. The path of the electrical center of the human heart determined from surface electrocardiograms. *Journal of Electrocardiology,* 1971, **4,** 29–33.

Baruch, D., Neufeld, H. N., Riss, E., et al. Human versus computer evaluation of electrocardiographic data in epidemiological studies. *Israel Journal of Medical Sciences,* 1969, **5,** 920–922.

Baumgarten, K., Frohlich, H., Seidl, A., et al. Continuous control of the fetal heart rate during labor and delivery by means of telemetric direct fetal electrocardiography. *Wiener Klinische Wochenschrift,* 1968, **80,** 307–310. (German)

Bernreiter, M. *Electrocardiography.* New York: Lippincott, 1963.

Blumenschein, S. D., Spach, M. S., Flaherty, J. T., et al. Exploratory electrocariography: Use of isopotential surface maps. *UCLA Forum in Medical Science,* 1970, **10,** 347–368.

Bojrab, M. J., Breazile, J. E., & Morrison, R. D. Vectorcardiography in normal dogs using the frank lead system. *American Journal of Veterinary Research,* 1971, **32,** 925–934.

Bolte, A., Unkel, B., & Wechselberg, K. Electrocardiographic follow-up investigations in newborn infants of various stages of maturity. *Zeitschrift für Kinderheilkunde,* 1968, **104,** 94–102. (German)

Borin, A. L., & Shketnik, A. A. On a method of synchronous recording of the electrocardiogram, phonocardiogram and motor activity of the fetus. *Akusherstvo i Ginekologiia (Moskva),* 1966, **42,** 20–22. (Russian)

Bovicelli, I., & Giardina, L. Direct and indirect cardiotachometry. On various technics of registration and derivation. *Minerva Ginecologica*, 1967, **19**, 919–922. (Italian)

Brandon, C. W., & Brody, D. A. A hardware trigger for temporal indexing of the electrocardiographic signal. *Computers and Bio-Medical Research*, 1970, **3**, 47–57.

Brody, D. A., Warr, O. S. 3rd, & Wennemark, J. R., et al. Studies of the equivalent cardiac generator behavior of isolated turtle hearts. *Circulation Research*, 1971, **29**, 512–524.

Castellani, L. Use of filters for EEG, ECG and ERG recording during electric anesthesia. *Minerva Anestesiologica*, 1967, **33**, 813–814. (Italian)

Chignon, J. C., Distel, R., Courtois, B., et al. Orientation of the analysis of electrical tracings regarding athletes. *Journal of Sports Medicine*, 1969, **9**, 241–244.

Cranfield, P. F., & Hoffman, B. F. (Eds.) *Electrophysiology of the heart*. New York: McGraw-Hill, 1960.

Ferrer, M. I. Instant electrocardiograms as a teaching aid. *Diseases of the Chest*, 1969, **56**, 344–349.

Fitzgerald, R. D., Vardaris, R. M., & Teyler, T. J. An on-line method for measuring heart rate in conditioning experiments. *Psychophysiology*, 1968, **4**, 352–353.

Gau, G. T., & Smith, R. E. The effect of electrode position on a modified frank electrovectorcardiographic lead system. *Mayo Clinic Proceedings*, 1971, **46**, 536–543.

Geddes, L. A., Rosborough, J., Garner, H., et al. Obtaining electrocardiographs of animals without preparing or penetrating the skin. *Veterinary Medicine and Small Animal Clinician*, 1970, **65**, 1163–1168.

Giardina, L., & Bovicelli, L. On a new recording method for fetal ECG. *Minerva Ginecologica*, 1967, **19**, 842–846. (Italian)

Goldman, M. J. *Principles of clinical electrocardiography*. Los Altos, California: Lange Med. Publ., 1970.

Hanish, H. M., Neustein, R. A., Van Cott, C. C., et al. Technical aspects of monitoring the heart rate of active persons. *American Journal of Clinical Nutrition*, 1971, **24**, 1155–1163.

Hoffman, I. (Ed.) *Vectorcardiography*. New York: Lippincott, 1965.

Holt, J. H., Jr., Barnard, A. C., Lynn, M. S., et al. A study of the human heart as a multiple dipole electrical source. I. Normal adult male subjects. *Circulation*, 1969, **40**, 687–696.

Kobayashi, T., Takeuchi, M., Koro, T., et al. Radio-telemetering of electrocardiogram and blood pressure. *Malattie Cardiovoscular I*, 1969, **10**, 129–141.

Kuaity, J., Wexler, H., & Simonson, E. The electrocardiographic ice water test. *American Heart Journal*, 1969, **77**, 569–571.

Kupfer, J. Technical description of a method for uninterrupted recording of electrocardiogram in electric perfusion of the heart in intact animals. *Biomedizinische Technik (Stuttgart)*, 1971, **16**, 109–115. (German)

Lemmerz, A. H. New aspects of routine electrocardiography. *Medizinische Klinik (München)*, 1970, **65**, 2060–2062. (German)

Milliken, J. A., Wartak, J., Lywood, D. W., et al. ECG data acquisition for computer analysis. *Canadian Medical Association Journal* 1970, **102**, 1281–1285.

Pare, W. P., Isom, K. E., & Reus, J. F. Recording the electrocardiogram from the squirrel monkey, *Saimira sciureus*. *Physiology and Behavior*, 1970, **5**, 819–821.

Pomerleau, O. F. A three electrode system for EKG recording and aversive shock. *Conditional Reflex*, 1969, **4**, 59–60.

Ramsay, D. A., Pomerleau, O. F., & Snapper, A. G. Two methods for obtaining electrocardiograms of chair-restrained monkeys. *Conditional Reflex*, 1968, **3**, 200–204.

Rippa, S., & Hulin, I. Causes of form changes of the ECG curve in dogs due to varying positions of the front limbs. *Zeitschrift für Kreislaufforschung*, 1969, **58**, 535–542. (German)

Ritota, M. *Diagnostic electrocardiography*. New York: Lippincott, 1969.

Sass, D. J. A method for recording myocardial ECG in animals during intense vibration. *Journal of Applied Physiology*, 1970, **28**, 361–364.

Van Petten, G. R., Evans, J. A., & Salem, F. A. A simple method for chronic measurement of the electrocardiogram and blood pressure in the conscious rat. *Journal of Pharmacy and Pharmacology,* 1970, **22,** 467–469.

Werth, G., & Wink, S. The electrocardiogram in healthy rats. *Archiv für Kreislaufforschung,* 1967, **54,** 272–308. (German)

Wolf, R. H., Lehner, N. D., Miller, E. C., et al. Electrocardiogram of the squirrel monkey, *Saimiri sciureus, Journal of Applied Physiology,* 1969, **26,** 346–351.

Electrodermal Activity

Burstein, K. R., & Epstein, S. Procedure for reducing orienting reactions in GSR conditioning. *Journal of Experimental Psychology,* 1968, **78,** 369–374.

Carpenter, F. A., Steinhaus, J. E., Turner, M. E., Jr., Parrott, T. S., Blumenstein, B. A., & Savino, M. F. Computer analysis of changes in electrodermal response due to "sedative" drugs. *Acta Anaesthesiologica Scandinavica Supplement,* 1966, **24,** 157.

Carson, V. G., Kado, R. T., & Wenzel, B. M. Method for recording foot pad impedance in freely moving mice. *Physiology and Behavior,* 1971, **6,** 77–79.

Epstein, A., & Schill, H. A. Electrodermal responses and tolerance thresholds for pure tones. *Acta Oto-Laryngolica (Stockholm),* 1968, **66,** 417–432.

Geer, J. H. A test of the classical conditioning model of emotion: The use of non-painful aversive stimuli as unconditioned stimuli in a conditioning procedure. *Journal of Personality and Social Psychology,* 1968, **10,** 148–156.

Greenblatt, D. J., & Tursky, B. Local vascular and impedance changes induced by electric shock. *American Journal of Physiology,* 1969, **216,** 712–718.

Grings, W. W., & Schell, A. M. Magnitude of electrodermal response to a standard stimulus as a function of intensity and proximity of a prior stimulus. *Journal of Comparative and Physiological Psychology,* 1969, **67,** 77–82.

Johns, M. W., Cornell, B. A., & Masterton, J. P. Monitoring sleep of hospital patients by measurement of electrical resistance of skin. *Journal of Applied Physiology,* 1969, **27,** 898–901.

Kajtor, F., & Halasz, P. Electroencephalographic, electrodermographic and pupillomotor responses under hexobarbital effect in man. *Acta Medica Academiae Scientiarum Hungaricae (Budapest),* 1968, **25,** 191–216.

Khalafalla, A. S., Turner, L., & Spyker, D. An electrical model to simulate skin dielectric dispersion. *Computers and Bio-Medical Research,* 1971, **4,** 359–373.

Lader, M. H. The unit of quantification of the GSR. *Journal of Psychosomatic Research,* 1970, **14,** 109–110.

Lang, A. H. On the physiological significance of the shape of the endosomatic galvanic skin reaction in the cat. *Acta Physiologica Scandinavica,* 1968, **74,** 246–254.

Lerner, J. The psychiatrist and the polygraph. *Journal of the American Geriatrics Society,* 1969, **17,** 979–984.

Maxwell, R. D., & Barton, J. L. The apparent skin resistance. *Acta Psychiatrica Scandinavica,* 1971, **47,** 92–105.

Pasquali, E., & Roveri, R. Measurement of the electrical skin resistance during skin drilling. *Psychophysiology,* 1971, **8,** 236–238.

Richter-Heinrich, E., & Lauter, J. A psychophysiological test as diagnostic tool with essential hypertensives. *Psychotherapy and Psychosomatics,* 1969, **17,** 153–168.

Vernet, E. Maury. Electrodermal responses to visual, auditory and olfactory stimulations in rats. *Journal de Physiologie Supplement (Paris),* 1970, **1**(62), 225–226. (French)

Electrogastrogram

Kahatsu, S. The current status of electrogastrography. *Klinische Wochenschrift,* 1970, **48,** 1315–1319.

Monges, H., & Salducci, J. A method of recording the gastric electrical activity in man. *American Journal of Digestive Diseases,* 1970, **15,** 271–276.

Roche, M., & Santini, R. Electrophysiologic study of the gastric motility. I. Comparative electrogastromyography. *Compte Rendu de la Société Biologie (France),* 1970, **164,** 1528–1534. (French)

Shiratori, T., Okabayashi, T., Harata, N., et al. Intraluminal lead of the gastric action potentials in human subjects: Use of a new suction needle electrode. *Surgery,* 1969, **66,** 483–487.

Sobakin, M. A., & Todorov, D. P. Implantation of non-polarizing electrodes for registering the ultra-slow potentials of the gastrointestinal tract in chronic experiments. *Biulleten Eksperimental'noi Biologii i Meditsiny,* 1968, **65,** 120–122. (Russian)

Author Index

Numbers in italics refer to the pages on which the complete references are listed.

A

Ackerman, P., 223, *227*
Adams, T., 241, 242, *267*
Adrian, E. D., 4, *60*, 139, *153*
Alvarez, W. C., 298, *304*
Anastasio, E. J., 291, *293*
Aoki, K., 89, *95*
Arai, T., 88, *94*
Arnold, J. E., 5, *61*
Aschan, G., 130, *131*
Aserinsky, E., 115, 129, *131*
Autrum, H., 66, *85*
Averill, J. R., 219, *227*
Ax, A. F., 286, *292*

B

Baeza, M., 138, 145, 147, *153*
Baker, L. E., 58, *61*
Baldridge, B. J., 113, *131*
Ban, P., 211, *228*
Ban, T., 174, *197*
Barber, T. X., 130, *131*
Barnett, A., 277, *292*
Barnothy, J. M., 110, *134*
Bartels, B., 214, 220, 222, *228*
Bartlett, M. S., 289, *292*
Bartoshuk, A. K., 213, 220, 226, *227*
Basmajian, J. V., 138, 142, 143, 145, 147, 152, *153*, *154*, 156, *162*
Baumel, M. H., 211, 215, *228*
Beebe-Center, J. G., 222, *227*
Beiser, G. D., 173, *199*
Benjamin, L. S., 213, 214, *227*, 290, *292*
Benson, H., 131, *135*

Berg, K. M., 221, *227*
Berg, R. L., 222, *227*
Berg, W. K., 218, 221, *227*
Bergeron, J., 277, *292*
Bergman, T., 129, *131*
Berlyne, D. E., 158, *162*
Bernstein, N., 157, *162*
Berry, F., 298, *305*
Bickford, R. G., 130, *134*
Billingsley, P., 173, *199*
Black, A. H., 161, *162*, 175, 177, 178, *197*
Blank, I. H., 281, *292*
Bloch, J. D., 292, *292*
Bloodworth, D. S., 143, *154*
Borrey, R., 285, *292*
Boyle, R., 222, *227*
Brauer, E. W., 244, *271*
Braun, J. J., 177, *198*
Braunwald, E., 173, *199*
Bray, C. W., 4, 5, 7, 8, *62*
Brazier, M. A. B., 277, *292*
Brener, J. M., 176, *197*, *198*
Bridger, W. H., 220, *227*, 292, *292*
Brommett, R. E., 26, *61*
Bronk, D. W., 139, *153*
Brown, B. B., 130, *131*
Brown, H. M., 83, *85*
Bruckner, A., 104, *131*
Bruno, L. J. J., 143, 151, *153*, *154*
Brusilow, S. W., 235, 245, 246, 248, 250, 256, 257, 258, *269*, *270*
Bullock, T. H., 84, *85*
Burch, N. R., 243, 244, 245, 252, *268*, 276, 277, 281, 285, *292*, *293*
Burger, R. E., 147, *154*
Burstein, K. R., 277, *292*

Bush, R. C., 191, *198*
Butler, R. A., 5, 25, *60*
Byford, G. H., 106, *132*

C

Cage, G. W., 235, 240, 247, 248, 253, *267, 268, 270*
Campbell, H., 206, 215, 216, *228*
Carmichael, L., 100, 128, *132*
Carpenter, D. O., 140, *154*
Carslöö, S., 147, *153*
Cazavelan, J., 254, *270*
Chai, C. Y., 174, *198*
Chase, H. H., 220, *227*
Chase, W. G., 191, *198*
Childers, D. G., 73, *86*
Childers, H. E., 285, *292*
Chism, R. A., 193, *198*
Christie, M. J., 232, 252, 260, 261, 262, 263, 267, *267, 271, 277, 292, 295*
Clifton, R. K., 204, 211, 221, *227, 229*
Cohen, D. H., 173, 174, *198*
Coles, E. M., 240, *269, 277*, 278, 280, 281, 284, *294*
Cooke, R. E., 245, 246, 248, 256, 257, 258, *270*
Cornsweet, J. C., 106, *134*
Cornsweet, T. N., 106, 110, *133, 134*
Covian, M. R., 174, *198*
Coyne, L., 277, *293*
Crawford, B. H., 76, *85*
Crescitelli, F., 79, *85*
Crowell, D. H., 220, *227*
Crowley, D. E., 11, *60*
Cullen, T. D., 277, *295*
Cutt, R. A., 27, *61*

D

Dalland, J. I., 16, 36, *60, 62*
Dallos, P., 49, *60*
Darrow, C. W., 240, 244, 245, 246, 250, 251, 254, 257, *268,* 282, 283, 289, *292*
Dauth, G., 177, *200*
Davidowitz, J. E., 143, 151, *153, 154*
Davis, H., 5, 7, 48, *60, 62*
Davis, J. F., 156, *162,* 280, 285, *294,* 298, *305*
Davis, R. C., 277, 285, *293,* 298, *305*

Day, J. L., 122, *134*
Dayton, G. O., 129, *132*
Dearborn, W. F., 100, 105, 106, 128, *132*
Deatherage, B. H., 7, *60*
Delebarre, E. B., 105, *132*
Dement, W. C., 129, *132*
DeRouk, A., 115, *132*
de Somer, P., 24, *60*
Dewan, E. M., 130, *132*
Dewar, J., 64, 65, *85,* 113, *132*
DeWick, H. N., 274, *294*
DeYoung, G., 212, *229*
DiCara, L. V., 173, 177, 193, *198, 199*
Djoleto, B. D., 190, *199, 200*
Dobson, R. L., 245, 248, 253, *267, 268*
Dodd, C., 211, 220, *228*
Dodge, R., 105, 106, *132*
Doving, K. B., 91, 93, *94*
Dowling, J. E., 66, *86*
Druger, R., 210, *229*
du Bois, R. E., 64, *85,* 113, *132*
Duffy, E., 158, *162*
Dunhoffer, H., 161, *162*
Dureman, I., 223, *227*
Durkovic, R. G., 174, *198*
Dykman, R. A., 173, *198,* 223, *227*

E

Edelberg, R., 122, *132,* 232, 233, 234, 239, 241, 242, 243, 244, 245, 246, 247, 250, 251, 252, 254, 255, 257, 258, 260, 261, 262, 263, 266, *268,* 274, 275, 276, 277, 278, 280, 281, 282, 284, 285, *293, 294*
Edfeldt, A. W., 147, *153*
Eimas, P. D., 202, *227*
Einthoven, W., 65, *85*
Eisenberg, R. B., 213, 221, *227*
Elden, H. R., 235, *268*
Eldredge, D. H., 5, 7, *60, 62*
Ellingson, R. J., 203, *227*
Elliott, D. N., 291, *293*
Ellis, R. A., 235, *269*
Ellsworth, D. W., 147, *153*
Elster, A. J., 174, *198, 200*
Engel, B. T., 131, *132,* 193, *198*
Enoch, D. M., 174, *198*
Epstein, S. E., 173, *199,* 277, *292*

Evans, C. R., 130, *133*
Evans, F. J., 277, 292, *294*
Evans, S. H., 291, *293*
Everett, M. A., 245, *269*
Eyssen, H., 24, *60*

F

Fabrigar, C., 138, 145, 147, *153*
Faw, T. T., 129, *132*
Feder, W., 120, *132*
Fedor, J. H., 298, *305*
Felton, G. S., 147, *154*
Fender, D. H., 100, 106, *132*
Fenz, W. D., 277, *292*
Féré, C., 122, *132*
Fernandez, C., 7, 27, 48, *60, 61, 62*
Fick, A., 283, *293*
Fields, C., 190, 195, *198*
Finer, B. L., 130, *131*
Finesinger, J. E., 281, *292*
Fisch, U., 8, *61*
Floyd, W. F., 260, 262, 263, *268*, 277, *293*
Forbes, T. W., 277, *293*
Ford, A., 121, 123, *132*
Fowles, D. C., 242, 243, 245, 247, 248, 250, 251, 255, 257, 260, 261, 264, *268*, 281, *293*
Fox, R. H., 249, *268*
Francois, J., 115, *132*
Freeman, G. L., 274, *293*
Frick, A., 235, 245, 255, *270*
Fröhlich, F. W., 65, *85*
Frömter, E., 235, 245, 255, *270*

G

Galambos, R., 27, *61*
Galindo, A., 178, *198*
Gantt, W. H., 173, *198*
Garafolo, L., 298, *305*
Garrison, P., 298, *305*
Gasser, H. S., 139, *153,* 173, 175, *198*
Gault, F. P., 298, *305*
Gavira, B., 277, *293*
Geddes, L. A., 58, *61*
Gelhorn, E., 178, *198*
Gellhorn, E., 161, *162*
Gerbrands, R., 282, 292, *293*
Germana, J., 157, 159, 160, 162, *162*

Gerstein, G. L., 189, *199*
Gesteland, R. C., 93, *94*
Giesige, R., 283, *293*
Gilson, A. S., 138, *153*
Goadby, H. K., 244, 253, *268*
Goadby, K. W., 244, 253, *268*
Goldberg, S., 214, 215, 217, 220, 222, *228*
Gordon, Jr., R. S., 235, 240, 247, *268*
Gotch, G., 65, *85*
Gouras, P., 77, *85*
Graham, D. T., 191, *198*
Graham, F. H., 191, *198*
Graham, F. K., 204, 210, 211, 212, 213, 218, 220, 221, 222, *227*
Granger, G. W., 76, *85*
Granit, R., 65, *85,* 100, *132*
Grastyan, E., 161, *162*
Gray, J. A. B., 88, *94*
Gray, M. L., 220, *227*
Green, A. M., 143, *153*
Green, E. E., 143, *153*
Green, H. D., 174, *199*
Green, J. D., 68, *85*
Greiner, T., 243, 244, 245, 252, *268*, 276, 281, *293*
Grieve, D. W., 188, *198*
Grings, W. W., 122, *132,* 277, 285, 292, *292, 293, 295*
Grinnell, A. D., 5, *61*
Grobstein, R., 221, *228*
Groves, P. M., 160, *162*
Guillemin, V., 110, *134*
Guinan, J. J., 13, *61*
Gulick, W. L., 9, 27, *61*
Gullickson, G. R., 244, 254, 257, *268*
Gustafson, L. A., 277, *295*
Guyton, A. C., 101, *132,* 178, *198,* 238, 247, 248, 262, *268*

H

Haapanen, L., 93, *94*
Hagbarth, K. E., 130, *131*
Hagfors, C., 284, 285, *293*
Haggard, E. A., 282, 291, 292, *293*
Haith, M. M., 108, 109, 129, *131, 132, 133*
Hale, L. J., 82, *85*
Hallman, S., 223, *229*

Hammond, L. J., 284, *296*
Hardyk, C. D., 147, *153*
Harris, L., 212, *229*
Harrison, J. B., 27, *61*
Harrison, V. F., 138, *153*
Hartline, H. K., 65, *85*
Harwitz, M., 211, 220, *228*
Hattle, M., 212, *229*
Hatton, H. M., 211, 218, 221, *227*
Hawkins, Jr., J. E., 5, *61*
Heath, H. A., 292, *293*
Heaton, W. C., 16, *61*
Hefferline, R. F., 143, 151, *153, 154*
Hegel, U., 235, 245, 255, *270*
Held, R., 100, *134*
Hellman, K., 235, *271*
Henneman, E., 140, *154*
Hepp-Raymond, M. -C., 11, *60*
Herman, P., 13, *62*
Herrick, C. J., 156, *162*
Hertzman, A. B., 243, *270*
Hess, W. R., 161, *162*, 174, *198*
Hicks, R. G., 283, *293*
Higashino, S., 88, 91, *94*
Hilton, S. M., 249, *268*
Hirsch, H. R., 277, *295*
Hirshenson, M., 108, *133*
Hnatiow, M., 176, *198*
Hoff, E. C., 174, *199*
Holmgren, F., 64, *85*
Holmstrom, V., 244, *271*
Holzgreve, H., 235, 245, 255, *270*
Hord, D. J., 292, *293*
Horridge, G. A., 84, *85*
Horvath, P. N., 248, 250, *270*
Hosoya, Y., 87, 93, *94*
Hothersall, D., 176, *197, 198*
Howard, J. L., 204, *229*
Hubel, D. H., 68, *85*
Hudson, W., 8, *61*
Huey, E. B., 105, *133*
Hupka, R. B., 277, *294*
Huttenlocher, J., 176, *200*
Hyman, A. B., 244, *271*

I

Iino, M., 89, *95*
Imai, K., 88, 91, 93, *95*
Ito, H., 66, 81, *86*

J

Jackson, J., 210, 212, 213, 222, *227*
Jacobs, A., 147, *154*
Jacobson, E., 113, *133*
Jacobson, J. H., 67, 77, 78, *86*
Janda, V., 143, 147, *154*
Janz, G. J., 120, *133*
Järvilehto, M., 83, *86*
Jennings, J. R., 219, *227*
Johnson, G., 257, *268*
Johnson, L. C., 203, *227, 292, 293*
Jolly, W. A., 65, *85*
Jones, A. E., 79, *86*
Jones, M. H., 129, *132*
Judd, C. H., 106, *133*
Jujimori, B., 277, *296*
Jung, R., 115, *133*
Jusczyk, P., 202, *227*

K

Kaada, B. R., 174, *199*
Kagan, J., 204, 206, 208, 212, 215, 216, 218, 225, *227, 228, 229*
Kahana, L., 27, *61*
Kahn, R., 65, *86*
Kahn, S. D., 143, *154*
Kalafat, J., 206, 215, 216, *228*
Kamiya, J., 130, *133*
Karpe, G., 65, *86*
Katzoff, E. T., 274, *293*
Keele, C. A., 260, 262, 263, *268*, 277, *293*
Keen, R. K., 220, *227*
Kerr, F. W. L., 174, *198*
Kerr, J., 190, *199, 200*, 209, 210, *228, 229*
Kessen, N., 129, *134*
Kessen, W., 108, 129, *133*
Khachaturian, Z. S., 190, *199, 200*, 209, 210, *228, 229*
Kimm, J., 150, *154*
Kimura, K., 91, *94*
Kitamura, H., 88, 91, 93, *95*
Kleitman, N., 113, 115, 129, *131, 132, 133*
Kligman, A. M., 245, 248, 249, *270*
Kline, T. S., 105, *132*
Koch, A., 247, 248, 262, *268*
Koefoed-Johnsen, V., 249, *269*
Koelle, G. B., 178, *199*
Kondo, M., 277, *296*

AUTHOR INDEX

Konishi, T., 7, *61*
Korner, A., 221, *228*
Korr, I. M., 242, 252, *270,* 284, *295*
Kramer, M., 113, *131*
Krasil'nikov, L. G., 298, 301, *305*
Kris, C. E., 100, 114, 115, *133*
Krogh, A., 175, *199*
Kruger, R., 209, 210, *228, 229*
Kuechenmeister, C., 285, *294*
Kuhne, W., 65, *86*
Kuno, Y., 122, *133,* 232, 233, 234, 235, 240, 245, 253, *269*
Kveim, K., 298, *305*

L

Lacey, B. C., 204, 208, 223, 225, *228*
Lacey, J. I., 192, *199,* 204, 208, 209, 213, 214, 215, 223, 225, 226, *228,* 289, 290, *294*
Lacey, O. L., 292, *294*
Lachin, J., 190, *199*
Ladd, W. L., 147, *154*
Lader, M. H., 239, *269,* 289, *294*
Lamansky, S., 104, *133*
Landis, C., 274, 277, *293,* 294
Lang, P. J., 176, *198*
Lazarus, R. S., 213, 219, *227, 228*
Lee, H. Y., 130, *135*
Legouix, J. -P., 48, *62*
Leibrecht, B. C., 146, *154*
Leiderman, P. H., 292, *295*
Leonard, J. L., 121, 123, *132*
Lettvin, J. Y., 93, *94*
Levinger, G., 277, *294*
Lewis, M., 206, 211, 214, 215, 216, 217, 218, 220, 221, 222, 223, 225, *228, 229*
Licklider, J. C. R., 33, *61*
Lindgren, P., 174, *199*
Lindsley, D. B., 138, *154*
Linhard, J., 175, *199*
Lipowski, Z. J., 298, *305*
Lippold, O. C. J., 156, *163*
Lipsitt, L. P., 220, *228*
Lipton, E. L., 209, 211, 220, *228*
Lissak, K., 161, *162*
Lloyd, A. J., 146, *154*
Lloyd, D. P. C., 239, 240, 248, 254, 259, 260, 263, *269,* 277, *294*
Lobitz, Jr., W. C., 235, *269*

Loevinger, R., 244, *271*
Loewenstein, A., 65, *86*
Longstreet, B., 285, *292*
Lordahl, D. S., 193, *199*
Lubin, A., 292, *293*
Luce, G. G., 113, *133*
Luther, B., 253, *269,* 291, *294*
Lykken, D. T., 121, *133,* 232, 233, 234, 236, 237, 240, 244, 245, 246, 250, 251, 253, 255, 258, 264, *269,* 277, 278, 279, 280, 281, 284, 285, 287, 288, 291, *294*

M

Mackworth, J. F., 107, *133*
Mackworth, N. H., 107, 129, *133*
MacLeod, D. P., 239, *269*
MacLeod, P., 88, *94*
MacNichol, E. F., 68, *86*
Malcuit, A., 221, *229*
Maley, M., 253, *269,* 291, *294*
Malmo, R. B., 158, *163,* 174, *199,* 280, 285, *294*
Mandarsz, I., 161, *162*
Mann, L., 129, *131*
Manning, A. A., 193, *199*
Marg, E., 114, *133*
Martin, I., 232, 233, 235, 239, 244, 245, 246, 254, 257, 260, 261, 262, *269, 271,* 277, 278, 279, 280, 281, 282, 285, *295*
Martin, L., 122, *135*
Massof, R. W., 79, *86*
Matsuo, K., 5, *62*
Maulsby, R. L., 276, *294*
Mazokhin-Porshnyakov, G. A., 66, *86*
McAllister, C. H., 106, *133*
McAuliffe, D. R., 7, *60*
McCall, R. B., 212, *229*
McDonald, T. F., 239, *269*
McElroy, W. D., 76, *86*
McGuigan, F. J., 147, *154*
McKendrick, J. G., 64, *85*
Meek, W. J., 173, 175, *198*
Meikle, M. B., 26, *61*
Melmon, K. L., 131, *132*
Metz, H. S., 130, *133*
Meyers, I. L., 113, *133*
Milkman, M., 190, *199, 200*
Miller, N. E., 113, *133,* 173, 177, 178, 193, *199*

Miller, R. D., 250, 251, *269*, 277, 280, 281, *294*
Mills, W. D., 138, *153*
Misrahy, G. A., 5, *61*
Moffett, A. R., 221, *229*
Montagna, W., 235, *269*
Montagu, J. D., 239, 240, *269*, 277, 278, 280, 281, 284, *294*
Moore, G. P., 189, *199*
Moore, P. A., 202, *229*
Morandi, A. J., 129, *133*
Mordkoff, A. M., 213, *228*
Morehead, S. D., 130, *135*
Morimitsu, T., 5, *62*
Morris, V. B., 80, *86*
Mortensen, O. A., 138, *153*
Morton, H. S., 298, *305*
Moss, H. A., 204, 208, 225, *228*
Mowrer, O. H., 113, *133*
Mozell, M. M., 92, *94*
Mulholland, T., 130, *133*
Munger, B. L., 235, 250, 256, *269*
Murphy, G., 143, *153*

N

Nakashima, T., 5, *62*
Neumann, E., 277, *294*
Newcomer, H. S., 139, *153*
Newhall, S. M., 104, *134*
Newman, J. S., 110, *134*
Niimi, Y., 260, 263, *269*, *270*
Nordquist, R. E., 245, *269*
Nunnally, J. C., 129, *132*
Nutbourne, D. M., 249, *269*, *270*

O

Obrist, P. A., 175, 176, *200*, 204, 223, *229*
O'Connell, D. N., 115, 117, 119, 120, 121, 122, 125, *134*, *135*, 277, 278, 279, 280, 281, 292, *295*
Offner, F. F., 70, *86*
Ogden, T. E., 83, *85*
Oken, D., 292, *293*
Olson, R. L., 245, *269*
Omeara, D., 130, *133*
Opton, E. M., 219, *227*
Orne, M. T., 120, 121, *134*, 278, 281, *295*

Oshry, E., 244, *271*
Osterhammel, P., 93, *94*
Oswald, I., 113, *134*
Ottoson, D., 87, 89, 91, 93, *94*

P

Paden, A. B., 298, *305*
Paintal, A. S., 291, *295*
Palin, J., 11, *60*
Papa, C. M., 245, 248, 249, *270*
Papadimitriou, M., 261, *270*
Pappas, B. A., 173, 177, *198*, *199*
Parmalee, A. H., 220, *229*
Peake, W. T., 13, *61*
Pearlman, J. T., 73, *86*
Peckham, R. H., 71, *86*
Peiss, C. N., 243, *270*
Perera, T. B., 151, *154*
Perez, R. E., 285, *295*
Perkel, D. H., 189, *199*
Perlman, H., 27, *61*
Perry, N. W., 73, *86*
Petajan, J. H., 147, *154*
Petersen, H., 4, *61*
Peterson, E. A., 11, 13, 16, 57, *60*, *61*, *62*
Petrinovich, L. F., 147, *153*
Philip, B. A., 147, *154*
Pierce, D. S., 147, *154*
Pitts, L. H., 173, *198*
Pitts, W. H., 93, *94*
Plutchik, R., 277, *295*
Polyak, S. L., 100, *134*
Ponzoli, V. I., 16, *61*
Powell, D. A., 174, *199*
Powers, W. R., 147, 152, *154*
Prince, J. H., 79, *86*
Pringle, J. W. S., 82, *86*

Q

Quilter, R. E., 280, *295*

R

Rahm, Jr., W. E., 9, 11, *61*, *62*
Randall, W. C., 243, *270*
Randall, W. D., 243, *270*
Ranson, S. W., 173, *199*

Raskin, C. D., 212, *229*
Rasmussen, W. C., 130, *134*
Ratliff, R., 106, *134*
Reeder, R. C., 178, *198*
Rice, E. A., 5, 25, *61*
Richmond, J. B., 209, 211, 220, *228*
Richter, C. P., 122, *134*, 239, *270*
Rickles, W. H., 122, *134*
Riggs, L. A., 65, 76, 77, *86*, 106, *134*
Riley, J. A., 58, *61*
Robinson, B. F., 173, *199*
Rose, R., 253, *269*, 291, *294*
Rosen, A., 174, *199*
Rosenberry, R., 250, 251, 260, *268*
Rosenblith, W. A., 27, *61*, 300, *305*
Rosenthal, R., 301, *305*
Ross, M. S., 244, *271*
Roth, N., 284, 285, *294*
Rothman, S., 232, 233, 234, 235, 240, 244, 245, 257, 260, 261, *270*
Rowland, V., 130, *135*
Roy, R. R., 261, *270*
Ruben, R. J., 8, *61*
Ruch, R. L., 113, *133*
Ruck, P., 66, *86*
Ruckmick, C. A., 274, *295*
Russell, R. W., 297, 298, 299, 302, *305*

S

Saaren-Seppälä, P., 223, *227*
Salapatek, P., 129, *133, 134*
Samaan, A., 173, *199*
Sandberg, H., 110, *134*
Sarkany, I., 242, *270*
Sarnoff, S. J., 175, *199*
Sato, M., 88, *94*
Sayer, E., 121, *135*, 260, 262, *271*, 277, 278, 281, *295*
Sayers, G., 174, *199*
Schachter, J., 190, *199, 200*, 209, 210, *228, 229*
Schlosberg, H., 292, *295*
Schneider, D., 93, *94*
Schneiderman, N., 174, 177, 192, 193, *198, 199, 200*
Schott, E., 113, *134*
Schulman, C. A., 220, *229*
Schulz, H. R., 221, *229*
Schulz, I., 235, 245, 249, 255, *270*

Schwartz, G. E., 131, *135*
Schwartzkopff, J., 16, *61*
Schwarz, B. E., 130, *134*
Scott, A. B., 130, *133*
Scully, H. E., 147, 152, *154*
Segundo, J. P., 189, *199*
Seliger, H. H., 76, *86*
Shackel, B., 100, 107, 114, 121, 122, 123, *134*, 278, 281, *295*
Shapiro, D., 131, *135*, 292, *295*
Shaver, B. A., 245, 246, 248, 256, 257, 258, *270*
Shibuya, T., 88, 91, 92, 93, *94*
Shinabarger, E. W., 5, 25, *61*
Shelley, W. B., 248, 250, *270*
Shimizu, K., 260, 263, *270*
Shorey, C. D., 80, *86*
Shuster, S., 242, 249, *270*
Siddons, G. F., 285, *293*
Sideroff, S., 174, *199*
Siegel, P. S., 292, *294*
Sillman, A. J., 65, 81, *86*
Silver, A. S., 235, *269*
Simard, T. G., 143, 145, 147, *153, 154*
Simmons, F. B., 9, *61*
Simons, D. G., 285, *295*
Singer, E. G., 291, *293*
Singh, H., 27, *61*
Siqueland, E. R., 202, *227*
Small, A., 5, *61*
Smith, C. A., 7, *60*
Smith, C. E., 277, *295*
Smith, O. C., 138, *154*
Smith, W. M., 110, *134*
Snow, Jr., J. B., 5, *62*
Sokolov, E. N., 204, *229*
Solomon, N., 174, 176, 177, *199*
Somjen, G., 140, *154*
Spaulding, S. J., 217, 218, *228*
Speisman, J. C., 213, *228*
Sperry, R. W., 156, *163*
Stammers, M. C., 242, *270*
Stanley, W. S., 292, *295*
Stark, L., 100, 110, 129, *134, 135*
Stecko, G., 138, 142, 143, *153*
Steele, W. G., 225, *229*
Steele, W. M., 106, *133*
Steinbach, H. B., 249, *270*
Steinback, M. J., 100, *134*
Steiner, J., 65, *86*

Steinschneider, A., 209, 211, 220, *228*
Stern, R. M., 298, 299, 300, 302, *305*
Stevens, J. K., 297, 298, 300, 302, *305*
Stewart, H. L., 130, *133*
Stout, G. L., 285, *293*
Strahan, R. F., 250, 251, *269*, 277, *294*
Strandberg, P., 174, *199*
Strother, W. F., 9, 11, 16, *61, 62*
Stuart, H. G., 106, *132*
Suchi, T., 245, 248, *270*
Suga, F., 5, *62*
Surwillo, W. W., 280, 292, *295*
Sutterer, J. R., 204, *229*
Sutton, D. L., 142, 150, *154*
Swihart, S. L., 82, *86*
Swinnen, M. T., 186, *200*

T

Tabowitz, D., 11, *60*
Tajimi, T., 260, 263, *270*
Takagi, S. F., 88, 89, 91, 92, 93, *94, 95*
Takahashi, T., 277, *296*
Takeuchi, H., 88, 91, 93, *95*
Tamiguchi, H., 120, *133*
Tarchanoff, J., 122, *134*
Tasaki, I., 5, 48, *62*
Terkildsen, K., 93, *94*
Thetford, P. E., 277, *293*
Thomas, L. E., 107, *133*
Thomas, P. E., 242, 252, *270*, 284, *295*
Thomas, R. C., 239, *270*
Thompson, R. F., 160, *162*
Thysell, R. V., 147, 149, *154*
Timo-Iaria, C., 174, *198*
Tinker, M. A., 128, *134*
Tobin, M., 190, *199, 200*, 210, *229*
Tomita, T., 65, 66, 81, *86*
Torok, N., 110, *134*
Tregear, R. T., 234, 264, *270*
Trehub, A., 254, *270*
Trevarthan, C., 123, *135*
Trowill, J. A., 177, *200*
Tucker, I., 254, *270*
Tursky, B., 115, 117, 119, 120, 121, 122, 123, 125, 131, *134, 135,* 277, 278, 279, 280, 281, 292, *295*

U

Ullrich, K. J., 235, 245, 255, *270*
Uno, T., 277, *295*

Ussing, H. H., 248, 249, 262, *269, 271*
Uvnas, B., 174, *199*

V

VanDercar, D. H., 174, 176, 177, 192, *198, 200*
Vanderwolf, C. H., 161, *163*
Van Dijck, P., 24, *60*
Varkarakis, M., 261, *270*
Vaughan, J. A., 241, *267*
Venables, P. H., 121, 122, *135,* 232, 233, 235, 236, 237, 239, 242, 243, 244, 245, 246, 247, 248, 250, 252, 254, 255, 257, 260, 261, 262, 263, 264, 267, *267, 268, 269, 271,* 277, 278, 279, 280, 281, 282, 284, 285, 287, 288, 291, *292, 293, 294, 295*
Vernon, J. A., 11, 13, 16, 26, 36, *60, 61, 62*
Verriest, G., 115, *132*
Vigorito, J., 202, *227*
von Békésy, G., 5, *62*
Vossius, G., 100, 110, 129, *134*

W

Wagman, I. H., 147, *154*
Wagner, H. G., 68, *86*
Wagner, H. N., 239, *271*
Wallace, R. K., 130, *135*
Walters, E. D., 143, *153*
Wang, G. H., 122, *135*
Wang, S. G., 174, *198*
Wartis, P. J., 110, *134*
Wasman, M., 130, *135*
Watanabe, T., 260, 263, *269, 270*
Watson, P. D., 122, *135*
Weale, R. A., 76, *85*
Webb, R. A., 175, *200,* 204, *229*
Weiner, J. S., 235, *271*
Weiss, O., 105, *135*
Welford, N. T., 208, *229, 285, 295*
Wendt, P. R., 107, *135*
Wenger, M. A., 277, *295*
Wenner, W. H., 221, *229*
Werblin, F. S., 66, *86*
Wesson, L. G., 239, 247, *271*
Westcott, M. R., 76, *200*
Wever, E. G., 4, 5, 7, 8, 9, 11, 16, 36, *62*
Whitman, R., 113, *131*
Wilcott, R. C., 122, *135,* 239, 240, 244, 245, 246, 251, 253, 257, 258, 259, 260, 264, *271,* 277, 284, *296*

AUTHOR INDEX

Wilder, J., 192, *200,* 213, *229,* 292, *296*
Williams, T. A., 190, *199, 200,* 210, *229*
Wilson, C. D., 211, 215, 218, 223, *228, 229*
Wing, N., 289, *294*
Witten, V. H., 244, *271*
Wohlbarsht, M. L., 68, *86*
Wood, D. M., 176, *200,* 223, *229*
Woodcock, J. M., 213, *229*
Woodruff, B. G., 239, *270*
Woods, R. H., 143, *154*
Wooten, B. R., 76, *86*
Worrall, N., 298, 300, 302, *305*
Wright, D. J., 122, *132*
Wruble, S. D., 16, *61*

Y

Yajima, T., 88, 89, *94, 95*
Yamazaki, K., 260, *269*
Yashida, H., 87, 93, *94*
Yasuno, T., 7, *61*
Yehle, A. L., 177, *200*
Yokota, T., 277, *296*
Young, L. R., 100, 110, 129, *134, 135*

Z

Zappalá, A., 147, *154*
Zettler, F., 83, *86*
Zilstorff, K., 93, *94*

Subject Index

A

Absolute scores, 208
Absolute threshold of hearing, 10
Acceleration component, 212
a.c. Cochlear potential, 4–5, *see also* Alternating cochlear potential
a.c. Impedance bridge, 282
Accrine sweat glands, 122
Acidosis, 178
Acoustic stimulus in cochlear recording, 31–43
   coupling to ear, 35–36
   pure tones definition, 30
   sound measurement, 36–43
   sound-producing systems, 30
   tone producing equipment, 31–36
     amplifier, 32
     attenuators, 32
     transducer, 33
Activation, 158
Activational peaking, 160
Activity cessation, 215
Affect, 203
Alpha training, 130
Alternating cochlear potential, 4, 8–58
   characteristics, 8–10
   data, 10–14
     frequency function, 10–13
     intensity function, 13
   physiological variables, 24–30
     accumulation of fluid in bulla, 27–30
       ossicular loading, 28
       shorting out of electrodes, 29
     anoxia, 25
     body temperature, 26–27
     in cat, 24
     in guinea pig, 24–25
     respiration, 25–26

Alternating cochlear potential (*contd.*)
   recording, 14–58
     anatomical considerations, 14
     apparatus and methods, 43
     artifacts, 55
     choice of experimental animal, 16
     chronic preparations, 23
     physiological variables, 24
     quantitative measurements, 51
     surgical procedures, 18
     acoustic stimulus, 30
Amplifiers
   for electrocardiogram, 184
   for electrodermal activity, 285
   for electrogastrogram, 299
   for electro-oculography, 114
   for electromyography, 140
Amplitude window, 187
Animals for electro-olfactogram, 93
Anion, 238
Anode, 280
Anoxia, 25
ANS, *see* Autonomic nervous system
Anxiety, 226
Apparent resistance, 278, 280
Arousal, 158
Arousal state, 226
Arrhythmia, 175
Arterial lumen, 172
Artificial respiration, 26
   overventilation, 26
   underventilation, 26
Assist mode, 176
A-to-D converter, 208
Atrial depolarization, 168
Atrial fibrillation, 171
Atrial premature beats, 171
Atrial systole, 169

SUBJECT INDEX

Atrioventricular bundle, 167
Atrioventricular node (AV), 167
Atropine 173, 189, 253, 260, 289
Attention, 203
Attentional response to visual stimuli, 217
  nature of stimulus, 221–222
  state, 220–221
Audio amplifiers, 141
Auditory bulla, 18
  surgical procedures, 18–23
    in cat, 23
    in guinea pig, 18–23
Auditory mode, 221
Auditory nerve action potentials, 7–8
Autocorrelation, 189, 300, 302, 304
Autonomic lability score, 290
Autonomic nervous system (ANS), 173, 203
Autonomic response stereotyping, 226
Autonomic–somatic integration, 159–161
AV, *see* Atrioventricular node

B

Baroreceptors, 173
Barrier layer, 234
Base-free units, 290
Beats per minute, 207
Behavioral regulation, 156
Bendover, 14, 27
Bias potential, 120, 279, 280
Biofeedback, 131
Biologically meaningful stimuli, 221
Biomechanics, 140
Biotelemetry, 182–184
  frequency modulation, 183
Biphasic response, 210
Bite board, 105
Blink artifacts, 125
Bony labyrinth, 14
Bony spiral lamina, 16
Bradycardia, 171
Broadcast band, 183
Bucking voltage, 282, 299, 300
Bulla, 12, 14, 17
Bundle of His, *see* Sinoatrial node

C

Cardiac cycle, 167–168
Cardiac output, 166, 172

Cardioaccelerator center, 175
Cardiotachometers, 186, 193, 195, 207
Cat, 17
Cathode, 280
Cathode ray tube, 107
Cation, 238
Central nervous system (CNS), 203
Chloriding process, 280
Ciliary muscles, 101
Classical conditioning, 176
Clinical application of electrogastrogram, 301
CNS, *see* Central nervous system
Cochlea, 19
  drilling, 19
  electrophysiology, 3–60
  functional check during surgery, 23
  of guinea pig, detailed surgical procedures, 20–21
Cochlear duct, *see* Scala media
Cochlear anatomy, 6
Cochlear microphonic, 4–5
Cochlear nerve, 16
Cochlear potentials, 4–8
  a.c., 4–5
  d.c., 5–7
    endolymphatic, 5
    summating, 7
Cognition, 203
Cognitive efficiency, 222
Color reception, 100
Component analysis, 218
Components of attention in infants, 219
Computer analyses of heart rate, 189
Computers in electrocardiogram data analysis, 190–191
Conditioned tachycardia, 173
Conditioning of autonomic responses, 193
Conductance, 289
Conjugate eye movements, 104
Conjugate negative variation, 130
Constant current, 282
Constant voltage, 282
Contact lenses in eye movement recording, 106
Continuous feedback, 196
Contractile force, 166
Control mode, 176
Convergence, 104
Coordination, 156
  of behavior and response, 156
  uncertainty, 156–158

Corneal reflection, 107
Corneoretinal potential, 113
Corneum, 232
  duct filling and hydration, 240–243
Corrections for intrasubject variations, 253
Coupled beat, 171
Cranial nerves, see individual nerve
Cross correlation, 189
Curarization, 175, 177, 178
Curve description, 210
Curve-fitting techniques, 210

D

Dark adaption, 100
Darrow bridge, 283
Data analysis of electrogastrogram, 299–300
Data reduction of heart rate, 188
d.c. Cochlear potentials, 5–7
  endolymphatic, 5
  summating, 7
Deceleration component, 212
Decelerative response, 215
Decibel scale, 36
Depolarization in heart muscle, 168
Dial®-urethane, 58
Diastole, 169
Difference scores, 208
Dilator pupillae, 104
Direction of gaze and tracking error measurements, 127
d-Tubocurarine chloride, 177
Ductal epithelia, 235
Duct filling model, 242
Dynamic eye tracking, 100

E

Eccrine sweat glands, 235, 239
ECG, see Electrocardiogram
Ectopic beats, 168
Ectopic foci, 168
EDR, see Electrodermal responses
EGG, see Electrogastrogram
EKG, see Electrocardiogram
Electrical activity of the skin, 122
Electrical radiation in cochlear recording, 56–57
Electrocardiograph, 207

Electrocardiogram (ECG), 166–197
  amplifiers and polygraphs, 184
  basic instrumentation, 179–193
  data analysis, 191
  description, 168–170
  detection, 186–188
  normal and abnormal patterns, 170–171, see also specific pattern
  operant conditioning and biofeedback, 193
  rationale, 166–167
  recording sites, 171
  regulation, 172–179
  response topography, 191
  specific pattern, 170–171
  usual method of recording, 185
Electrode(s)
  for cochlear potentials, 49–51
    active, 50
    chronic, 51
    ground, 51
  for electrocardiogram, 179–180
    implanted, 180
    nonpolarizing, 179
    pins, 180
    polarizing, 179
    sites, 179
  for electrodermal recording, 277–281
    artifacts, 278
    bipolar, 278
    configurations, 278
    materials, 278
    monopolar, 278
    polarization, 279
    shape, 278
  for electrogastrogram, 298–299
  for electro-oculography, 115, 120–121
    preparation, 125
    silver–silver chloride sponge, 120
    silver–silver chloride suction cup, 121
  for electro-olfactogram, 92
  for infant heart rate recording, 204–205
  for single motor unit training, 142–143
Electrode paste, 115, 205
Electrode placement anxiety, 205
Electrode polarization, 180
Electrode–skin junction, 277
Electrodermal activity mechanisms, 231–267
  summary, 265

SUBJECT INDEX

Electrodermal effector system components, 254
Electrodermal phenomena, 235–237
  basal, 235
  endogenous, 237
  exogenous, 235
  tonic, 235
Electrodermal recording, 273–292
  amplification, 285
  constant current, 282
  constant voltage, 284
  electrodes, 277–281
  input signal conditioning, 281–282
  measurement units, 286–292
  physical variables, 276
  registration, 286
  terminology, 274
Electrodermal responses (EDR), 232, 237, 274
Electrogastrogram (EGG), 297–304
  clinical application, 301
  data analysis, 299–300
  data interpolations and examples, 300
  history, 298
  problems and caveats, 301–304
  recording technique, 298–299
  subject restraint, 299
Electrogenic sodium pump, 239
Electrolyte, 281
Electrolyte medium, 180
Electromechanical counter, 186
Electromyogram, 139
Electromyographs, 140
Electromyography (EMG), 137–153, 155–162, *see also* Single motor unit training
  general techniques, 141–143
    electrodes, 142–143
    equipment, 141–142
  psychophysiological significance, 156
Electroneurogram, 139
Electro-oculography (EOG), 100, 111–129
  a.c. versus d.c. recording, 119
  electrode skin circuit stability, 120–122
    problems of recording electrodes, 120
    preparation of skin, 122
  head movement, 123
  history, 113
  instrumentation, 114–120
  recording problems, 114–128

Electro-olfactogram (EOGs), 87–94
  animals, 93
  electrodes, 92
  as generator potential, 88–92
  history, 87–88
  recording equipment, 92
  stimuli, 93
  types of responses, 88–89
    negative off, 88
    negative on, 88
    positive off, 88
    positive on, 88
Electroretinogram (ERG), 63–85, 113
  clinical uses, 77
  definition, 64
  history of techniques, 64–66
  measurements and interpretation, 83–84
    amplitude, 83
    latency, 84
  recording
    electrode configuration, 70–71
      differential, 70
      single-ended, 70
    preparations, 76–83
      arthropods, 81–83
        head and eye, 82
        whole animal, 81
      birds, 79
      humans, 76
      lower vertebrates, 80–81
        excised eye, 80
        perfused retinal, 81
        whole animal, 80
      mammalian, 78–80
        cats, 79
        ground squirrels, 79
        monkeys, 79
        rabbits, 79
        rats, 79
      other, 83
    signal amplification, 71–73
      averaging devices, 72
      d.c. amplifiers, 72
      resistance–capacitance coupled amplifiers, 71
    techniques and electrodes, 66–70
      contact lens, 67
      glass capillary, 67
      solid core, 68–70

Electroretinogram (ERG) (contd.)
  suction, 68
  wick, 66
 stimulus, 73–75
  d.c. light sources, 74
  filters, 74
  light guides, 75
  monitoring, 75
  photostimulators, 73
  shutters, 74
EMG, see Electromyography
Endodermal phenomena, 273
Endolymph, 15
EOG, see Electro-oculography
EOGs, see Electro-olfactogram
Epidermal duct, 247
Epidermal hydration, 260
Epidermal membrane, 245, 263
Epidermal membrane theory in skin conductance responses, 246
Epidermis, 232
Ergotropic, 161
ERG, see Electroretinogram
Etching solutions, 68
Evaporative water loss (EWL), 241
Evoked potential, 8
Evoked responses of electrodermal phenomena, 274
EWL, see Evaporative water loss
Exercise and heart rate, 173
Experimenter effect, 304
Extrasystoles, 171
Eye blink, 175
Eye function, 100
Eye movement(s), 100–104
 human, see Human eye movements
 musculature, 101
 neuromuscular control mechanisms, 100–104
 recording methods, 104–113
  direct observation, 104
  electro-oculography, 111
  mechanical coupling, 104
  other recent methods, 110–111
  photoelectric recording, 110
  photography and corneal reflection, 105
   history, 105–108
   infant, 108
 saccadic jumps, 100
 smooth pursuit, 100
 types, 102–103
Eyelids, 100–101

F

Féré effect, 273
Fiber optic cables, 108
Field effect transistor, 181
Finger withdrawal, 175
Fire etching of tungsten wire, 69
Fixation time, 215
Floating reference level detector, 195
FM, see Frequency modulation
Fourier analysis, 30, 200, 304
Frequency modulation (FM), 183

G

Gallamine triethiodide, 177
Galvanic skin potential, 125
Galvanic skin response (GSR), 202, 232, 274, see also Electrodermal recording
Gastrointestinal activity (GI), 297
General anesthesia, 18
 diallylbarbituric acid, 18
 pentobarbital sodium, 18
 urethane, 18
GI, see Gastrointestinal activity
Granular layer, 232
Ground electrode, 205
GSR, see Electrodermal recording, Galvanic skin response
Guinea pig, 17

H

Habituation, 160, 212
Hair cells, 16
Half-cell potentials, 279
Half-cell voltage, 180
Head holder, nontraumatic, 18–19
Head rotation measurement, 123
Heart, 167–172
 cardiac cycle, 167
 structure, 167
Heart rate(s) (HR), 166, 202
 descriptors, 211–213
 in infants, variability, 225–226
 recording in infants, 201–226
  data analysis, 208–211
  infant sample record, 206
  initial level effects, 213–215
  laboratory considerations, 204–207
  measurement, 207–208

Heart rate(s) (HR) (*contd.*)
  response, 203
    and cognitive functioning, 222–225
    developmental issues, 219–222
    in infants, studies, 215–219
    in various mammals, 169–170
Helicotrema, 16
Hippocampal theta activity, 161
Homeostatic mechanism, 213
Homeostatic process, 210
Homoscedasticity, 289
Horizontal eye movement measurement, 125
Horny layer, 232
HR, *see* Heart rate
Human eye movement(s), 100–131
  applications, 128–130
  as artifact, 130
  behavioral measure, 129
  control system, 129
  developmental studies, 129
  recording, 100–131
  reading ability, 128
  sleep and dreaming, 129
  states of consciousness, 129
Hybrid computers, 286
Hydration model, 242
Hydraulic capacitance, 241
Hydrophilic, 232
Hyperthyroidism, 175
Hypnosis, 130
Hypotension, 173
Hypothalamus, 174
Hypothyroidism, 174

I

Impedance, 185, 282
Impedance matching, 285
Implantable transmitter, 183
Inadequate ventilation, 178
Independence of means and variances, 289
Individual variability, 210
Infancy, physiological recording, 201–203
  approaches, 202
  rationales, 203
Infant behavior, 202
Infrared filters, 75
Initial response, 211
Input–output function, *see* Alternating cochlear potential, data
Instrumental conditioning, 176

Instrumentation for electro-oculography, 114–120
  amplifiers, 114
  recording systems, 114
  stable d.c. recording systems, 115
  time constant, 115
Insulating metal electrodes, 68–69
Interatrial septum, 168
Interpolated premature beat, 171
Interstimulus interval, 192
Interval histogram, 189
Intraretinal electrodes, 65

K

Kangaroo rat, 17
Kinematic analyses, 156

L

Laser techniques, 131
Latency-to-peak magnitude, 211
Latency-to-trough magnitude, 211
Law of initial value (LIV), 192, 213, 290
Lead I, 170
Lead II, 170
Lead III, 170
Levator palpebrae superioris muscle, 100–101
Level detectors, 128
Light adaption, 100
Light guides, 75
Linearity of regression, 289
LIV, *see* Law of initial value
Locus of electrogastrogram, 301–302
Low torque potentiometer, 123
Lumen negative potentials, 248

M

MAP, *see* Muscle action potentials
Mean response measure, 208
Measurement units of electrodermal activity, 286–292
Medulla in cardiovascular regulation, 173
Methylene blue, 244
Membrane location in skin conductance responses, 244
Membrane potentials, 237
Membrane responses in skin conductance responses, 243

Mho, 289
Microcosmic model, 138
Micromhos, 289
Miniature galvanometers, 141
Miniature skin electrodes, 205
Mirror galvanometer, 110
Modiolus, 16
Molar behavioral level, 203
Molecular behavioral level, 203
Motoneuron, 138
Motor units, 138
Motor unit potentials, 139–140
　amplitude, 139
　duration, 139
Movement artifact, 205
Mucous layer, 232
Mueller's muscle, 101
Multiphasic response, 209
Multivariant approach, 218
Multivariate analyses, 226
Muscle action potentials (MAP), 159
Myoelectric prostheses, 138

N

NaCl reabsorption in skin, 248
Narrow band-pass filters, 75
Neocortex limbic system, 174
Neonatal period, 204
Neonates, 204
Nernst equation, 238
Neuromuscular control mechanisms of eye movements, 100–104
Neutral density filters, 74
Noise, 125
Nonlinearity, 284
Normal cardiac cycle, 169
Normal sinus rhythm, 171
Nystagmus, 113

O

Orbicularis muscle, 100
Off effect, 65
Ohm, 289
One-shot multivibrator, 187
Organ of Corti, 4, 5, 16
Orienting, 212
Orthoses, 138

Ossicles, 14
　incus, 14
　malleus, 14
　stapes, 14

P

Paced respiration, 176
Palpebral fissure, 100
Parasympathetic nervous system, 173
Paroxysmal tachycardia, 171
Peak detector, 187
Peak magnitude, 211
Perilymph, 15
Peripheral resistance, 172
PGR, see Psychogalvanic reflex
Photoemitting diode, 196
Photomultipliers, 110
Photoreception, 64
Point, 209
Positive-pressure respiration pump, 26
Poststimulus time histogram, 189
Propanolol, 173, 189
P–R segment, 169
Psychogalvanic reflex (PGR), 274
Psychophysiological control systems, 138
Psychophysiological time, 209
Pulse former, 186
Pupil diameter control, 78
Pupil size, 101
Pure cone electroretinogram, 77
Purkinje fibers, 167
P wave, 169

Q

QRS complex, 169
Quantitative measures of alternating cochlear potentials, 51–55
　oscilloscope, 55
　wave analyzer, 52

R

Radioactive phosphate, 244
Range ratio procedure, 291
Rapid eye movement (REM), 115
Reaction time paradigm, 222
Real time, 209
Real time photography, 107

SUBJECT INDEX

Recording
  of alternating cochlear potential, 43–51
    amplifier, 46–48
      differential, 46
      single ended, 48
    differential electrode method, 48–49
    electrical noise, 44
    electrodes, 49–51
    shielding, 45
  equipment for electro-olfactogram, 92
Recti muscles, 101
Reducing skin impedence, 122
Reduction of cable artifacts, 181
Reference detector, 195
Regression coefficients, 213
Regulation of cardiac responses, 172–179
  cardiac–somatic coupling, 175
  curarization, 177–179
  heart rate and respiration, 175
  neural and hormonal control, 172
  relationship of heart rate to other cardiovascular responses, 172
REM, see Rapid eye movement
Resistance, 282
Respiratory acidosis, 178
Response parameter, 212
Response uncertainty, 157
  effects on behavior, 158–159
Ringing, 33
Round window, 8, 22
R–R interval, 207
Running histogram, 195
R waves, 188, 207

S

Saccadic eye movements, 116
Saline bridge, 280
Sampling procedures, 208, 212
SA, see Sinoatrial node
Scala media, 5, 15
Scala vestibuli, 15
Scali tympani, 14
Schmitt trigger, 186, 193
SC, see Skin conductance
SCL, see Skin conductance level
SCR, see Skin conductance response
Secretory theory in skin conductance responses, 244
Secretory epithelia, 235

Semipermeable membrane, 237
Sensitivity functions, see Alternating cochlear potential data, frequency function
Serial correlation, 189
Sequential histogram, 189
Seventh cranial nerve, 100
Shaping procedure, 194
Shutters, 74
Signal decay time, 35
Signal rise time, 35
Single motor unit training (SMUT), 137–153
  covert operant conditioning, 151
  history, 138
  muscles used, 145
  personality factors affecting control, 152
    handedness, 152
    internal states, 153
    previous training and skill, 152
  preliminaries, 143
  psychophysiologic techniques, 147
  reaction-time experiments, 147
  sample routines, 145
Sinoatrial node (SA), 167
Sinus arrythmia, 171
Sixth cranial nerve, 101
Skin and sweat gland anatomy, 232–235
Skin conductance (SC), 274
Skin conductance level (SCL), 235, 251–253
  intrasubject variations, 252
  nonsudorific pathway, 251
  sweat gland pathway, 252
Skin conductance response (SCR), 239–251
  duct filling versus membrane, 250
  evidence of sweat gland activity, 239
  for two types, 250
  membrane theory, 243
Skin conductance response production, 240–243
Skin function, 232–234
Skin impedance (SZ), 274
Skin potential (SP), 274
Skin potential level (SPL), 235, 259–267, 274
  epidermal component, 260–265
  sweat gland component, 259–260
Skin potential response (SPR), 235, 253–259, 274
  diphasic or positive, 257–259
  monophasic negative, 253–257
Skin preparation for eye movement recording, 125

Skin resistance (SR), 274
Skin resistance level (SRL), 235
Skin resistance response (SRR), 235
Skin water permeability, 233–234
Slow positive drift, 263
Slow shifts in eye position, 115
Smoked drum kymograph, 105
SMUT, see Single motor unit training
Sodium pump, 255
Sodium reabsorption, 255
Sound cannula description, 34–35
Sound measurement, 36–43
  calibration, 42–43
  methods, 39–42
    probe tube, 40
    substitution, 38
  reference levels, 36
Sphincter pupillae, 104
Spiral ganglion, 16
Spontaneous responses of electrodermal phenomena, 274
SP, see Skin potential
SPL, see Skin potential level
SPR, see Skin potential response
SR, see Skin resistance
SRL, see Skin resistance level
SRR, see Skin resistance response
Standing potential, see Corneoretinal potential
Stapedius, 14
State as arousal continuum, 214
Statistical bias, 212
Statistical control for prestimulus levels, 214
Stimuli for electro-olfactogram, 93
Stimulus relevance, 221
Stranger anxiety, 205
Stratum compactum, 233
Stratum corneum, 232
Stratum germinativum, 232, 245
Stratum granulosum, 232
Stratum lucidum, 232
String galvanometer, 113
Succinylcholine, 177
Surgical procedures, cochlea, 18–24
  in cat, 22–23
  chronic, 23–24
  electrode immobilization, 24
  in guinea pig, 20–21
Sweat, 281
Sweat gland anatomy, 232–235

Sweat-produced hydration, 242
Sweat reabsorbtion, 247–248
Sympathetic nervous system, 173
Sympathectomy, 173, 253
Sympathetic ganglionectomy, 239
Systole, 169, 172
SZ, see Skin impedance

T

Tachycardia, 171
Tail-end attenuator, 32
  schematic diagram, 58
Tarchanoff effect, 273, see also Endodermal phenomena
Tensor tympani, 14
Third cranial nerve, 101
Thorium-X, 244
Tibialis anterior muscle, 138
Time constants in electro-oculography, effects on several types of eye movements, 115
Time-locked signals, 188
Tracheal cannula, 18
Trough magnitude, 211
Trophotropic, 161
T wave, 169, 261

U

Unity gain voltage follower, 181

V

Vagal stretch receptors, 175
Varying skin potential, 120
Vasomotor center, 173, 174
Velcro tape, 124
Ventricular ectopic beats, 171
Ventricular excitation, 169
Ventricular fibrillation, 171
Ventricular repolarization, 169
Ventricular systole, 169
Vestibular nystagmus, 130
Video disk, 108
Videotape, 108
Visible sweating, 240
Visual field, 107
Voltage divider, 282

W

Water barrier of skin, 233–234
Water reabsorbtion
  mechanisms in skin, 248
  reflex, 247, 258
  response, 249

Wever–Bray potential, 5
Wheatstone bridge, 124, 282

Z

Zero deceleration, 212

DATE DUE

MY 24 '78